ECONOMIC SPACES OF PASTORAL PRODUCTION AND COMMODITY SYSTEMS

Ashgate Economic Geography Series

Series Editors:
Michael Taylor, Peter Nijkamp, and Tom Leinbach

Innovative and stimulating, this quality series enlivens the field of economic geography and regional development, providing key volumes for academic use across a variety of disciplines. Exploring a broad range of interrelated topics, the series enhances our understanding of the dynamics of modern economies in developed and developing countries, as well as the dynamics of transition economies. It embraces both cutting edge research monographs and strongly themed edited volumes, thus offering significant added value to the field and to the individual topics addressed.

Other titles in the series:

Economic Spaces of Pastoral Production and Commodity Systems

Markets and Livelihoods

Edited by

JÖRG GERTEL
Leipzig University, Germany

and

RICHARD LE HERON
The University of Auckland, New Zealand

Routledge
Taylor & Francis Group

LONDON AND NEW YORK

First published 2011 by Ashgate Publishing

2 Park Square, Milton Park, Abingdon, Oxon OX14 4RN
711 Third Avenue, New York, NY 10017, USA

Routledge is an imprint of the Taylor & Francis Group, an informa business

First issued in paperback 2016

British Library Cataloguing in Publication Data
Economic spaces of pastoral production and commodity
    systems : markets and livelihoods. -- (Ashgate economic
    geography series)
    1. Pastoral systems. 2. Pastoral systems--Economic
    aspects. 3. Livestock--Cooperative marketing. 4. Economic
    anthropology. 5. Economic geography.
    I. Series II. Gertel, Jörg. III. Le Heron, Richard B.
    338.1'76-dc22

Library of Congress Cataloging-in-Publication Data
Gertel, Jörg.
    Economic spaces of pastoral production and commodity systems : markets and
livelihoods / by Jörg Gertel and Richard Le Heron.
        p. cm. -- (Ashgate economic geography series)
    Includes bibliographical references and index.
    ISBN 978-1-4094-2531-1 (hardback)
    1. Pastoral systems--Economic aspects. 2. Exchange. 3. Commodity exchanges 4.
Human geography. 5. Economic anthropology. 6. Economic geography. I. Le Heron,
Richard B. II. Title.

GN407.7.G47 2011
306.3'4--dc23

                                                                        2011025756

ISBN 978-1-4094-2531-1 (hbk)
ISBN 978-1-138-26130-3 (pbk)

Contents

List of Figures

List of Contributors

Hans-Georg Bohle is Professor and Head of the Research Group of Development Geography at the University of Bonn, Germany.

Ingo Breuer is Research Associate at the Institute of Oriental Studies at the University of Leipzig and a project leader at the DFG-funded Collaborative Research Centre 'Difference and Integration' (SFB 586) at the University of Leipzig, Germany.

Jörg Gertel is Professor at the Institute of Orientale Studies at the University of Leipzig and Director of the DFG-funded Collaborative Research Centre 'Difference and Integration' (SFB 586) at the University of Leipzig, Germany.

Andreas Gruschke is Research Assistant at the DFG-funded Collaborative Research Centre 'Difference and Integration' (SFB 586) at the University of Leipzig, Germany.

Barbara Harriss-White is Professor of Development Studies at the Wolfson College and Director of the School of Interdisciplinary Area Studies Programme at the University of Oxford, UK.

Jörg Janzen is Professor of Geography at the National University of Mongolia, Ulaanbaatar and staff member at the Institute for Human Geography at the Freie Universität of Berlin, Germany.

Guma Kunda Komey is Assistant Professor of Geography at Juba University, Sudan and a researcher at the DFG-funded Collaborative Research Centre 'Difference and Integration' (SFB 586) at the University of Halle-Wittenberg, Germany.

David Kreuer is Research Assistant at the DFG-funded Collaborative Research Centre 'Difference and Integration' (SFB 586) at the University of Leipzig, Germany.

Hermann Kreutzmann is Professor of Human Geography and Director of the Zentrum für Entwicklungsländerforschung (ZELF) at the Freie Universität Berlin, Germany.

Erena Le Heron is a postdoctoral researcher at the School of Environment at the University of Auckland, New Zealand.

Richard Le Heron is Professor of Geography at the School of Environment at the University of Auckland, New Zealand.

Chris Lloyd is Industry Development Manager at EBLEX (English Beef and Lamb Export), a division of Agriculture and Horticulture Development Board (AHDB), UK.

Hussein Mahmoud is Lecturer at the Department of Geography at the Egerton University of Njoro, Kenya and Postdoctoral Research Fellow at the Max Planck Institute for Social Anthropology, Halle/Saale, Germany.

Nikola Rass is Associate Programme Officer at the United Nations: Convention to Combat Desertification (UNCCD) in Bonn, Germany.

Florian Stammler is Senior Researcher in Anthropology at the Arctic Centre, University of Lapland in Rovaniemi, Finland and associated to the Scott Polar Research Institute at the University of Cambridge, UK.

Martin Wiese is Senior Program Specialist at the Ecohealth Team of the International Development Research Centre (IDRC) in Ottawa, Canada.

Fred Zaal is Assistant Professor at the Amsterdam Institute for Metropolitan and International Development Studies (AMIDSt) at the University of Amsterdam and Senior Advisor for the Area of Sustainable Economic Development at the Royal Tropical Institute (KIT) in Amsterdam, Netherlands.

Acknowledgements

'*Stability, gentlemen, is the one thing we can't deal with*' (in Krebs, A.V. The Corporate Reapers, 1992, 305) – this statement by a trader at the Chicago Board of Trade can easily be applied to nomadic peoples and to the pastoralists alike. But in contrast to traders who handle foreign commodities and foreign money, pastoral livelihoods are directly involved, affected and exposed to changing environments and risks. Flexibility and mobility is inscribed in their livelihoods and their relation to markets, particularly since sedentary knowledge systems increasingly prescribe human-nature relations.

This book originates from discussions at the Universities of Leipzig and Halle-Wittenberg about the interrelations between mobile and settled people that started with the inception of the Collaborative Research Centre 'Difference and Integration' in 2001 (Sonderforschungsbereich 586, www.nomadsed.de). As part of a series of colloquia and workshops, the conference on 'Pastoralists and the World Market: Problems and Perspectives' in Leipzig in April 2006 prompted the focus on markets and livelihoods as inextricably linked everyday realities. For funding this project, and for offering the possibility to progress our research over a lengthy period, we thank the German Research Foundation (DFG) for its financial support.

As editors we wish to thank our respective institutions for actively supporting agri-food research and scholarship over many years. We discovered our mutual interests in pastoralism in 2003 when Jörg was a visitor at the then School of Geography and Environmental Science, the University of Auckland. This was further cemented through Richard's participation in the 2006 conference and Jörg's revisits to Auckland and New Zealand in 2009 and 2010. We both owe much to research stimulation that springs from our involvement in the Australasian Agri-Food Network and its annual conferences and the collegiality and intellectual exchange that characterises the New Zealand Marsden funded 'Biological Economies: Knowing and making new rural value relations' project (2010-2013).

A number of people played key roles in ensuring the completion of the book. We are grateful to Katy Crossan and Carolyn Court at Ashgate for shepherding the project along and assisting us with queries. We thank Sonja Ganseforth from the University of Leipzig for her valuable suggestions that helped get the book project started. Igor Drecki from the School of Environment at the University of Auckland handled all our wishes and aspirations for visualisation with great care and admirable patience. He professionally crafted the figures in the book. As the book neared completion Marie-Christine Hakenberg from the University of Leipzig spent long hours in skilfully editing and meticulously checking the text and figures. We are indebted to the dedication of Erena Le Heron from the School of Environment at the Auckland University for her invaluable input as an external referee on the argument of all chapters, her critical overview of the book as it evolved, her insistence on clarity of style and meaning, and her attention to detail on every aspect of the book. We also thank the Ashgate reviewers for their initial incitements regarding the emphasis of the book and for their later encouraging and helpful suggestions.

Finally we would like to express our gratitude to the contributors of this book for their patience and support of our requests throughout the publication process.

Jörg Gertel and Richard Le Heron
14 June 2011

PART I
Pastoralists in the Market

Chapter 1

Introduction: Pastoral Economies Between Resilience and Exposure

Jörg Gertel and Richard Le Heron

This book attempts to interrogate early twenty-first century pastoralism and pastoral economies by putting into tension ideas about markets and livelihoods. As such it is a project in the spirit of recent research and scholarship in economic geography, agri-food and development studies dedicated to producing knowledge on a planetary scale. The book asks three questions that draw on and point to new research imaginaries in a globalising world. The questions are:

- What does it mean to try to make a livelihood from pastoralism in different parts of the world?
- How are markets implicated in the livelihoods of pastoralists, wherever they are to be found?
- How far does or should, livelihood-informed thinking and action that emerges in particular geographic contexts necessarily mean engaging in market-making relations?

Our questions are designed to ground the book's enquiry in the remarkably rich resource of field-centred ethnographic research on pastoralism and pastoral economies that is available, and to unsettle the categories of livelihood and markets that form very different points of entry into any research on socio-economic relations. They spring from a growing awareness and concern that across every sphere of economic activity market fundamentalism and triumphalism are sweeping away questions that challenge its market assertions, claims and promises. They are in fact unashamedly questions that prioritise and foreground human concerns of everyday surviving, living and hoping, questions that prompt reflection on the bodily placement of pastoralists as they strive to handle their exposure to multiple influences and seek to create conditions of some stability and resilience. In asking these questions we are forced to examine how situated knowledges about pastoral economies accumulated in different parts of the world might be framed into new and deeper understandings. In this way our open ended questions embrace wider intellectual concerns such as 'where do markets or livelihood practices come from' and 'how are they performed into existence, by and for whom?'

The questions initially began to surface during our dialogue at the April 2006 conference on 'Pastoralists and the World Market: Problems and Perspectives' at the University of Leipzig, Germany. The conference formed part of the major German funded research programmes on 'Difference and Integration – Interdependencies between Nomadic and Sedentary People' (SFB 586) and 'Critical Junctures of Globalisation' (GRK 1261). Being able to bring together researchers whose pastoral economy interests spanned much of the globe afforded an unprecedented opportunity to probe issues relating to globalising processes through the lens of territorially distributed empirical and longitudinal studies. The great social and economic thinker, Karl Polanyi, who has influenced many contributing to the book, might have described the conference as a proto-project in comparative economy, given its focus on empirically informing discussion of pastoral economies and positioning them relative to their others. We should add, however, that while our questions may be ambitious, the book is more modest in its aspiration and what it achieves. Our step, in seeking to elucidate new styles of knowing pastoralism and pastoral economies is, we hope, to have used empirics and theorising creatively and differently.

When we began to outline our individual and collectively diverse research trajectories to each other at the conference we soon realised we were doing a global mapping of pastoralism and particular experiences of differing pastoral economies. The research trajectories were more than case studies put in a row. They rather reflected insights from very different situations all over the world. Indeed, the presenters were speaking from and about many contexts, that when considered through the eyes of the German and French academy, have been the focus of ongoing and heavily funded research programmes over many years. The Collaborative Research Centre 'Difference and Integration' at the Universities of Leipzig and Halle (SFB 586, 2001-2012) is focusing, for example, on interdependencies between nomadic and sedentary people, the Max-Planck-Institute for Social Anthropology in Halle is analysing integration and conflict in Asian and African societies, the Maison Méditerranéenne des Sciences de l'Homme in Aix-en-Provence has a long standing research interest in North African mobilities, and the Collaborative Research Centre Akazia in Bonn (SFB 389, 1996-2008) investigated arid climate, adaptation and cultural innovation in Africa. The experiences recounted all added up to a highly variegated landscape of human experience and knowledge relating to pastoralism and pastoral economies.

As editors, we reflected on the most strategic intervention that would allow the research programmes of the mainly European participants to be performative. By this we aimed to place the wealth of research into a new framing that elicited fresh and globally useful knowledge. Driven by the conference design, the provocation of putting markets and livelihoods into tension was very effective. The responsiveness to this framing on the part of participants was challenging. In their experience, the worlds of production and pastoral economies could not be reduced to one or the other of these categories. Moreover, the inseparability of markets and livelihoods struck a clear chord with the researchers present. The connections and relationships between markets and livelihoods dominated the discussion.

The keynote speakers at the conference, Barbara Harriss-White and Hans-Georg Bohle, directly challenged the hidden privileging of markets in the global commodity chain, global supply chain and global value chain and livelihood literatures as well as the market centred discourses of national and international policy communities. As successive presenters articulated the specifics of the pastoral economies they were studying we began to appreciate the remarkable revelatory experience we were sharing. The conference priority of assembling a group of researchers investigating pastoralism in Africa, Asia, Europe, Australia and New Zealand, regardless of academic traditions of declaring strong messages about correct ways to represent pastoralism, led to an equivalence of stature of pastoralism from any place with other places, and an equivalence of contribution amongst the presenters regardless of their background and longevity of field research. This was felt at the time. No recounted experience of pastoralism was any more or less relevant, pertinent or illustrative of pastoral economy experiences. No presentation had more or less merit because its narrative was tied to particular theoretical categories about or to which participants had more or less exposure or commitment. These conditions of engagement led to an openness of exchange around shared and up-to-date accounts with carefully framed critiques of the crucial categories of markets and livelihoods. Started by the conference and furthered by ongoing empirical and conceptual work in the collaborative research centres we established a discursive field around knowledge production on economic spaces of pastoral economies (see for example: www.nomadsed.de).

Situating Pastoralism and Pastoral Economies

The following five observations seek to establish a number of broad contours of pastoralism and to situate research on pastoral economies within the broader field of geographical knowledge production. First, pastoral activities are the reality of many people's livelihood practices around the globe. Extensive pastoral production takes place on some 25 per cent of the world's land area, it comprises herds of nearly a billion head of camel, cattle and smaller livestock and provides 10 per cent of global meat production. Livestock production is crucial for the livelihoods and well-being of up to 200 million households (Rass 2007). Until recently the largest share of benefits generated by pastoralism was obtained from grazers on marginal lands, where other economic activities provide lower returns or were not viable at all. But land use and market systems are changing at a rapid pace. Globalisation, growing international interdependencies, and integration into foreign exchange systems are fostering new pressures but are also creating new opportunities for pastoralist producers.

Second, there is little recognition of the existence of pastoral activities. The diverse characteristics of pastoralism and pastoral economies are commonly misrepresented, its importance understated and their diversity and enduring nature discounted. Statistical knowledge, especially counting, and more precisely the translation of social relation systems into numerical relation systems, is a major part of continually constituting the world of modern power that nomadic people inhabit. Pastoral and nomadic societies are probably among the most important groups that escape the counting practises of sedentary powers. Pastoral production provides multiple products, such as meat, hides, milk, hair, blood, and manure, while livestock also perform additional roles such as transport and draught power, food storage, and capital reserve, as cash buffer and a hedge against inflation. Yet, knowledge concerning pastoral production and marketing remains weak, however, they are part of the 'diverse economies' (Gibson-Graham 2008) – characterised, for example, by non-market transactions, unpaid labour, and non-capitalist enterprises. Official statistics barely capture economic values associated with pastoralism (Breuer 2005). Movement and mobility mean pastoralists and their herds (or in more sedentary arrangements, livestock flows amongst farms, en route to meat or other processing plants and from country to country) are hard to contain within national statistical systems. Many goods and services provided by pastoralist systems, particularly in Africa and Asia, are statistically unknown, due to their 'informal' activities (classified as such by sedentarised urban powers), the lack of systematic data collection and the failure to disaggregate national datasets into others. As a consequence, the economic and social contribution of nomadic people and pastoral societies remains structurally underestimated. In the same line of argument the framing of data deficits pops up in neoliberal contexts of 'integration', when international 'development' agencies complain, for example, about the lack of rigorous data on possible transmission of animal disease and viruses, and lament the constraints in guaranteeing reliable and sustained commodity supply for export markets.

Third, until now analyses of pastoral commodity and livestock marketing systems have been restricted to regional perspectives. Zaal (1999) focuses on two African countries (Kenya and Burkina Faso), Schlee's (2004) edition on ethnicity and livestock markets concentrates on West Africa, while McPeak and Little (2006) investigate pastoral livestock marketing in eastern Africa. A complementary regional focus is provided by Kerven's (2003) collection that reveals the effects of large-scale privatisation in Kazakhstan and Turkmenistan (see Bruun 2006 for Mongolia). Recent insights for Latin America are provided by Westreicher et al. (2007), Grandia (2009), van Ausdal (2009), and Walker et al. (2009). So far, only Chang and Koster (1994) offer a wider geographical range of case studies, their approach, however, remains largely historical.

Fourth, beyond this background there are four significant dynamics that are constitutive of the broad contours of world markets and the particular articulations of pastoral economies: (a) Pastoral production for world export markets is regionally concentrated. The global sheep meat market is dominated by Australia and New Zealand, and so is the wool export market. Excluding intra-EU trade, 88 per cent of the world trade in sheep meat is sourced from them (see Fig. 16.2). In contrast, almost 50 per cent of the world beef exports originate from four countries in Latin America (Brazil, Argentina, Uruguay, and Paraguay) (Brown 2009). Large national livestock production, however, does not necessarily correspond with high volumes of export. China, although a major producer of meat and wool, is a net importer. (b) Nomadic pastoral production in Africa and Asia – though increasingly commercialised – seldom meets international standards, particularly for meat; marketing is either protected (by tariffs) or restricted (by frontiers and regulations) to national markets, trans-border and proximate regional trade. However, millions of pastoral households depend on their livestock to make a living in rural areas. Yet, while meat from beef, sheep, goats and yak rarely reach affluent customers around the world, niche products for international export markets, particularly from Asia – such as cashmere from Mongolia, reindeer antlers from Siberia or caterpillar fungus from Tibet – constitute far reaching economic spaces. (c) The processes and patterns of organisations linking producers and consumers constantly anew can be conceptualised as constitutive of economic spaces made up of relations near and far. Sheep meat, for example, is supplied from Western Australia as live sheep to Arab Gulf countries and competes in Saudi Arabia with live sheep from Somalia or Sudan and with frozen meat from New Zealand. Driven by urbanisation, oil wealth, realigning retail structures, and concerns of animal welfare consumption patterns change with crucial economic impacts. The global *halal* market, for example, amounts to US$ 2.1 trillion and is growing at an annual rate of US$ 500 million (ACIL Tasman 2009). (d) Pastoral livelihoods, both in industrialised countries (i.e. Australia, New Zealand) and also in least developed countries (i.e. Somalia, Sudan) are often linked. They might even depend on the very same (external) decisions, political regimes and economic regulations that are often beyond their reach. The transport of expensive premium lamb meat cuts from New Zealand to European supermarkets takes advantage of New Zealand's guaranteed sheep-meat quota into the EU of 227,854 tonnes. This in return limits market opportunities for proximate North African pastoralists.

Finally, from the point of view on market and livelihood encounters experienced by nomads, pastoralists and farmers, exchange relations are always embedded in societal structures. Markets and livelihoods are inextricably linked, and are continuously reproducing and constituting (new) economic spaces. As Massey (1999: 28) reminds us, space is, first of all, a product of interrelations, constituted through interactions. Space is also a sphere of multiplicity, in which distinct trajectories coexist. As interrelations are necessarily embedded material practices which have to be carried out, space is always in a process of becoming – it is never finished, never closed. 'This relationality of space together with its openness means that space also always contains a degree of the unexpected, the unpredictable' (Massey 1999: 37). Subsequently, our framing of economic spaces conceptualises them as being constituted through interactions, as spheres of multiplicity, as historically superimposed trajectories, as unfinished, and always in the process of being made. Moreover, economic spaces are inextricably constituted by economic knowledge (Mitchell 2008). But there is no simple divide between a virtual world of economic theory and a real world outside it. Every economic project hence involves – as Mitchell puts it – 'multiple arrangements of the simulated and that to which it refers' (Mitchell 2008: 1116). This feature has much bearing on how understandings of pastoralism and pastoral economies are shaped. Often simplistic assumptions of the state of things are contested by lived realities.

These five aspects are in a manner of speaking tests of lived realities, connections back to the presence of undervalued life. They register pastoralism and pastoral economies into framings that make broadly visible their hidden dynamism. But this increased visibility is hardly a sufficient investigative end. It is stepping off point and a basis for analysis. A preliminary look at exchange relations provides the beginnings of a re-orientation, in which it becomes possible to learn from the subject matter of experience about relations around which politics and ethics of survival may form.

Conceptual Approaches to Markets and Livelihoods

Informed by economic anthropology and its contextual knowledge of exchange relations the book is unreservedly a study of economic geographies, involving social and economic relations of agriculture and food, in developing trajectories. One of the more perplexing features of contemporary development theorising is that despite the consensus that production-consumption relations are imbricated in the social reproduction of groups and societies, two rather divergent theoretical traditions have come to dominate much thinking in recent years. These are the global commodity chain, global supply chain, global value chain (hereafter GCC/GSC/GVC) framework and the livelihood framework, respectively production to consumption relations and relations of reproduction. Realising that the two traditions are relatively unconnected in the literature invites a re-thinking, and a re-linking. This is the emphasis of the book. But first, we outline what is commonly understood as markets and livelihoods in these traditions. Our commentary is very selective and aims to provide some background on the traditions to give a sense of why bringing the categories of markets and livelihoods together is the guiding strategy of the book.

A primary thrust of the GCC/GSC/GVC literature is the examination of the interrelations that are uncritically assumed to exist and for the most part only to be market exchange relations. What results is a silence about the boundaries of theoretical enquiry, which springs from a specialisation of investigation into in-chain developments. This is often accompanied by a narrowness of thinking about how producers from developing countries might integrate in global markets which van Grunsven (2009) attributes to western centrism. The idea of a chain is often imposed on realities, thereby leading to a view that the study of chain relationships forms a study of market relations. The chain is considered to be the market. This said, the tradition recognises the challenges of ensuring value circulation and realisation throughout the chain and how any interruption to such value motion puts in doubt business viability and household welfare.

The identification of three variants of the GCC/GSC/GVC literature reflects different interests and emphases on the part of researchers. Two recent reviews (van

Grunsven 2009, Hassler 2009) provide an excellent introduction to where chain-type thinking came from. The earliest tradition, strictly speaking a commodity chain rather than a more fully developed GCC focus, has been especially concerned with investment: labour relations and of households in these relations. The early work drew heavily on the world-systems framework. GSC studies that have arisen with the appearance of supply chains in many industries are mainly concerned with organisational strategies to bind production-consumption relations inside the chain (Le Heron et al. 2001). The GVC variant, possibly the most recent, focuses on where and how value is generated and appropriated, and is increasingly an interest of business researchers and scholars (Gereffi et al. 2005).

Overlapping these discernible threads are longstanding attempts to place core, semi-peripheral and peripheral economies into understandable relationships, especially by making more explicit the spatiality of commodity chains (Hughes and Reimer 2004, Leslie and Reimer 1999). Inspired by and partly derived from world systems thinking, some agri-food researchers have examined the emergence of food regimes and extended the investigation of the geography of commodity chains (Fold and Pritchard 2005). These relatively stable structural arrangements are not static, developing under historical and geographical influences. Food Regime theorists identify three regimes to date. Initial work by Friedmann and McMichael (1989) has been revisited, in the early 2000s (Friedmann 2005, McMichael 2005) and again at the end of the 2000s (Burch and Lawrence 2009, Campbell and Dixon 2009, McMichael 2009).

In short: Commodity, supply and value chain approaches intersect in their focus on commodity flows. They are primarily concerned with the system of activities and the networks of businesses involved in the production, distribution and delivery of products or services to customers. The conditions of social reproduction and livelihood security are, however, beyond the scope of these approaches. The notion of a 'chain' rather invites, as an analytical device, a simplistic reading that amalgamates the distinct spheres of social networks (connecting, for example, producers with processors, and so on), the manifold ways of financial transactions and credit schemes (connecting, for example, electronic with human assessments of risks), and the specific – space and time related – social context of commodities into an allegedly unified chain. As economic relations are embedded within society (Granovetter 1985), the understanding of a specific commodity (like meat or wool) can, conceptually speaking, however, not be reduced to its mere physical presence. It is always also part of a (usually western) property and social system and thus interwoven with specific conditions of production, exchange, consumption and reproduction. Hence, alongside economic transactions and the spatial and temporal flow of commodities, local nomadic and pastoral livelihoods and their capabilities, based on rules and resource, are shaped, and vice versa are structuring commodity, supply and value chains.

What is significant for pastoralism and pastoral economies research is that while issues that are actively scrutinised by livelihood researchers also constitute the object of enquiry in the GCC/GSC/GVC literature, they are framed very differently. The above theoretical emphases tended for many years to write out political and ethical concerns. Over the past decade a number of pioneer and insightful studies of different commodity systems have tried to prioritise the links between chain dynamics and producer livelihoods. Very few of the studies, however, draw upon the livelihood literature familiar to most researchers of pastoralism and pastoral economies.

If the GCC/GSC/GVC literature amounts to a top-down framing of socio-economic and institutional relations, then the livelihood literature, with its overt consideration of subsistence production, reproduction crises and resource-portfolios is best regarded as a bottom-up perspective. This detailed analysis of the social practices of the so-called survival economy, vulnerability, and livelihood security gathered momentum in the 1980s.

The 'economy for survival' approach assumes that the 'section of the insecure' apply different strategies in order to secure survival (Evers et al. 1984) by spreading (economic) risks rather than maximising profits. Several economic activities may generate revenues, and revenues from different family members are often pooled in a joint household. Subsequently, risks of social reproduction are redistributed among generations. The young and the old, who can either not work yet, or not work anymore, as well as women and children, who often perform unpaid tasks, render indispensable assistance, albeit not directly participating in earning of monetary revenue. The elderly may receive transfer incomes, since they may have access to state welfare services or even receive pensions from foreign countries. From here we learned that even individuals can produce with both subsistence and market orientation, that they are integrated into different social networks and are able to act in a spatially mobile fashion.

The concept of vulnerability results from Chambers' critique of the notion of poverty within the development discourse (1989). He emphasises that the concept of poverty has degenerated to a catchall term, concealing social differences, and perpetuating stereotypes about the poor. Chambers rather differentiates between several dimensions of deprivation. Aside from poverty, he enumerates isolation, powerlessness as well as vulnerability. In so doing, Chambers highlights that vulnerability is not the same as poverty. It means not lack or want, but defencelessness, insecurity, and exposure to risk, shocks and stress (1989: 1). He explains: 'Vulnerability here refers to exposure to contingencies and stress, and the difficulty in coping with them. Vulnerability has thus two sides: an external side of risks, shocks, and stress to which an individual or household is subject; and an internal side which is defencelessness, meaning a lack of means to cope without damaging loss (1989: 1).' Hence, the group of the poorest may not correspond with the most vulnerable and vice versa (Bankoff et al. 2004, Watts and Bohle 1993).

The livelihood approach associated with vulnerability research first appeared in the early 1990s and like the vulnerability concept mainly originates from the work of the Chambers-School at the Institute of Development Studies (IDS) (Chambers and Conway 1991). The concept treats livelihood as the 'capabilities, assets (including both material and social resources) and activities required for a means of living' (Carney 1998). Disposable assets from different kinds of capital (natural, financial, social and human capital (DFID 2000)) are a major explanatory dimension in how sustainable a livelihood finally is (Devereux et al. 2011, Ellis 2000, Scoones 1998). Gertel's critique (2007) stresses the theoretical weaknesses of the concept. He notes the absence of any insights into the causes of the risk generation, the reasons for economic inequality, and its limited analytical scope on economic possibilities. This understanding reduces livelihoods conceptually and empirically to the monetary level. Other approaches, such as Giddens' rule-resource-complexes and Bourdieu's concept of capital offer a way out of the utilitarian input emphasis of conventional livelihood writing. A notion of resources as capability of doing things allows linkages to biology (e.g. bodies ultimately die), to labour (e.g. accumulated income from the past can be transferred into present livelihoods, enabling access to education) and to property rights and thus to the discursive construction of rules (e.g. encoded in the discourse about private ownership vis-à-vis the global commons). The concept of resources offers the possibility of linking the notion of uneven development with the status of the physical body of a person, and it also provides insights into the social causes of reproduction problems and thus opens a new perspective to understand the social dimension of economic spaces.

The Book

The preceding discussion raises questions about the constitution of economic spaces of pastoral production and commodity systems, and about the vulnerability and resilience of the economic practices of pastoralists and their communities. As already indicated the contributors largely draw on two bodies of knowledge: the agri-food perspective of global commodity and value chains, focusing on the movement of pastoral products; and the livelihood perspective with its insights into vulnerability and risks for households. While markets are comprehended as social and political institutions that are crosscutting borders and boundaries, the empirical studies – from post-colonial Africa, post-socialist Asia, and a range of capitalist economies – represent various exchange systems and different frames of state-market relations. This framing reveals three principal dynamics, namely transformations from subsistence to market production, transitions from state to market production, and integrative relations in and of markets. However, the spatial situatedness of these dynamics and their implied evolutionary causalities is sometimes challenged and sometimes affirmed by the empirical findings. The mobility of livestock related commodities and market information is expanding. Commodity and value chains are stretching over wide distances, bridging national borders and not only foster the transformation of market dynamics interpenetrating the chain but also are linking producer and consumer constantly anew. These are loaded distinctions. In the 2010s we are trying not to read the dynamics as suggestive of progressive directionality, always away from subsistence relations and always towards market relations. Instead, we critically privilege the descriptions and analytics of the researchers and the voices of the pastoralists, whose experiences in the field and in life, suggest much indeterminancy, and much evidence of connection and disconnection, to market and livelihood relations. We finally invite the reader to situate herself or himself in this debate and, more particularly to reflect, if our arguments should be restricted to pastoral economies or rather extended to the wider society. Assuming that this kind of framing requires context the book starts with provocations about markets and livelihoods.

Pastoral Production and Commodity Systems: Provocations About Markets and Livelihoods

Barbara Harriss-White (Chapter 2) criticises conventional economic approaches to the conceptualisation of markets. She explores different perspectives – from economic sociology, political economy, social structures of accumulation to the focus on commoditisation – and stresses the crucial importance of a theoretical plurality to study markets. She argues that markets have to be studied specifically rather than generally as is conventional. The chapter challenges the simplistic concepts of supply and demand that continue to shape and fuel market discourses. In contrast, it rather retains a focus on processes, relationships and interactions. Harris-White explores how economic markets are vehicles for the exercise of forms of social authority and arenas for struggles between political interests. She considers markets as bundles of institutions and as fields of accumulation. Her critical overview provides framings of engagement with economic processes.

Taking up a position very much dedicated to the specificities of markets and livelihood systems and viewing how increasing market integration entails many new risks to pastoral systems and households, Hans-Georg Bohle (Chapter 3) develops a framework for the assessment of market risks and vulnerabilities. He identifies key elements of vulnerability – stress, exposure, sensitivity, response and outcome – and analyses their operationalisation in the sustainable livelihood approach. Combining these insights with the concept of real markets, seeing vulnerability of households in the contexts of socio-political structures and power-relations, Bohle's integrative framework maps not only market risks, but also the economic, social, and institutional spaces of pastoral production and commodity systems. His chapter hence provides an analytical tool for the discussion of the problems and perspectives that pastoralists face in the process of market integration, both in terms of their social vulnerability and human security.

Post-Colonial Africa: From Subsistence to Market Production

Nomadic people in Africa have been trading and marketing animals for a very long time. Economic spaces for pastoral products were shaped by various phases and missions of European colonialism, by colonial boundaries constructions, by inventing, naming, and delineating tribes and their territories in respect to tax collection and for means of physical control, and also by land use policies, infrastructural developments and various direct interventions and regulations including diseases and price controls. In some cases colonial powers actively hampered pastoral trade, while in others they were effectively promoting pastoral efforts to market their stock. For cases in Kenya, Nigeria and Niger, Kerven (1992) argues in her historical reassessment of pastoral livestock markets that pastoralists are price-responsive, attuned to market fluctuations and not inclined to withhold saleable animals from the market. She concludes that government interventions such as price controls, licensing of traders, and implementation of taxes can seriously hamper the efforts of pastoralists to market their livestock (Kerven 1992). While European colonialism incorporated and imprinted Africa largely in the image of the British and French empires, political independence of the newly emerging nation states in the twentieth century did not include an independent economic restart. Export markets and commodity chains originating in Africa stayed oriented towards France and Great Britain, while local and regional livestock markets prevailed.

Until recently there have been limited opportunities for smallholder livestock producers in Africa to access international markets. Major market opportunities for livestock products in most countries of the continent unfold at the domestic and regional levels. Five different market chains can be distinguished: First, subsistence production and self consumption of animal products within households and local communities, where almost no price building takes place. Second, regional, and national trade where the products are sold in nearby villages and cities, increasingly for cash; they often work as collection centres, both to provision local (urban) demand and to build up a surplus to be marketed beyond the region (see Wiese, and Komey chapters). Third, cross-border trade; it is predominately oriented towards larger urban centres like Nairobi, Abidjan or Casablanca (see Mahmoud, Zaal, Breuer and Kreuer, in this volume). Here, various intermediate traders are involved and price building starts to be complex, and detached from a single market place. Fourth, live exports of animals to the Arab Gulf countries, a trade that Somalia and Sudan (El Dirani et al. 2009) are particularly involved in. Fifth, industrial market systems of vertically integrated chains; they are linking production, processing and the movement of pastoral products from the producer to the consumer (Behnke 2008). This often implies crossing national and continental borders to provide consistent year round supply, and to meet international quality and safety standards (Perry and Dijkman 2010).

Regional differences are inscribed into the distribution of these market chains between the pastoral economies in North and East Africa, and in southern Africa (Behnke 2008, Dutilly-Diane 2007, Kocho et al. 2011, McPeak and Little 2006). The northern countries are often poor, at least until recently, before large scale oil exploitation started in Sudan and Chad. They are composed of weak states and are run by governments that aim to control nomadic people (Casciarri 2009). This situation is different in southern Africa, where governments support or even may subsidise livestock producers. Comparing traditional pastoral production systems and ranching in Africa Scoones (1995) reveals, however, that the value of communal area cattle production by far exceeds returns from ranching; in Zimbabwe ten times and in Botswana three times. For countries like Mozambique, South Africa, Tanzania, Uganda, Ethiopia and Mali traditional production systems are comparable with commercial herds but do also offer multi product outputs and hence provide higher overall returns (Davies and Hatfield 2007).

The following chapters focus on the situation in north and east Africa. In contrast to Asia, where the marketing of wool and fibre products are of crucial importance, African pastoral economies are largely linked via meat markets to the wider economy. In response to the emergence of globalising markets and the rapid transformation of commodity chains, the African livestock sector is undergoing profound changes. The social and environmental consequences are important since small-scale livestock producers especially in Sub-Saharan Africa are increasingly marginalised. However, pastoral livelihood systems offer the drylands, as Martin Wiese (Chapter 4) argues, a considerable and potentially sustainable form of production. In Chad, pastoral livestock production remains an important pillar of the national economy even after the start of oil exportation in 2003. This is remarkable since the facilities for livestock export from Chad are rudimentary. Pastoral societies, like the Dazadaga and Juhayna, have shown a remarkable resilience during the latest pan-Sahelian droughts and the civil war in Chad. Nevertheless, they face an increasing encroachment of day-to-day livelihood security. Not market failures but – as Wiese argues – political crises are threatening these pastoral livelihoods.

The destructive power of the state vis-à-vis pastoral livelihood systems is also discernible in Guma Komey's (Chapter 5) analysis of the driving forces, the function and the spatial pattern of the war-born markets in the context of the pastoral-sedentary relations in the Nuba Mountains, Sudan. Komey examines the survival strategies and coping mechanisms deployed by the involved parties in a situation characterised by extensive insecurity and mounting risk. The arrangement of new forms of local markets (informal and smuggler markets) by the pastoral Baqqāra and the sedentary Nuba is discussed as part of the limited choices of survival strategies in response to the war situation. In the externally induced and state driven civil war, he argues, market relations operate as a starting point in a wider set of trade chains that systematically subjugate local economies to national and global markets. This implies that the multifaceted dynamics taking place in local markets across the different social fields are not merely local dynamics. Rather, they are, in most cases, local manifestations of national and globalisation processes with all their social, political and economic dimensions.

Inclusion of traditional pastoral livestock systems in globalising meat markets has progressed to the point that functional specialisation has developed in various pastoral livelihood systems within pastoral society. Though traditional pastoral production still exists successfully, commercial capital-based and labour intensive systems have developed, including an intricate division of labour, linked to a flexible and hierarchical, both culture-based and commercial, international meat value chain. Fred Zaal (Chapter 6) investigates traditional pastoral livestock systems and the history of trans-border livestock trade in North Tanzania and southern Kenya. In the analysis of the impact of developing meat value chains directed towards markets in urban areas like Nairobi, five major household strategies are identified and the solidification of social and economic inequality is observed; more successful households directly participate in the new commercialised value chain, and less successful households are indirectly involved through the commoditisation of their labour and other resources.

Targeting a central market in Nairobi as in the study above, Hussein Mahmoud (Chapter 7) examines strategies of risk management among North Kenyan cattle traders. Facing extremely high risks from political conflicts, violence, credit insecurity and degenerated infrastructure, this chapter examines innovations in livestock marketing. Livestock trading in northern Kenya is one of the toughest and most risk-prone jobs in the region, yet successful livestock traders have not only been able to transform the ways in which trading is conducted, but have also become resilient to risks. Disconnecting the market chain of trading live animals from the circuits of capital is revealed as the core mechanism, but it is trust embedded in social networks that allows for the adoption of risk minimising strategies.

The first section of the book ends by assessing the impact of neoliberal policies and the shifting market spaces within a peripheral pastoral region in eastern Morocco. Ingo Breuer and David Kreuer (Chapter 8) investigate how the integration of the local population into international markets and commodity chains affect their livelihoods that, until recently, depended almost entirely on pastoral livestock production. They argue that both meat and fodder markets are undergoing profound changes, with severe consequences for eastern Moroccan pastoralists. Free trade schemes and the implementation of the plan for a 'Green Morocco', designed by international consultancy McKinsey for the Moroccan government, are transforming agricultural production systems, threatening local market chains and will largely undermine local livelihoods. Hence, new market spaces will expand the social spaces of insecurity.

Post-Socialist Asia: From State to Market Production

Although pastoral systems in Central Asia present a rather heterogeneous picture with their geographical environments, varying social contexts, and differing political development trajectories in the last two decades, they still share a number of characteristics. Most of these pastoral production systems had been integrated into the centralised apparatus of socialist economic production in the Soviet Union or in communist China. Especially in the Soviet Union, livestock production was collectivised and converted to a state-organised high-input and heavy-capital industry. Traditional nomadic pastoralists were often forced into a settled existence in large-scale production units with only limited seasonal movement, while the provision of highly subsidised inputs, the processing of produce and the controlling of marketing channels encompassing an entire continent were state responsibilities (Kerven 2006).

After the collapse of the Soviet Union, the loss of state subsidies, the breakaway of established distribution channels, agricultural reform and privatisation measures profoundly changed the conditions of pastoralist production. The degree and speed of restructuring and privatisation has been different in the newly emerging countries, with high-speed reformers like Kazakhstan and Kyrgyzstan experiencing a dramatic decline of livestock and devastating ramifications especially for pastoralists in peripheral areas. Turkmenistan, proceeding on a slower pace of privatisation and restructuring, has been able to preserve some degree of stability (Kerven 2003). Even for the Soviet satellite state Mongolia, which was not fully incorporated into centralist Soviet economy, and not as dependent on external inputs, the breakdown of the Soviet Union and the loss of Soviet meat markets still had severe consequences (Janzen and Enkhtuvshin 2008, Marin 2008). China and some other countries, in contrast, were able to avoid such devastating effects, having started gradual agricultural reform at an earlier point and having greater flexibility in their agricultural production forms and scales (for Afghanistan see Callahan 2007, Kreutzmann 2007, de Weijer 2007; for Siberia see Habeck 2005, Intigrinova 2010, Stammler 2005; for Tibet see Gruschke 2009).

In most cases, national processing facilities were not maintained under privatisation, and while exports were increasingly shifting to raw products, institutional barriers as well as quality impingements continually hindered the exploration of new meat markets. In particular, remote mountainous regions are often practically cut off from market access, even though a certain amount of informal or even illegal cross-border trade can be assumed. Exports of livestock produce from post-socialist countries are now dominated by wool, leather and other non-meat products (these have always played a much greater role in Asia than in Africa). Even the wool market, however, has been depressed since the 1990s, especially for non-processed, low quality wool. In recent years the production of cashmere has proven quite lucrative and increased in several Central Asian countries. Other groups have been exceptionally successful with certain niche products, as is the case with ingredients for traditional Chinese medicine such as reindeer velvet antler or caterpillar fungus discussed in case studies in this volume.

Hermann Kreutzmann (Chapter 9) contributes an historical perspective on pastoral life in peripheral mountain regions of Central Asia. Geo-political interferences, boundary-making, socio-economic reforms and revolutionary movements had strong and for the most part detrimental effects on the livelihoods of nomadic people. Kreutzmann demonstrates how territories in contemporary countries like Afghanistan, Pakistan, Tajikistan, and the People's Republic of China have been divided since the late nineteenth century by international boundaries as the result of an imperial 'Great Game', resulting in a varied spectrum of legislation, infrastructure development and regional planning. These dividing lines disrupted traditional migratory paths of seasonal nomads and created an arena of continuous contestation in the Pamirs, the Hindukush and the Himalayas, while the economic value of these mountain areas as a transit region for herders and traders was lost. As Kreutzmann reveals, Soviet modernisation projects further dislodged pastoral life through forced settlement of migratory nomads. After the end of the Cold War, however, market-oriented reforms are aiming to replace collective strategies and privatising collective production, while some cross-border trade is revitalised by pastoralists.

Jörg Janzen (Chapter 10) explores the upheaval of the export-oriented livestock production system in Mongolia after the end of the Soviet Union. After the privatisation of cooperative production, the loss of state subsidies, and the collapse of all marketing facilities, the economy for livestock export production has been reduced to a subsistence-oriented system, fostering large scale social differentiation within the pastoral population. With the destruction of most processing factories, Mongolia's export industry has been reduced to exporting primarily raw materials such as wool and hide. Combined with overstocking, accelerating desertification, and the impact of severe climatic conditions this resulted in a massive loss of animals and the deterioration of production and living conditions for mobile livestock keepers. Impoverished pastoralists are increasingly forced to give up herding and populate new yurt quarters on the outskirts of Ulaanbaatar. These pressures are forcing pastoralists to reshape their strategies and negotiate a new presence in the Asian markets. Today the sale of cashmere hair from goats is the major source of income for Mongolian herders.

Andreas Gruschke's study (Chapter 11) of pastoral livestock keeping in the Yushu Region on the Tibetan highlands is an example of successful integration into the world market for traditional Chinese medicine. Even though pastoral livestock keeping is under increasing pressure from deteriorating markets for livestock and decreasing pasture areas, the larger part of Yushu is nevertheless a region where pastoral livestock keeping persists. Livestock, however, is not necessarily the income source for these households. A niche economy built on a single resource allows integration into the world market, capitalising on a unique product: the caterpillar fungus. The marketing of this fungus makes up 50-80 per cent of households' cash income. Gruschke argues that this relationship allows many Tibetan pastoralists to subsist with their livestock rather than give up animal husbandry, especially after local pastoralists were recently able to assert greater participation in the marketing chain of this commodity.

Another product targeting the world market for traditional Chinese medicine is presented in Florian Stammler's contribution (Chapter 12). Using the velvet antler trade between nomads and Far Eastern businesspeople as an example, he analyses how tundra people welcome international trade. Their involvement in a rather recent economic activity shows how they cleverly manoeuvre to make the most of opportunities for diversified income. Nenets nomads make explicit distinctions between different spheres of production, as a result of which the nature of money becomes personalised: the source of income determines the sphere of spending. The argument is, in this and other cases, that involvement in the global economy does not have to replace but rather supplements a solid basis of traditional subsistence or domestic market production.

Industrialised Commodity Systems: From Commercialised Production to Integrated Markets

Based on the commitments of the Millennium Development Goals to eradicate extreme poverty and hunger, the Food and Agricultural Organization (FAO) launched the Pro-Poor Livestock Policy Initiative in 2001. In this wider frame of western development policy Nikola Rass (Chapter 13) analyses the impact of drought on markets and the associated response of pastoralists to the risk of drought. As this chapter largely draws on cases from Africa it could also have been placed in part one of the book, but on the other hand it tackles western instruments of interventions. Rass outlines different income generating strategies and increasingly commercialised production objectives of pastoralists in order to understand the pastoralists' response to markets. There are several reasons pastoralists often do not follow market incentives to sell livestock, but rather follow a long-term strategy of herd expansion. Droughts often have extremely disruptive effects on markets, while at the same time commercialisation uproots traditional household strategies. The adaptation of traditional strategies to new contexts is discussed as a desirable goal. In this light, Rass debates opportunities of the modern techniques of Early Warning Systems and evaluates examples of market-focused interventions.

Chris Lloyd (Chapter 14) examines the UK sheep industry that is mainly focused on meat production and, with 16 million breeding ewes, is the largest producer in Europe. A great variety of climatic and natural conditions in the UK has contributed to the historical development of a large stratified breeding structure, linking farms from the mountains to the more productive lowlands and exploiting the beneficial production characteristics of the local breeds and hybrid vigour. The UK is also one of the largest markets for pastoral products and, because of seasonality, a main importer for lamb, mainly from counter-seasonal New Zealand. Cheap market prices as well as changing consumer demands in recent years have been challenging the UK sheep industry. Changes in consumer preferences and eating habits, societal changes and new health concerns have led sheep producers to devise new brands and new marketing strategies, boasting health benefits of leaner meat and ecological soundness.

Under neo-liberalising reforms in New Zealand the country's pastoral sectors have undergone major transformations. New Zealand's pastoral commodity systems are now re-connected into the globalising world food and fibre economy on distinctively different bases to the arrangements that prevailed some two decades ago. Richard Le Heron (Chapter 15) articulates a trajectory of economic and institutional investment approach to situate New Zealand-based pastoral activities – dairy and sheep (meat and wool) production – in the wider (and changing) international political economy of GATT, WTO and other market-forming and market-shaping institutions and to outline the re-alignment of New

Zealand's pastoral commodity systems into world markets. The chapter explores changing enterprise and organisational relations to reveal connections between market-making relations and livelihood implications in the New Zealand context.

Around 3.5 million Australian sheep are exported to the Middle East annually, representing the largest transfer of live animals between continents. It has also been a relationship fraught with multiple and markedly different perceptions about and interpretations of sheep, in their life and death. Jörg Gertel, Erena Le Heron and Richard Le Heron (Chapter 16) juxtapose the dynamics of the trade industry and its relation to the regional economy in Western Australia (with an economic value of about AUD$ 323 million in export earnings in 2009) with the arguments from animal welfare organisations (and their valued-concerns about the cruelty of this trade), and finally with the cultural practices in the Gulf countries (about the religious notions of fresh sheep meat consumption). They argue that there are not only economic values but cultural and social values attached to sheep, that enter into the making and unmaking of market relations, and in the linking of livelihoods across continents.

The last chapter reviews the book's achievements. This chapter considers how different and new styles of situated narratives about engaging in exchange relations might be developed and put into circulation in policy, aid, community, and business circles. Richard Le Heron and Jörg Gertel (Chapter 17) suggest such narratives are likely to be more compelling if they are embodied accounts of real risks encountered and lived in fields of exchange relations, and explicitly recognise socio-economic spaces, in which they are constituted and experienced.

Conclusion

This book thus brings together experiences and information ranging from integrated industries to those of subsistence producers. We reveal that integration into international commodity chains and distant markets entails risks, often generated far away from the sphere of production, while the consequences are always reflected in local livelihoods. This holds especially true for marginalised pastoralists, who enter market and commodity chains in relatively powerless positions. State regulations and war or recent boundary-making can have devastating effects on pastoral trade routes, markets, and livelihoods. Jumps in energy prices also expose farmers and pastoralists, who depend on selling their products in distant markets. These problems are exacerbated in areas where insecurity (be it in the form of cattle theft, contested land rights or civil war) and increasing environmental degradation and droughts pose constant threats.

Of great interest are the innovative solutions pastoralists apply in order to cope with risks, to adapt to changing conditions and, in some cases, even to turn them to their advantage. The book's evidence documents heartening strategies. In civil war Sudan, warring tribes sustained their interdependent livelihoods through the establishment of secret illegal 'peace or smuggler markets'. In northern Kenya, cell phones facilitate informal cash transfer systems. And in Siberia, helicopters transporting supplies are turned into an economically viable means for smuggling large quantities of velvet antlers, a substance important in Chinese medicine, to bigger cities to be sold to Far Eastern traders. The ingenuity of Tibetan pastoralists, making use of a caterpillar fungus, another valuable ingredient for Chinese medicine, that grows exclusively on their lands and is experiencing high demand in the last years, also refutes the generalising assumption that integration into global markets necessarily has detrimental effects on traditional pastoralist livelihoods. Similarly Siberian reindeer herders only participate in the global trade with velvet antlers in order to generate cash for the purchase of luxury articles, while continuing to maintain traditional – and sustainable – pastoralist lives.

Finally, we contend the book's exploration of pastoralism and pastoral economies allows some preliminary re-imaginings of research questions and research methodologies. In particular the book's positioning work, that is, its empirical exposure of the world's pastoralism and pastoral economy experiences, facilitates a different level of engagement with the formative relationships and interactions of market-making and livelihood-making.

Acknowledgement

Erena Le Heron provided extremely helpful suggestions relating to the ideas and messages of this introductory chapter.

References

ACIL Tasman. 2009. *The Value of Live Sheep Exports from Western Australia.* [Online]. Available at: http://www.rspca.org.au/assets/files/Campaigns/ACIL Tasman/ACILTasmanValueofLiveSheepExports2009.pdf [accessed: 26 March 2011].

Ausdal, S. van. 2009. Pasture, profit, and power: An environmental history of cattle ranching in Colombia, 1850-1950. *Geoforum*, 40(5), 707-719.

Bankoff, G., Frerks, G. and Hilhorst, D. (eds). 2004. *Mapping Vulnerability. Disasters, Development and People.* Sterling: Earthscan.

Behnke, R.H. 2008. The economic contribution of pastoralism: Case studies from the Horn of Africa and southern Africa. *Nomadic Peoples*, 12(1), 45-79.

Breuer, I. 2005. Statistiken oder wie werden 'Nomaden' in Marokko gemacht? in *Methoden als Aspekte der Wissenskonstruktion. Fallstudien zur Nomadismusforschung*, edited by J. Gertel. [Online]. Available at: http://www.nomadsed.de/fileadmin/user_upload/redakteure/Dateien_Publikationen/Mitteilungen_des_SFB/owh8breuer.pdf [accessed: 29 March 2011].

Brown, R. 2009. *Global Developments in the Red Meat Markets and Industries (Beef Sector Emphasis).* (AMIC Meat Industry Conference, Sept. 25th 2009,

Girafood). [Online]. Available at: http://amicconference.com/Richard_Brown. pdf [accessed: 14 June 2010].

Bruun, O. 2006. *Precious Steppe: Mongolian Nomadic Pastoralists in Pursuit of the Market.* Lanham: Lexington Books.

Burch, D. and Lawrence, G. 2009. Towards a third food regime: Behind the transformation. *Agriculture and Human Values*, 26, 267-279.

Callahan, T. 2007. The Kyrgyz of the Afghan Pamir ride on. *Nomadic Peoples*, 11(1), 39-48.

Campbell, H. and Dixon, J. 2009. Introduction to the special symposium: Reflecting on twenty years of the food regimes approach to agri-food studies. *Agriculture and Human Values*, 26(4), 261-264.

Carney, D. (ed.). 1998. *Sustainable Rural Livelihoods. What Contribution Can We Make?* London: Department for International Development.

Casciarri, B. 2009. Between market logic and communal practices: Pastoral nomad groups and globalization in contemporary Sudan (Case studies from central and western Sudan). *Nomadic Peoples*, 13(1), 69-91.

Chambers, R. 1989. Vulnerability, Coping and Policy. *IDS Bulletin*, 20(2), 1-7.

Chambers, R. and Conway, G. 1991. *Sustainable Rural Livelihoods: Practical Concepts for the 21st Century.* (IDS Discussion Papers, 298). Brighton: Institute for Development Studies.

Chang, C. and Koster, H.A. (eds). 1994. *Pastoralists at the Periphery. Herders in a Capitalist World.* Tucson: University of Arizona Press.

Davies, J. and Hatfield, H. 2007. The economics of mobile pastoralism. A global summary. *Nomadic Peoples*, 11(1), 91-116.

Devereux, S., Sabates-Wheeler, R. and Longhurst, R. (eds). 2011. *Seasonality, Rural Livelihoods and Development* (with a Foreword by Robert Chambers). London: Earthscan, forthcoming.

DFID (Department for International Development). 2000. *Sustainable Livelihoods Guidance Sheets.* [Online]. Available at: http://training.itcilo.it/decentwork/staffconf2002/presentations/SLA%20Guidance%20notes%20Section%202. pdf [accessed: 18 April 2011].

Dutilly-Diane, C. 2007. Pastoral economics and marketing in North Africa: A literature review. *Nomadic Peoples*, 11(1), 69-90.

El Dirani, O.H., Jabbar, M.A. and Babiker, I.B. 2009. *Constraints in the Market Chains for Export of Sudanese Sheep and Sheep Meat to the Middle East.* Nairobi: ILRI.

Ellis, F. 2000. *Rural Livelihoods and Diversity in Developing Countries.* Oxford: Oxford University Press.

Evers, H.-D., Claus, W. and Wong, D. 1984. Subsistence reproduction: A framework for analysis, in *Households and the World Economy*, edited by J. Simth et al. (eds). Beverley Hills: Sage Publications, 23-36.

Fold, N. and Pritchard, B. (eds). 2005. *Cross-Continental Food Chains.* London: Routledge.

Friedmann, H. 2005. From colonialism to green capitalism: Social movements and emergence of food regimes, in *New Directions in the Sociology of Global Development.* (Research in Sociology and Development, 11), edited by F. Buttel and P. McMichael. Oxford: Elsevier, 227-264.

Friedmann, H. and McMichael, P. 1989. Agriculture and the state system: The rise and decline of national agricultures, 1870 to the present. *Sociologia Ruralis,* 29, 93-117.

Gereffi, G., Humphrey, J. and Sturgeon, T. 2005. The governance of global value chains. *Review of International Political Economy,* 12, 78-104.

Gertel, J. 2007. Mobility and insecurity. The significance of resources, in *Pastoral Morocco,* edited by J. Gertel and I. Breuer. Wiesbaden: Reichert, 11-30.

Gibson-Graham, J.K. 2008. Diverse economies. Performative practices for 'Other Worlds'. *Progress in Human Geography,* 32(5), 613-632.

Grandia, L. 2009. Raw hides: Hegemony and cattle in Guatemala's northern lowlands. *Geoforum,* 40(5), 720-731.

Granovetter, M. 1985. Economic action and social structure: The problem of embeddedness. *American Journal of Sociology,* 91, 481-510.

Grunsven, L. van. 2009. Global commodity chains, in *International Encyclopedia of Human Geography,* edited by R. Kitchin and N. Thrift. Amsterdam: Elsevier, 539-547.

Gruschke, A. 2009. *Ressourcennutzung und nomadische Existenzsicherung im Umbruch. Die osttibetische Region Yushu (Qingahi, VR China).* PhD Thesis (unpublished), University of Leipzig.

Habeck, O. 2005. *What It Means to Be a Herdsman: The Practice and Image of Reindeer Husbandry Among the Komi of Northern Russia.* Münster: LIT.

Hassler, M. 2009. Commodity chains, in *International Encyclopedia of Human Geography,* edited by R. Kitchin and N. Thrift. Amsterdam: Elsevier, 202-208.

Hughes, A. and Reimer, S. (eds). 2004. *Geographies of Commodity Chains.* London: Routledge.

Intigrinova, T. 2010. Social inequality and risk mitigation in the era of private land: Siberian pastoralists and land use change. *Pastoralism,* 1(2), 178-197.

Janzen, J. and Enkhtuvshin, B. (eds). 2008. *Present State and Perspectives of Nomadism in a Globalizing World.* Ulaanbaatar: Admon Press.

Kerven, C. 1992. *Customary Commerce. A Historical Reassessment of Pastoral Livestock Marketing in Africa.* London: ODI.

Kerven, C. (ed.). 2003. *Prospects for Pastoralism in Kazakstan and Turkmenistan. From State Farms to Private Flocks.* London: Routledge.

Kerven, C. 2006. *Review of the Literature on Pastoral Economics and Marketing: Central Asian, China, Mongolia and Siberia.* (Report Prepared for the World Initiative for Sustainable Pastoralism). Odessa Centre Ltd., UK: IUCN EARO.

Kocho, T., Abebe, G., Tegegne, A. and Gebremedhin, B. 2011. *Marketing Value-Chain of Smallholder Sheep and Goats in Crop-Livestock mixed farming System of Alaba, Southern Ethiopia.* (Small Ruminant Research 2011). [Online: ILRI].

Available at: http://dx.doi.org/10.1016/j.smallrumres.2011.01.008 [accessed: 29 March 2011].

Kreutzmann, H. 2007. Afghanistan and the opium world market – Poppy production and trade. *Iranian Studies*, 40(5), 605-621.

Le Heron, R., Penny, G., Paine, M., Sheath, G., Pedersen, J. and Botha, N. 2001. Global supply chains and networking: A critical perspective on learning challenges in the New Zealand dairy and meat commodity chains. *Journal of Economic Geography*, 1, 439-456.

Leslie, D. and Reimer, S. 1999. Spatializing commodity chains. *Progress in Human Geography*, 23(3), 401-420.

Marin, A. 2008. Between cash cows and golden calves: Adaptations of Mongolian pastoralism in the 'Age of the Market'. *Nomadic Peoples*, 12(2), 75-101.

Massey, D. 1999. *Power-Geometries and the Politics of Space-Time. Hettner-Lecture 1998.* Heidelberg: Department of Geography.

McMichael, P. 2005. Corporate development and the corporate food regime, in *New Directions in the Sociology of Global Development.* (Research in Rural Sociology and Development, Vol. 11), edited by F. Buttel and P. McMichael. Amsterdam: Elsevier, 265-299.

McMichael, P. 2009. A food regimes analysis of the 'world food crisis'. *Agriculture and Human Values*, 26(4), 281-295.

McPeak, J.G. and Little, P.D. (eds). 2006. *Pastoral Livestock Marketing in Eastern Africa. Research and Policy Challenges.* Bourton: ITGD Press.

Mitchell, T. 2008. Rethinking economy. *Geoforum*, 39, 1116-1121.

Perry, B. and Dijkman, J. 2010. *Livestock Market Access and Poverty Reduction in Africa: The Trade Standards Enigma.* (PPLPI Working Paper No. 49). Rome: FAO.

Rass, N. 2007. *Policies and Strategies to address the Vulnerability of Pastoralists in Sub-Saharan Africa.* (PPLPI Working Paper No. 37). Rome: FAO.

Schlee, G. (ed.). 2004. *Ethnizität und Markt. Zur ethnischen Struktur von Viehmärkten in Westafrika.* Köln: Rüdiger Köpke Verlag.

Sconnes, I. 1995. *Living with Uncertainty. New Directions in Pastoral Development in Africa.* London: Intermediate Technology Publications.

Scoones, I. 1998. *Sustainable Rural Livelihoods. A Framework for Analysis.* (IDS Working Paper, 72). Brighton: Institute for Development Studies.

Stammler, F. 2005. *Reindeer Nomads Meet the Market. Culture, Property and Globalisation at the 'End of the Land'.* Berlin: Lit Verlag.

Walker, R., Browder, J., Arima, E., Simmons, C., Pereira, R., Caldas, M., Shirota, R. and Zen, S. de. 2009. Ranching and the new global range: Amazônia in the 21st century. *Geoforum*, 40(5), 732-745.

Watts, M.J. and Bohle, H.-G. 1993. The space of vulnerability. The causal structure of hunger and famine. *Progress in Human Geography*, 17, 43-69.

Weijer, F. de. 2007. Afghanistan's Kuchi pastoralists: Change and adaptation. *Nomadic Peoples*, 11(1), 9-37.

Westreicher, C.A., Mérega, J.L. and Palmili, G. 2007. The economics of pastoralism: Study on current practices in South America. *Nomadic Peoples*, 11(2), 87-105.

Zaal, F. 1999. *Pastoralism in a Global Age. Livestock Marketing and Pastoral Commercial Activities in Kenya and Burkina Faso*. Amsterdam: Thela Thesis.

Chapter 2

Theoretical Plurality in Markets Conceived as Social and Political Institutions

Barbara Harriss-White

There is no science in the study of markets unless their relevant institutions are incorporated (Penny 1985).

To conceive of economic phenomena as embedded is not to renounce theory, certainly not, it is to start theorising differently (Caille 1994).

Introduction

The small minority of students of economics who wonder how supply is supplied and demand demanded is asking a question to which the answer certainly cannot be found within the discipline – and a fully satisfactory answer may not be found at all. It is not that the tools of economics have not been used to address institutions. It is that economic markets are vehicles for the exercise of forms of social authority, the origins of which lie outside markets and which operate outside markets as well as inside them. In the same way, economic markets are one of the arenas for struggles between political interests. Markets do not perform 'subject to' institutions, they *are* bundles of institutions and are nested in others. In this essay, mindful of the fact that there can be no 'theory of everything', we will consider some of the ways in which markets have been 'theorised differently' as social and political constructs, and will consider the significance for their functioning of social forms of regulation and of continual political contestation.

Markets and Institutions

All theories are social constructs. All theories about the economy – not excluding neoclassical ones – embody assumptions about forms of social organisation. The default concept of the competitive market has long been known to embody strong and implausible assumptions about society, needing several means by which large numbers of buyers and sellers are equally endowed with (accurate) information and can transact rapidly in one (virtual) space. While some have concluded that economics is 'under-socialised', others have found that its depiction is socialised all right, but in ways inconsistent with that of real, observed economies.

How Neoclassical Economics Handles Real-Economy Institutions

Theories have come into being under particular economic and political conditions and perform political roles. In the last quarter of the twentieth century, neoclassical economics has used its assumptions of an historical methodological individualism and of rational, maximising behaviour to colonise three areas of society and politics – the household, the state and the firm (and the contracts it makes). In the new home economics (Folbre 1994, Haddad et al. 1997), the allocation of household labour between productive activity outside the home using land, labour and other markets and reproductive activity inside the home (for example the preparation of meals, cleaning or childcare, which are modelled as the production of 'z'-goods) is theorised as an efficient response to the existence of prices for the labour of the different genders, and to the *difference* between those prices, on markets outside the household, subject to constraints (such as a irreducible minimum of z-good production by one gender or a fixed leisure requirement for one or both of the one-generation parents). Whether one parent is a benevolent dictator or whether and how each individual's interests can be subordinated to a collectively agreed goal is a bone of contention in this approach.[1] In the new political economy (not so new, and unhistorically political) the state has been modelled as the sum of individual maximising behaviour by its agents, with parties or policies taking the place of goods, voters proxying for consumers and votes replacing currency.[2] In the new institutional economics (NIE), the (corporate) firm has been cast as a bundle of labour contracts, under continual renegotiation in the light of performance, the institutional boundaries to which are conditioned only by supervision and measurement costs (Hodgson 1988: 178). Further, since spot contracts are rare, especially in developing economies, a variety of non-spot contracts have been theorised (either by means of analysing their existence conditions or by means of pair-wise comparison and contrast) as efficient solutions to the problems of asymmetrical information and differentials in the costs of useful knowledge, of transactions costs (the costs of search, negotiation and enforcement of contracts) and of protection against opportunistic behaviour, moral hazard and adverse selection.[3] The rational solutions to coordination problems in economic landscapes in which such concepts and tools are deployed result in outcomes which cannot otherwise be explained in economics. The sharecropping contract, actually rather rare, has received such varied attention that it can be studied as a microcosm of the NIE approach to the micro-economics of development (Byres 1983, Majid 1994).

---

1   See the critical reviews in Folbre (1994), Haddad et al. (1997); plus the important debates in Jackson and Pearson (1998), Young et al. (1981).

2   Bates (1981) is paradigmatic for developing countries.

3   See Bardhan (1989), Basu (1994), Harriss, Hunter and Lewis (1995), Subramanian (1992).

Problems with this Approach

These acts of theoretical colonisation are applied to highly stylised households, states, firms and contracts, overlooking a larger problem for explanation which immediately confronts any student of 'actually existing' households, states, firms and markets in developing economies. This is the simultaneous co-existence of a great range of types of institution, of contracts (and of 'technologies') each of which has non-trivial internal variation.[4] This theoretical colonisation also isolates institutions and wrenches them from their contexts of property distribution and power.[5] Power is not necessary to explanations couched in transactions or information costs; at best it is reduced to 'opportunism' or 'self interest with guile' and then generally considered to be practised by employees or agents rather than by owners or principals. Allocative and/or contractual behaviour is assumed to have to be devised by agents who are accepted as being imperfectly informed, with limits to their cognitive competence and boundaries to their rationality. Such solutions are theorised as efficient, as minimising transactions costs, because transactions costs minimisation is their objective. Such reasoning is not only tautological, it also flies in the face of evidence both for the persistence of inefficient technologies and institutions (North 1990), and for the role of institutions in protecting people from a superabundance of information – rather than from its scarcity.

In fact, it is not possible to evaluate the efficiency of actually existing institutions because there are no actually existing de-institutionalised alternatives with which to compare them. Institutions are also inadequately theorised as a constraint on maximising behaviour. They are actually the way societies work – a society cannot exist without them – they are facilitating and enabling. While in the theory, factors driving change in institutions are exogenous and/ or assumed (typically these factors are changes in price, technology, population and/or 'subjective preference'), in practice institutions are continually changing. They evolve from within, through the action of individual agents (for a range of motivations), through the consequences of the unavoidable use and degradation of energy (Martinez-Alier 1987), and from the coercive and competitive creation and defence of wealth and property (Harriss-White 1996). Furthermore, the matrix of institutions through which the economy is regulated and controlled replicates itself and evolves over time in interaction with its components in ways that make time, space and society unique (Harriss-White 2003). Unfortunately, history is therefore essential to theory. The new home, political and institutional economics tend to dehistoricise markets and all the non-market institutions they touch. Their political role is, first, to reduce capitalism to the interplay of supply and demand,

---

4   See Patnaik (1994), for an attempt to theorised this variation in agricultural production. See Rogaly (1996), for the co-existing diversity of labour contracts and the range of variation within any given contract.

5   Bhaduri (1983) and Subramanian (1992), are rare attempts to theorise power.

second (subject to some specific qualifications), to reduce markets to prices and third, therefore, to naturalise markets.

In *Agricultural Markets from Theory to Practice*, I reviewed the use of frameworks derived from industrial organisation theory and from theories of value chains/*filieres* for the evaluation of the economic efficiency of markets (Harriss-White 1998). Although these approaches are concerned with assets structures, competitive conditions, price formation and performance, they do not allow for the non-economic structuring of markets. Alternative, critical, historical ways of conceiving of and analysing markets in developing economies recognise that markets are social and political phenomena. However, they are all also in some ways unsatisfactory. We will consider four here: the approaches of economic sociology, of the politics of markets, of social structures of accumulation and of commodification.

Economic Sociology

At the heart of this approach is the insight that the economy is not the sum of individual maximising motivation but is a socially embedded phenomenon. Karl Polanyi, one of the most influential exponents of economic sociology,[6] argued that there were three economic principles. These are reciprocity (where price is the product of custom), redistribution (where price is the product of command) and market exchange (where price is the product of supply and demand) (Polanyi 1957). Underdeveloped economies are regulated in various combinations by these three principles, in such a way that economic transactions cannot be understood outside their social relations. For instance, it is through relations of kinship or religion that occupation may be regulated;[7] it is through gender relations that the terms of participation – firm size, activity, credit, the divisions of task and labour relations – are negotiated or thrashed out.[8] The moral character of goods may affect the structure and performance of markets;[9] and the network may be the means whereby jobs are secured and commodities distributed.[10] In many societies, and until recently, market exchange has been literally marginal. Its sites – marketplaces – may have evolved from a healthy (but economically irrational) distance from the edge of a town; its practitioners may have been migrant, and/or stigmatised and/or have kept themselves socially separate in order to reduce the heavy social

---

6   The great founding father of which is Max Weber (1922 (1978)).

7   See van Ufford (1999), on Benin's cattle trade.

8   See Pujo (1997), for its role in Guinee's rice economy.

9   See Harriss-White (1996), for the moral status of garlic, onion and tobacco in the rural Indian economy.

10   See Granovetter (1995), for labour markets in the US; see Meagher (2003), for informal petty commodity production and marketing in southern Nigeria.

obligations that generally come with wealth.[11] As economies evolve, not only do social arrangements cede regulative authority to political, legal procedure (custom gives way to contract) but market exchange also comes to dominate other principles of economic regulation. Polanyi argued that society is thereby transformed to suit the interests of the self regulating market. The direction of causality reverses – from the economy being embedded in social arrangements to society being embedded in market-serving arrangements. Polanyi realised that, at its extreme, his argument gave rise to a contradiction. For markets are not only principles of allocation driven by supply and demand, they have destructive properties. Not only do they destroy other principles of economic allocation, but also by themselves they destroy human life because they cannot protect it. As Sen put it more recently, markets are consistent with any income distribution, including one where some have no income at all.[12] Markets are so inherently destructive that a pure market society (aka pure capitalism) cannot exist. Society and states intervene to protect markets from destroying society. In all markets, there are thus continual political tensions between more and less regulation, between social and state regulation and between market regulation and politically necessary socially protective redistribution.

So, it is not surprising that empirical work in the most *advanced* market economies has revealed that the most developed market exchange is far from being socially disembedded. Davis, in his study of markets in the UK, discovered 'at least' 80 different forms of non-market exchange – mixtures of price formation that did not owe themselves purely to the interaction of supply and demand (Davis 1992). In the UK for example, festive occasions or the rhythm of vehicle registration, not price reductions, are the reasons for huge surges in demand. This increase in demand is not reflected in systematic price hikes. Emler's work on social information (gossip) showed that 90 per cent of all corporate transactions are personalised.[13] Hodgson has stressed the foundational importance in complex systems of production of the routine (or the 'meme') (Hodgson 2002). Gender, age, educational history, region and class of origin, religion (if any) and participation in civic associations shape the structures of production, trade and consumption. Habituated behaviour does not necessarily minimise costs, even transactions costs; but it makes production possible. All this work confirms that forms of power originating outside the economy have enduring impacts on economic performance. Furthermore, the spheres of public service and of the production, distribution and consumption of public goods are not only celebrations of a variety of market-limiting values but are also part of the political shield protecting society from market-created casualties (Leys 2001).

It is one thing to accept that even modern markets are embedded and even to seize the nettle of defining these social institutions and to attempt to analyse and measure their impacts. It is quite another matter to recognise that exactly

---

11   Evers and Schrader (1994); see Clough (1995), for Hausa Nigeria.
12   Sen (1981), quoted in Mackintosh 1990.
13   Prof. Emler (1999), personal communication.

the same institutions have been theorised as examples of different economic principles. Moral-economic activity may be interpreted substantively in terms of norms of reciprocity, generosity, the expression of social rules about the siting of spatial arrangements, the sharing of work, cooperative access. But it is also open to formalist interpretation in terms of long term self-interest, individual risk-minimisation and insurance or solutions to problems of lack of trust (Granovetter and Swedberg 1992). For the substantivist economic sociologist, the economy is moulded by non-economic factors while for the formalist (for that school of economic sociology that is influenced by the new institutional economics) the economy is shaped by distributions of information, uncertainty and relative factor scarcities. What both approaches have in common is their reliance on norms, norms of social embeddedness in one and norms of individual self interest in the other.

The Politics of Markets

Markets are not just socially embedded phenomena, they are sites of the exercise of power in ways which are not reducible either to the tool kit of the new institutional economics or to norms of social embeddedness. Following Max Weber, market exchange is 'always the resolution of conflicts of interest' (Weber 1922 (1978)) and it is this pursuit of interest that we define as politics, rather than the narrow conception of party – or electoral – power. It is not only the variety of real world markets, it is also the repeated experience of adverse and apparently unintended consequences of that sort of development policy which uses a deinstitutionalised conception of markets that has required the development of a framework for the political analysis of markets. Basic markets in developing countries have been found not to function well and also to change the way they function according to the seasons, as well as in extreme circumstances.[14] Attempts to regulate them – through consensus, or through democratic governance, or through the narrow legal prescription of proper contract – have often foundered because the interests at play in markets are highly unequal (Harriss-White 1995b).

Market power, the capacity of one agent to direct the action of another – a capacity which is intrinsic to transactions – is situated in a structure of power which determines the choices available to participants: the choices between tactics and between objectives. What are these structures of power? There appear to be four:[15]

First, there is *state power*. State and market are not separate but in practice densely intertwined. In turn, the state's involvement in markets takes two forms. One is direct participation – a form of politics well known to have taken a severe battering

---

14   See Crow (2001), for seasonal changes in Bangladesh; see Cutler (1988), Keen (1994), and Ravallion (1987), for market behaviour in famine conditions in Ethiopia, Sudan and Bangladesh respectively; see Palaskas and Harriss-White (1993, 1996), for analysis of price behaviour over the medium term in North East and South East India.

15   This discussion owes much to White (1993).

from the Bretton Woods financial agencies and from market fundamentalists. Yet the state is bound to be an active player in any market which needs creating and protecting under conditions which deter private capital (for example in remote regions, or where marketed surplus is small and sporadic, or where sellers are unable to borrow money against their products); or when a combination of roles are required which the private sector cannot play (for example short term profit maximising trade together with the long term security and redistribution of food and essential commodities). The second kind of intervention involves the *political regulation* of markets. There are then several layers or kinds of political regulation. One is where the state exercises *parametric* power in markets in order to correct distortions or achieve developmental goals (Often this kind of politics materialises around *development projects* – in credit to producers or traders, in infrastructure, in the deliberate social targeting of certain products [especially grain], in production conditions [irrigation/land reclamation/tenurial reform] and so on). Another is where state power *pervades* markets through the policies which make markets work (the legal definition of property, sites of trade, licensing laws, the calibration of weights and measures, the regulation of money and contracts, the existence of institutions of adjudication and enforcement of these pervasive laws). Third, redolent of Foucault's notion of the capillary nature of power, states may *saturate* markets, 'even' determining the hours of sale, the description of contents, the proximity of goods to one another, the environmental quality of packaging, the display of brands and prices and so on. These examples are confined to the point of sale. The labelling on a Coke bottle makes the point, but the bureaucratic politics of state saturation behind this labelling is far less visible.

If the complex roles of markets are to be understood, if the effectiveness of markets is to be appreciated, then we have to understand this institutional patterning, because, as markets develop, these are exactly the kinds of regulation which transferred from custom to contract, from social institutions to legal ones. In this process the law becomes a political resource – the object of determined attempts at capture, of evasion, or of manipulation alongside customary norms in legally pluralist regulative systems (von Benda-Beckmann and van Meijl 1999, van Ufford 1999).

At the same time the regulators cannot be assumed only to regulate. They also protect their own interests and may themselves embody conflicts of interest.[16] The result is markets which are structured so as not to be socially neutral. The analyst must also be aware of differences between procedure and practice – of the existence of incomplete, inconsistent and/or inconsistently amended law, and of the varying scope for improvisation in the practice of regulation, such that regulation takes on a local character moulded by the interests of political and social elites (Bavinck 2003).

---

16 For a common example: when regulators discipline *and* represent *and* promote those regulated (Monbiot 2001).

The second dimension of the politics of markets is therefore that driven by *association*, by means of which (some) participants act collectively in their own interests, in ways which may be antagonistic to others. The resolution of conflicts of collective interest often leads to endogenous regulation. The latter takes several commonly observable forms, theorised as association, network and hierarchy. *Association* defines formal organisations (trades unions, trade associations and consumer groups). Sometimes lumped together as 'civil society', these groups have formed to create and protect rents and thus evade competition, also to guarantee the collective preconditions for market competition and/or sometimes paternalistically to represent or to control labour.[17] The developmental outcome of the politics of association will depend on the type of rent created (Khan and Jomo 2000). *Networks* are symbols for repeated interactions which counteract the working of competitive markets and are often portrayed theoretically as the manifestation of relations of trust (Castells 2000, Meagher 2003). Regarded as developmentally positive, networks are also exclusive and can operate as 'conspiracies against the public' (Smith 1995). A wide range of socio-economic relationships have been reduced by network theory to (layers of) nodes and flows. This is as gross a reductionism as that of the individualism of neo classical economics. The *hierarchy* is manifest in the firm. Far from being a consensual unit, an 'island of coordination in a sea of market relations' (Hodgson 1988) the firm is a governance structure bristling with micro-politics; 'a combat unit designed for doing battle in the market' (White 1993), with hierarchical controls to maintain internal discipline in ways which benefit the owners. Yet, highly dynamic and synergistic inter-relations have now been researched within and between clusters of firms in third world markets (Nadvi 1999, Schmitz and Nadvi 1999).

The third dimension of markets as political institutions concerns the politics of *economic structure*. Here, the distribution of endowments shapes the exchange between individual elements and affects the relative returns to market engagement (Bardhan 1991). In both advanced and developing economies industrial organisation analysis has revealed that the impacts on performance of complex competitive conditions (of which monopoly and competition are but the extremes) are inconclusive (Harriss 1979a, 1979b). In developing countries under conditions of non-voluntaristic commercialisation (induced by the need to repay debt or pay taxes), market exchange is better understood not in terms of allocative efficiency but rather as a mechanism of extraction of surplus by one class from another (Crow 2001). The function of exchange is not to clear the market but to gain advantage of producers, a role of markets which Bhaduri has termed their 'class efficiency' (Bhaduri 1986). Market transactions will then be an expression of the relative power of dominant and subordinate classes. In agriculture, for instance, interlocking and more complex (triadic) contracts in markets for labour, credit, raw materials, products, water and transport may be manipulated to give propertied

---

17   See Harriss-White (1993, 2003), for explorations of this type of market-driven politics in India.

classes the capacity to benefit from unequal sets of choices. Interlocked contracts have been used to replace the elements of non-contractual obligation in contracts under circumstances where they have been challenged.[18]

In the fourth dimension of the politics of markets, markets are arenas for the expression of forms of *social authority and status* derived outside the economy. The distinction between this dimension as conceived of here and as seen in economic sociology turns around the 'politics of markets' insistence on analysis not merely of social embeddedness but also of the exercise of social power. Ethnicity is one obvious form (see van Ufford and Zaal 2004, for its operation in African cattle markets). Patriarchy is another. Gender relations regulate market exchange through restrictions on task and work, by screening access to labour markets, through ideologies of subordination, through rules of market participation prejudicial to women (see Jackson and Pearson 1998, Pujo 1997, Robson 2002). As a result of the way markets are gendered, economic growth may be constrained, competition suppressed and the social piety of the women in the business family may affect the creditworthiness of the family business (Laidlaw 1995). Religious authority is another. Divine authority expressed in the economy can be the basis of the formation of occupationally specialised social groups. Such groups may supply the preconditions for competition (information, skilling, contacts, access to finance, collective insurance and even livelihood guarantees). They may also be the foundation of apparently secular corporatist regulative institutions. The existence of such groups can be explained in terms of the minimisation of information and transactions costs but the point is that such groups are never merely groups and their purpose never merely confined to the economy. Religion (or sect, denomination or *biradari*) can and do determine the spatial arrangement of residence and marketplace. In developing economies, laws deemed 'personal' or 'customary' are ways in which divine authority regulates property, individual rights to property and the distribution of property on inheritance, marriage or the partition of a business. Religion has also been found capable of defining the rules of transaction with co-religionists and to differentiate them from those with 'others' (Harriss-White 2003). Social power may also find fields of play affecting the behaviour of markets in the media, in education and in political movements emerging in civil society (Harriss-White and Harriss 2007).

Not only will a politics of markets consolidate and challenge these four fields, it will also operate in contestation across the 'boundaries' of these dimensions. Not only will a politics of markets operate within a nation state, it will also operate internationally and even globally.

---

18   Janakarajan (1993). Agricultural labour contracts involved obligations to maintain irrigation infrastructure. Land sales between upper and lower castes has led to low caste agricultural labour's refusing to carry out these non-contractual obligations. This change in contractual content has contributed (together with the proliferation of private open wells) to the collapse of the tank irrigation system.

Evidently markets achieve order by means other than the purely economic and by forms of politics not restricted to parties or states. These arenas of power are not to be presumed to be developmentally beneficial. They can be the base for market exclusion.

While it should now be evident that markets are political institutions and that their politics can be extremely complicated and multilayered, the approach outlined here is a framework rather than a theory. It is greedy of evidence of a sort that is hard to come by, particularly in official statistics on markets which have very limited purposes (defining eligibility for tax, tracking price fluctuations and the standard of living). The 'rich description' to be had from the use of such a framework will be specific to place and time. So higher-order statements must be *expected* to be falsified.

Social Structures of Accumulation (SSA)

This is an approach to market analysis and theory which has emerged from a series of insights from economic history. These include the evident lack of equilibria anywhere on the surface of the planet, the long waves of the business cycle and the debates over their causes (are they due to bunched cycles of innovation, or to the life cycles of capital goods and infrastructure, or to contingent factors such as wars or new technologies?) (Gordon, Edwards and Reich 1982, Kotz, McDonough and Reich 1994). The SSA school has theorised business cycles as being due to the structural nature of the cradle of social institutions which support accumulation (or the creation of productive wealth). The relationships between the elements of this matrix of institutions are not fixed but are thought to follow regular and predictable courses. Their unravelling will herald the end of a stable phase of a business cycle and their re-configuration will consolidate another. The SSA thus has elements which are social, political and even ideological – enabling investment. But the SSA does not simply help to minimise investment risk, it also regulates contradictions and conflicts and reduces insecurity over the long term so that profit levels can be maintained and sustained.

If we are to understand markets we have to understand their SSA. What are the key social structures? According to Kotz, they are those governing the control of raw materials, the labour process, consumption and demand, money and credit. They draw our attention away from politics *per se* towards certain formal institutions hitherto not much discussed in this exploration of markets: the legal constitution of firms, labour laws, the banking system, ideas (especially those institutions which weaken conflict and control the 'unruly tendencies' of labour) (Kotz 1994).

This list has been criticised as arbitrary. In fact it follows from the idea that capitalist development takes place in conflict: that between financial, industrial and mercantile capital; and conflict between firms in specific markets, both of which are conflicts over the distribution of value. Conflict also erupts between capital as a whole and labour and between capital and peasants.[19] Such conflict belongs to the category of contradiction, by which is understood institutions with opposing interests which nonetheless cannot function without one another. Social structures of accumulation are all those institutions which enable the regulation of such conflict and contradiction. Other aspects of SSA theory are disputed. The rules governing the start, the breakdown or the succession of SSAs have not been established; the debate over the cause of crises of capitalism (whether due to the dynamic of capital itself, or due to the dynamic of the SSA or due to the relationships between capital and its SSAs) is unresolved. The timing of the development of key SSAs have not been shown to accord with the long swings of the macro-economy. Some components, particularly markets for finance, are inherently unstable (Fitzgerald 2002). The notion of structure has been criticised as mechanical and essentialist; though, while some essentialism is unavoidable, it is analytically useful to examine an element of a SSA in isolation prior to considering its inter-relationships. But clearly, the potential of the SSA approach is that the SSA can be explored not only historically in relation to the macro-economy but also in the contemporary period – unrelated to crises of regulation or growth – and in the micro-economy too.

Commodification

> It is a law, based on the very nature of manufacture ... that the transformation
> of the social means of production and subsistence must keep extending (Marx
> 1999 (1867): 222).

The commodity markets which proliferate in developing economies do so under capitalism. Capital has to grow. It does not only 'keep extending' according to a logic of expanded reproduction, which requires a continual increase in consumption and investment – leading to both the concentration and the centralisation of capital. Capital does not only extend due to constant competition which requires the extension of labour hours or the deterioration of labour conditions for the same or less pay, or increases in the productivity of labour with new technology and/or newly invented commodities. The sociologist of labour, Ursula Huws (2003) also suggests that capital also 'keeps extending' by non-stop commodification. So it is necessary to understand the specifically *capitalist* institutional dynamics of *commodity* markets.

The process of commodification involves a sequence of relations. First, un-valorised productive and reproductive tasks are carried out for their essential usefulness – their use value. They are then replaced by craft work for sale and/

---

19   Despite the discussion of market-driven politics in the section on the politics of markets, it must not be assumed that there cannot be conflict between capital and parts of the state.

or by paid services. In turn these are replaced by mass produced commodities in conditions which generate economies of scale – cost-reducing technological change. In this process workers are deskilled and displaced. New mechanisms of control 'taylorise' the production of these goods. These commodities in turn require new *services* – typically transport, information, repair and maintenance, advertising and marketing, finance and insurance. These are also in turn industrialised, involving further de-skilling along with the managerialisation of services. Each wave of commodification is accompanied by new technology. This new technology is *not* immiserising or labour displacing, because of the creation of new commodities including commoditised forms of services which require labour. This demand continually stimulates the labour market. The manner in which commodification stimulates labour or strips itself of low profit activity and shifts it back to the consumer and the territory of use value is an empirical and specific question.

Research into the commodification of staple food in Africa and Asia shows that new technology displaces jobs disproportionately allocated to women (Harriss-White 2005, Pujo 1997). It leads to the degradation of work (due to the reductive nature of commodity production: physically repetitive, mentally restricting and constraining social relations). It also leads to conditions in which it is very difficult to organise labour, and to job losses (as outsourcing is spatially and socially relocated).

While the theoretical focus of Huws' analysis is the working household under advanced capitalism, Huws herself makes global connections between the twin processes of commodification and accumulation and the search for cheap labour, the outsourcing of production and the transfer of work to women of different classes (as in highly educated call centre labour in India) and/or different stages of life (as with unmarried women in Bangladesh and China) and in conditions where rights at work and rights to social security are absent (as in sweatshops everywhere).

The domestic sphere is not the only one being commodified. As Leys (2001) has demonstrated, international capital mobility has enabled state-provided services everywhere to be attacked, commodified and transformed into fields of accumulation. Where states have actively participated in markets, the institutions through which state-trading takes place are being dismantled and privatised. Residual state intervention is subsequently subjected to a continual attack. The political process is on-going, 'evolutionary', and complex. As Szlezak (2006) has shown for Argentina, it may be one involving contingent coalitions between (or convergent interests of) state policy elites and capital rather than the open capture of the state by capital. In this process, services and/or rights have to be redefined in commodity form. Users have to be re-constituted as consumers and their needs have to take the economic form of demand. Employees have to be shed and re-employed under circumstances which enable surplus value to be extracted from them. This is a risky process which the state ultimately is forced by capital to protect or risk damaging economic and/or political crises. Through tax concessions, changes in the structure of subsidies, residual infrastructure provision and social security, the state, itself acting increasingly like a corporation, underwrites the risk to capital of this proliferation of markets. The impact of market-driven forms of regulation on democratic politics is to replace party politics with countervailing political influence through individual and collective corporate leverage on elected governments, through the co-option of business in consultation and policy advice, through the key positioning of corporately motivated individuals inside state bureaucracies, through the capture of regulatory bodies by the interests regulated such that states are driven to the competitive appeasement of corporate interests and/or leave them to self-regulation through unenforceable voluntary codes of conduct (Gupta 2002, Leys 2001). Corruption is only one component of an armoury of political tactics (Guhan and Paul 1997, Khan and Jomo 2000). Samir Amin has called this 'low intensity democracy' (Amin 2002). In societies where democratic politics prevails or is being established, not only does the active scope of regulative policy become undemocratically accountable to (international and national) capital rather than (local) voters, thereby undermining the collective values associated with the non-market sphere and its goods and services, but even the very process of policy making and implementation is prey to commodification.

The making of policy agendas is subcontracted to private knowledge or information services, the making of laws is outsourced to the proliferating legal management consultancy sector, access to state resources is subcontracted to more or less commodified quangoes or service agencies and evaluation and enforcement services are performed by specialist agencies. Residual state capacity to define the public interest or mediate between conflicting interests is critically compromised. Individuals move between positions in business and government without charges of public conflict of interest and without cognitive dissonance.

Conclusion

Each of the approaches that have been developed to systematise our understanding of markets on the ground has evolved so as to privilege certain institutions at the expense of others. The NIE is centred upon contracts, firms, farms and households. Economic sociology focuses upon networks, labour markets, corporations and the state. The politics of markets requires analyses of the state as participant and regulator, of collective institutions, of assets and their relation to tactics of competition or collusion, of the social power in which markets are embedded – and of each dimension to the others. The social structure of accumulation school has revealed the importance of the legal regulation of each stage of transfer of property rights in the process of production, distribution and consumption. In developing economies this last approach requires extension so as to establish the roles of non-state, social structures which regulate the economy and stabilise accumulation – such as gender, religion, ethnicity, region, age or life cycle status – and even language – though language has never to date been researched as a SSA. The new theoretical territory of the politics of commodification has to date focussed on party politics, the process of open democratic debate leading to market regulation and the far less transparent – Szlezak calls it 'hidden' – architecture of market-driven politics. The processes of the politics of commodification are secretive and obscure (and possibly dangerous) to research and to date the politics of resistance to this process – by labour, by civil society – have also been residualised.

Each approach privileges certain modes of explanation at the expense of others – information and transactions costs in the case of the NIE, social norms, types of exchange and network in economic sociology, market-driven politics in the approaches considering markets as political phenomena, the interconnection of state (and non-state) regulative structures in the SSA and the political roles of (global and national) capital, the state and the 'domestic' sphere of social reproduction in the dynamics of commodification. While the NIE does not need reference to time or space in order to develop theory, in all the others, the insights of history and geography are crucial to any attempt at explanation.

It is hard for the economist to evade the proposition that markets are social and political constructs whose performance is affected and continually changed by relations of authority. These are established outside the economy and act both inside and outside it. The diversity of meso-level theory already generated by the observation of actually existing markets testifies to the complexity of markets. Interdisciplinary work is necessary to understand them.[20] This work must celebrate local specificities. It must explore the impact on market structure, regulation and behaviour that is brought about by political and social ideologies as well as the material conditions in which exchange takes place.

Acknowledgments

This is a reworked version of a chapter entitled 'On Understanding Markets as Social and Political Institutions in Developing Economies' published in Ha-Joon Chang. 2003. *Rethinking Development Economics*. London: Anthem, 481-97, and of a paper 'Commercialisation, Commodification and Gender Relations in Post Harvest Systems for Rice in South Asia' published in 2005 in *Economic and Political Weekly*, June 18-24, Vol. XL (25), 2530-2542.

---

20   In the last four decades, there have been two big waves of interest in rural markets in developing economies. In the 1960s and 70s, industrial organisation methods were used to evaluate the competitiveness of local markets. If they were concluded to be competitive (which by and large they were, even when the detailed ethnographic evidence showed otherwise) there was then no case for extensive state intervention. The second wave of field research was provoked in the 1990s by the failure of structural adjustment conditions to have their predicted impacts upon domestic markets. See Harriss (1979a, 1979b), for reviews of the first phase and de Alcantara (1993), Bryceson (1993), Crow (2001) and Dorward, Poulton and Kydd (1998), for reviews of the second phase.

References

Alcantara, C.H. de. 1993. *Real Markets: Social and Political Issues of Food Policy Reform*. London: Frank Cass.

Amin, S. 2002. Economic globalisation and political universalism: Conflicting issues?, in *Globalisation and Insecurity*, edited by B. Harriss-White. London: Palgrave.

Bardhan, P. 1989. *The Economic Theory of Agrarian Institutions*. Oxford: Clarendon Press.

Bardhan, P. 1991. On the concept of power in economics. *Economics and Politics*, 3(3), 265-277.

Basu, K. (ed.). 1994. *Agrarian Questions*. Delhi: Oxford University Press.

Bates, R. 1981. *Markets and States in Tropical Africa*. Berkeley: University of California Press.

Bavinck, M. 2003. The spatially splintered state: Myths and realities in the regulation of marine fisheries in Tamil Nadu, India. *Development and Change*, 4(34), 633-657.

Benda-Beckmann, F. von and Meijl, T. van (eds). 1999. *Property Rights and Economic Development*. London: Kegan Paul.

Bhaduri, A. 1983. *The Economics of Backward Agriculture*. New York: Academic Press.

Bhaduri, A. 1986. Forced commerce and agrarian growth. *World Development*, 14(2), 267-272.

Bryceson, D. 1993. *Liberalising Tanzania's Food Trade: The Public and Private Faces of Urban Marketing Policy, 1939-1988*. London: James Currey.

Byres, T.J. 1983. *Sharecropping and Sharecroppers*. London: Routledge.

Caille, A. 1994. D'une économie politique qui aurait pu être. *Pour une autre économie/Revue du mouvement anti-utilitariste dans les sciences sociales Paris*, 3(1), 153-160.

Castells, M. 2000. *The Rise of the Network Society*. Oxford: Blackwell Publishing.

Clough, P. 1995. *The Economy and Culture of the Talakawa of Marmara*. PhD Thesis, Oxford University.

Crow, B. 2001. *Markets Class and Social Change: Trading Networks and Poverty in Rural Asia*. London: Palgrave.

Cutler, P. 1988. *The Development of the 1983-5 Famine in Northern Ethiopia*. PhD Thesis, London University: London School of Hygiene and Tropical Medicine.

Davis, J. 1992. *Exchange (Concepts in Social Thought)*. Minneapolis: University of Minnesota Press.

Dorward, A., Poulton, C. and Kydd, J. 1998. *Smallholder Cash Crop Production Under Market Liberalisation: A New Institutional Economics Perspective*. Wallingford: CAB International.

Evers, H.D. and Schrader, N. 1994. *The Moral Economy of Trade: Ethnicity and Developing Markets*. London: Routledge.

Fitzgerald, E.V.K. 2002. The security of international finance, in *Globalisation and Insecurity*, edited by B. Harriss-White. London: Palgrave.

Folbre, N. 1994. *Who Pays for the Kids: Gender and the Structures of Constraint.* London: Routledge.

Gordon, D., Edwards, R. and Reich, M. 1982. *Segmented Work, Divided Workers.* Cambridge: Cambridge University Press.

Granovetter, M. and Swedberg, R. (eds). 1992. *The Sociology of Economic Life.* Boulder: Westview.

Granovetter, M. 1995. *Getting a Job: A Study of Contracts and Careers.* Chicago: University of Chicago Press.

Guhan, S. and Paul, S. 1997. *Corruption in India.* New Delhi: Vision Pub.

Gupta, S. 2002. *Corporate Capital and Political Philosophy.* London: Pluto.

Haddad, L. et al. (eds). 1997. *Intrahousehold Resource Allocation in Developing Countries.* Baltimore: Johns Hopkins.

Harriss, B. 1979a. There's method in my madness, or is it vice versa? Measuring agricultural market performance. *Food Research Institute Studies*, XVI(2), 40-56.

Harriss, B. 1979b. Going against the grain. *Development and Change*, 10(3), 368-384.

Harriss J., Hunter, J. and Lewis, C. (eds). 1995. *The New Institutional Economics and Third World Development.* London: Routledge.

Harriss-White, B. 1993. The collective politics of foodgrains markets in South Asia. *Bulletin, Institute of Development Studies*, 24(3), 54-63.

Harriss-White, B. 1995b. Order ... order ... Agro-commercial micro-structures and the state: The experience of regulation, in *Institutions and Economic Change in South Asia*, edited by B. Stein and S. Subrahmanyam. Delhi: Oxford University Press, 275-314.

Harriss-White, B. 1996. *A Political Economy of Agricultural Markets in South India.* New Delhi: Sage Publications.

Harriss-White, B. (ed.). 1998. *Agricultural Markets from Theory to Practice.* London: Macmillan.

Harriss-White, B. 2003. *India Working.* Cambridge: Cambridge University Press.

Harriss-White, B. 2005. Commercialisation, commodification and gender relations in post harvest systems for rice in South Asia. *Economic and Political Weekly*, 40(25), 2530-2542.

Harriss-White, B. and Harriss, E. 2007. Unsustainable capitalism: The politics of renewable energy in the UK, in *Coming to Terms with Nature*, edited by L. Panitch et al. London: Merlin Press, 72-101.

Hodgson, G. 1988. *Economics and Institutions.* London: Polity.

Hodgson, G. 2002. *Is Social Evolution Lamarckian or Darwinian.* Paper to the International Workshop on Evolutionary Economics (University of Herts, UK).

Huws, U. 2003. *The Making of a Cybertariat.* London: Merlin.

Jackson, C. and Pearson, R. (eds). 1998. *Feminist Visions of Development: Gender Analysis and Policy.* London: Routledge.

Janakarajan, S. 1993. Triaidic exchange relations: An illustration from South India. *Bulletin, Institute of Development Studies*, 24(3), 75-82.

Keen, D. 1994. *The Benefits of Famine*. Princeton: Princeton University Press.

Khan, M. and Jomo, K.S. 2000. *Rents, Rent Seeking and Economic Development in Asia*. Cambridge: Cambridge University Press.

Kotz, D.M. 1994. The regulation theory and the social structures of accumulation approach, in *Social Structures of Accumulation: The Political Economy of Growth and Crisis*, edited by D. Kotz, T. McDonough and M. Reich. New York: Cambridge University Press, 85-98.

Kotz, D.M., McDonough, T. and Reich, M. (eds). 1994. *Social Structures of Accumulation: The Political Economy of Growth and Crisis*. New York: Cambridge University Press.

Laidlaw, J. 1995. *Riches and Renunciation: Religion, Economy and Society Among the Jains*. Oxford: Clarendon Press.

Leys, C. 2001. *Market-Driven Politics*. London: Verso Books.

Mackintosh, M. 1990. Abstract markets and real needs, in *The Food Question*, edited by H. Bernstein et al. London: Earthscan Publication Ltd, 43-53.

Majid, N. 1994. *Contractual Arrangements in Pakistan's Agriculture: A Study of Share Tenure in Sindh*. PhD Thesis, Oxford University.

Martinez-Alier, J. 1987. *Ecological Economics: Energy Environment and Society*. Oxford: Blackwell Publishing.

Marx, K. 1999 (1867). *Capital: An Abridged Edition*. Oxford: Oxford University Press.

Meagher, K. 2003. *Informalisation and Embeddness in South Eastern Nigeria*. PhD Thesis, Oxford University.

Monbiot, G. 2001. *Captive State: The Corporate Take-over of Britain*. London: Pan Macmillan.

Nadvi, K. 1999. Shifting ties, social networks in the surgical instruments cluster of Sialkot. *Development and Change*, 30(1), 143-77.

North, D. 1990. *Institutions, Institutional Change and Economic Performance*. Cambridge: Cambridge University Press.

Palaskas, T. and Harriss-White, B. 1993. Testing marketing integration: New approaches with case material from the West Bengal food economy. *Journal of Development Studies*, 30(1), 1-57.

Palaskas, T. and Harriss-White, B. 1996. Identification of market exogeneity and market dominance by tests instead of assumptions: An application to Indian material. *Journal of International Development*, 8(1), 111-123.

Patnaik, U. 1994. Tenancy and accumulation, in *Agrarian Questions*, edited by K. Basu. Delhi: Oxford University Press, 155-202.

Penny, D. 1985. *Starvation: A Political Economy*. Canberra: Australian National University.

Polanyi, K. 1957. *The Great Transformation*. Boston: Beacon Press.

Pujo, L. 1997. *Towards a Methodology for the Analysis of the Embeddedness of Markets in Social Institutions: Application to Gender and the Market for Local Rice in Eastern Guinee.* PhD Thesis, Oxford University.

Ravallion, M. 1987. *Markets and Famines.* Oxford: Clarendon Press.

Robson, E. 2002. *Gender, Households and Markets: Social Reproduction in a Hausa Village, Northern Nigeria.* PhD Thesis, Oxford University.

Rogaly, B. 1996. Agricultural growth and the structure of 'casual' labour-hiring in rural West Bengal. *Journal of Peasant Studies*, 23(4), 141-160.

Schmitz, H. and Nadvi, K. 1999. Clustering and industrialisation: Introduction. *Special Issue on 'Industrial Clusters in Developing Countries'. World Development*, 27(9), 1503-1514.

Sen, A.K. 1981. *Poverty and Famines.* Clarendon: Oxford.

Smith, A. 1995. *An Enquiry into the Nature and Causes of the Wealth of Nations.* London: Pickering.

Subramanian, S. 1992. Appendix: A model of triadic power relations in the interlinkage of agrarian markets, in *Trade and Traders, The Making of the Cattle Market of Benin*, edited by S. Subramanian, P.Q. van Ufford. PhD Thesis, Amsterdam University, 190-201.

Szlezak, P. 2006. *The Political Economy of Pension Privatisation in Argentina, 1990-2005.* PhD Thesis, Oxford University.

Ufford, P.Q. van 1999. *Trade and Traders: The Making of the Cattle Market in Benin.* Amsterdam: Thela Thesis.

Ufford P.Q. van and Zaal, F. 2004. The transfer of trust; ethnicities as economic institutions in the livestock trade in West and East Africa. *Journal of the International African Institute*, 74(2), 121-145.

Weber, M. 1922 (1978). *Economy and Society: An Outline of Interpretive Sociology.* Berkeley: University of California Press.

White, G. 1993. Towards a political analysis of markets. *IDS Bulletin*, 24(3), 4-11.

Young, K. et al. 1981. *Of Marriage and the Market.* London: CSE Books.

Chapter 3

Social Vulnerability and Livelihood Security: Towards an Integrated Framework for Market Risk Assessment

Hans-Georg Bohle

Spaces of Risk and Vulnerability

The making of global markets in terms of pastoral commodity systems and the integration of pastoral production systems into global markets constitute considerable risks for the pastoral actors involved. It is therefore the main objective of this chapter to propose an integrated framework for market risk assessment in conjunction with the making of a global pastoral market. The social spaces of pastoral production and commodity systems are embedded in institutions that frame the transformation process of market relations in risky environments. In this transformation process, the entire system of entitlements, power relations and livelihoods of pastoral communities is at stake. Processes of privatisation, commercialisation and new competition from different market actors pose high degrees of risks for pastoral societies that have to undergo profound shifts in political and institutional regulations. New types of conflicts emerge when pastoral livelihoods are reorganised along new production and commodity regimes, and this also implies major transformations in the power positions of the various market participants. The new market dynamics require innovative forms of risk-taking behaviour from all market actors, particularly from the pastoral groups. New types of uncertainties including processes of disentitlement and disempowerment can change the entire fabric of social relations that determine access and control to the pastoral modes of production and distribution.

It is against this background that this chapter seeks to develop an integrated framework for pastoral market risk assessment that is based on the concepts of social vulnerability and livelihood security. In the first section, five key components of vulnerability will be identified that, together, constitute a baseline for any vulnerability assessment. A second section will then demonstrate how disciplinary approaches in vulnerability research generally focus on one or two of these components only. New frameworks for vulnerability assessment that seek to link all these components in an integrated manner are presented in a third section. Fourthly, the UK Department for International Development's (DFID) Sustainable Livelihoods Framework is discussed as an example for integrated vulnerability assessment that focuses on issues of social vulnerability. Finally, in a fifth section, the foregoing discussion will be used to propose an integrated framework for market risk assessment. The chapter seeks to provide an analytical tool for the discussion of the problems and perspectives that pastoralists face in the process of market integration, both in terms of their social vulnerability and human security.

Key Components of the Framework

The first step for the proposition of an integrated market risk assessment framework is to identify the key components of such a framework. For this purpose, the basic concept of vulnerability will have to be discussed. Then, vulnerability has to be operationalised and the key components for vulnerability assessments identified. In a next step, the components of a framework for market risk assessment will have to be linked in an integrated way. Finally, the most basic components of the framework will have to be put into the various strands of discussion that the social sciences have in conjunction with social risk analysis.

What is Vulnerability?

Vulnerability has been defined as the degree to which a system is likely to experience harm due to stress (Kasperson 2001). Social vulnerability, therefore, is the degree to which a social system is likely to experience harm, and livelihood vulnerability can be defined as the degree to which a livelihood system is likely to experience harm due to shocks, stress and perturbations. In a more refined way, household vulnerability has been defined by the World Bank's Social Protection Unit (SPU 2001) in the following way:

> A household is said to be vulnerable to future loss of welfare below socially accepted norms caused by risky events. The degree of vulnerability depends on the characteristics of the risk and the household's ability to respond to risk. Ability to respond to risk depends on household characteristics – notably their asset-base. The outcome is defined with respect to some benchmark – a socially accepted minimum reference level of welfare (e.g. a poverty line) (Alwang et al. 2001: 4).

How to Operationalise Vulnerability?

Three key components of vulnerability can be identified: the characteristics of risk; the ability to respond to risk; and the outcome in respect to some minimum level of reference. Special attention has to be paid to the characteristics and behaviour of the systems components, particularly in terms of coping and adaptive capacities. The vulnerability of a pastoral system that is being integrated into new types of market relations is therefore a function of the character, magnitude and rate of market risks to which a system is exposed, its sensitivity, and its adaptive capacity.

What are the Key Components of Vulnerability Assessment?

The basic components of any vulnerability assessment (Bohle 2001) can be grouped into three major sections. The first section focuses on the stress and stressors to which a production and distribution system is exposed. This section includes risky events, shocks, perturbations and disturbances. The second section comprises of three elements: exposure, sensitivity and response. This requires the identification of the exposure units on which stress and stressors impact; the sensitivity of the respective systems elements and the types of responses including coping and adaptation. The third section consists of potential harms or the outcomes of risky impacts on the system. These outcomes can be positive or negative, but have to be classified according to a given socially accepted reference level.

How Does the Literature Treat Vulnerability Components?

It is clear that even the simplest systems are far too complex to account for all variables, processes and disturbances. A literature analysis on how various vulnerability components are treated in the vulnerability literature, which was undertaken by the Social Protection Unit of the World Bank (Alwang et al. 2001: 24-25), has shown that the focus is either on risk, on response or on outcomes (Fig. 3.1).

While the vulnerability literature on poverty dynamics and food security focuses particularly on outcomes (in terms of poverty lines; food production, distribution and consumption figures), literature on environmental vulnerability concentrates on the respective risks and exposures (for example in terms of environmental stressors, climate change and so on). Literature on social and anthropological perspectives of vulnerability, including asset-based approaches and sustainable livelihoods, however, puts its focus on the treatment of response mechanisms. Here, issues of human agency, of coping capacities and adaptability

Vulnerability Literature	Treatment of		
	Risk	Response	Outcome
Poverty dynamics	0	0	++
Asset-based approaches	0	++	0
Sustainable livelihoods	+	++	0
Food security	+	+	++
Disaster management	+	+	++
Environment	++	0	++
Sociology / anthropology	0	++	+
Health / nutrition	0	0	++

Figure 3.1   How the literature treats vulnerability components
Source: Alwang et al. (2001: 24-25).

are brought up. It is argued in this chapter that an integrated framework for market risk assessment that seeks to link all the major components of vulnerability in an integrated manner is still missing. The widely discussed Sustainable Livelihoods Framework, however, can be regarded as a model that can serve as a base for framing market risk assessments.

Sustainable Livelihoods Security: An Integrated Vulnerability Assessment Framework

According to DFID (1999: 1.1) 'a livelihood system comprises the capabilities, assets (including both material and social resources) and activities required for a

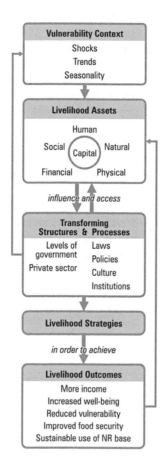

Figure 3.2   The Sustainable Livelihoods Framework
Source: DFID (1999).

means of living'. In order to help understand and analyse the livelihoods of the poor, the Sustainable Livelihoods Framework (SLF) has been developed by Scoones (1998: 4) and made a central part of the Sustainable Livelihoods Guidance Sheets by DFID (1999: 1.1). The framework which is presented in schematic form in Fig. 3.2 views poor people as operating in a context of vulnerability. Within this context, they have access to certain assets ('capitals') or poverty reducing factors. These gain their meaning and value through the prevailing social, institutional and regulatory environment. This environment also influences the livelihood strategies – ways of combining and using assets – that are open to people in pursuit of beneficial livelihood outcomes that meet their own livelihood objectives (Bohle 2009).

The core principles of any livelihood analysis are as follows (DFID 1999: glossary):

- Effort should be devoted to identifying and understanding the livelihood circumstances of marginalised and excluded groups.
- Analysis should take into account important social divides that make a difference to people's livelihoods. For example, it is often appropriate to consider men, women, different age groups, etc. separately. It is not sufficient to take the household as the sole unit of analysis.
- The sustainable livelihood (SL) approach seeks to build upon people's strengths and resourcefulness. When conducting analysis it is important to avoid thinking only about need.
- The SL approach embraces the idea of dynamism. Avoid taking one-off snap shots and instead think about change over time, including concerns about sustainability.
- There will never be a set recipe for certain methods to use under certain circumstances. Flexibility is the key. Equally, it is not necessary to produce one definitive 'map' of livelihoods. Different 'maps' may be appropriately used for different purposes.

When we look closer at the components of vulnerability in the Sustainable Livelihoods Framework, the key elements that were discussed above can be identified at the intersections of the various components (Fig. 3.3). Stressors for livelihood systems, for example, are located at the interface between vulnerability contexts and livelihood assets. The exposure side of vulnerability focuses on the linkage between livelihood assets and transforming structures and processes. The sensitivity of the livelihood system is largely determined by the interaction of transforming structures and processes with livelihood assets, on the one side, and livelihood strategies, on the other. Response mechanisms are located in-between transforming structures and livelihood strategies. The outcomes in terms of livelihood vulnerability or human security are found at the interlinkage between livelihood strategies and livelihood outcomes.

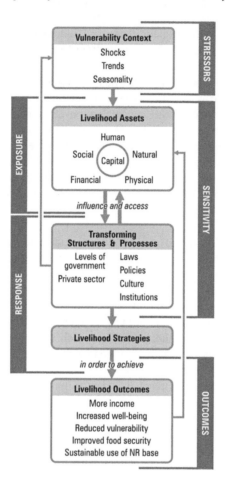

Figure 3.3   Components of vulnerability in the Sustainable Livelihoods Framework

Source: Author.

Towards an Integrated Framework for Market Risk Assessment

The task of integrating the various components that constitute a framework for market risk assessment requires, in a first step, to focus on the relational perspectives of vulnerability assessments. The Sustainable Livelihoods Framework, for example, is a typical example for 'box and arrows' conceptualisations (Fig. 3.2). While the boxes are presented as closed entities and operationalised in the analytical process, the arrows remain largely undetermined and vague. It is argued here that it is the dynamic linkages between the various components of risk assessment frameworks that should be at the centre of attention of any assessment practice. Therefore, the focus of the assessment has to shift away from structural to relational perspectives to come to terms with the risky nature of market relations.

A second requirement for integrated risk assessment practices is to deconstruct and theorise the 'black boxes' in the frameworks that have been proposed so far. Turning back to the Sustainable Livelihoods Framework, the entire component of livelihood assets has largely remained under-theorised in terms of the underlying assumptions. When livelihood assets are addressed, the arrows proposed in the framework have to be theorised, for example, in terms of the entitlements that determine the access to the various assets that make up livelihoods. When we then address transforming structures and processes, a new focus has to be on power relations and processes of empowerment/disempowerment. Finally, the livelihood strategies that characterise coping and adaptation behaviour have to be analysed in terms of the agency of the actors involved, their agendas and the arenas of livelihood struggles. In short, any attempt to identify the spaces of pastoral production and commodity systems under the impact of risky forms of market integration has to identify and map the spaces of entitlement, empowerment and agency. It is argued that the concept of 'real markets' can be best utilised to integrate entitlement approaches, political economies of empowerment and perspectives of agency into market risk assessments.

Determining Pastoral Market Risks Through the Concept of 'Real Markets'

The concept of 'real markets', as introduced by Mackintosh (1990), starts from various propositions. It is argued that the most dramatic outcomes of market integration, such as food crises, famines and disasters, can be caused by the 'normal' working of markets. This proposition rejects the neoliberal assumption that such disasters are basically the results of markets working 'badly'. Hunger and famine are therefore not the result of market failures, but, on the contrary, of the basic principles of markets that respond to demand backed by cash and not by the needs of the vulnerable. Market integration, from the perspective of 'real markets', forces risk-taking upon those who are integrated into market mechanisms on a vulnerable and subaltern base. Sen (1981: 112), when analysing the 1974 famine in the Wollo region of Ethiopia, remarks that: 'the pastoralist, hit by the drought, was decimated by the market mechanisms'. Such an entitlement perspective on market risks requires the analysis of the dynamics of specific types of markets at specific times. The spaces of 'real markets' operate in the contested contexts of a politicised world.

Following from these propositions, a number of strategic issues emerge that have to be examined when assessing pastoral market integration. 'Real market' approaches have to study markets as sets of social relations structured by institutions, interests and power. A particular emphasis has to be put on the terms by which people come to market; these can be dominant, hegemonic terms which meet with subaltern and vulnerable terms of market participation. Furthermore,

Stressors →	Entitlements →	Power Relations →	Agency →	Harm
Privatisation Commercialisation	Impacts on direct entitlements	Livelihood elements politicised	New dynamics of risk taking behaviour	Loss of opportunities, choices
New competition Shifting political and institutional regimes	Impacts on market-based entitlements	Conflicting interests in organising livelihoods	Coping / adapting under uncertainty	Disentitlement, disempowerment Human insecurity
Natural hazards Health risks	Impacts on institutional entitlements	Shifts in power positions of market participants	Changing fabric of social relations determining access and control	Livelihoods resilience undermined

Figure 3.4   Framework for market risk assessment
Source: Author.

any study of 'real markets' has to examine how entitlements (direct, market-based and institutional) are created, contested, lost and won in real markets. Therefore, the study of 'real markets' has to include the class structure of trade and the ways in which markets are abrogated rather than developed. Finally, non-market relations have to be integrated which surround and structure all markets and also shape the economic spaces of pastoral production and exchange.

An Integrated Market Risk Assessment Framework: Reconciling Livelihoods Analysis with 'Real Market' Approaches

The preceding discussion on integrated vulnerability analysis and the concept of 'real markets' has revealed that entitlements, power relations and human agency are the basic building blocks of market risk assessments. Accordingly, pastoral production and exchange systems, and their vulnerability towards the risk of market integration are framed in terms of the stressors that impact on entitlements, power relations and human agency and the (positive or negative) outcomes (Fig. 3.4).

Looking at the stressors, the integrated risk assessment framework points to processes of privatisation, commercialisation and new competitions. Shifting political and institutional regimes in conjuncture with market integrations have to be analysed. Additional disturbances such as natural hazards and health risks have to be taken into account. In conjunction with pastoral entitlements, the impacts of market integration on direct, market-based and institutional entitlements have to be scrutinised. Their effects on the power relations within the 'real markets' of the pastoral system have then to be analysed. The politicisation of livelihoods, conflicting interests in organising them and shifts in power positions of market participants will be major perspectives. When addressing human agency, new dynamics of risk-taking behaviour, coping and adaptation capacities under conditions of uncertainty and the changing fabric of social relations that determine access and control of basic pastoral resources have to be examined. Lastly, on the outcome side of the assessment process, loss of opportunities and choices, processes of disentitlement and disempowerment, but also issues of human security and livelihoods resilience will have to be determined. This framework does not only provide an analytical tool for the discussion of the problems and perspectives that pastoralists face in the process of market integration, but it also serves to map the economic, social and institutional spaces of pastoral production and commodity systems.

References

Alwang, J., Siegel, P.B. and Jorgensen, S.L. 2001. *Vulnerability: A View from Different Disciplines*. (Social Protection Discussion Paper Series, No. 0115). New York: The World Bank, Human Development Network, Social Protection Unit.

Bohle, H.-G. 2009. Sustainable livelihood security. Evolution and application, in *Facing Global Environmental Change: Environmental, Human, Energy, Food, Health and Water Security Concepts*. (Hexagon Series on Human and Environmental Security and Peace, Vol. 4), edited by H.G. Brauch et al. London: Springer-Verlag, 521-528.

Bohle, H.-G. 2001. Vulnerability and Criticality: Perspectives from Social Geography. *IHDP Update, Newsletter of the International Human Dimensions Programme on Global Environmental Change*, 1-7.

DFID (Department for International Development). 1999. *Sustainable Livelihoods Guidance Sheets*. London: DFID.

Kasperson, R.E. 2001. Vulnerability and global environmental change. *IHDP Update*, 2, 2-3.

Mackintosh, M. 1990. Abstract markets and real needs, in *The Food Question: Profit Versus People*, edited by H. Bernstein, B. Crow, M. Mackintosh and C. Martin. London: Earthscan Publication, 43-53.

Scoones, I. 1998. *Sustainable Rural Livelihoods. A Framework for Analysis*. (IDS Working Paper 72). Brighton, Sussex: Institute of Development Studies.

Sen, A. 1981. *Poverty and Famines: An Essay on Entitlement and Deprivation*. Oxford: Oxford University Press.

PART II
From Subsistence to Market Production: Post-Colonial Africa

Chapter 4

Livestock Production and Pastoral Livelihood Security in Western Chad

Martin Wiese

Introduction

The global livestock sector is undergoing profound changes in response to globalisation and rapidly growing demand for animal food products, especially in developing countries. New global players in livestock production have recently emerged in developing regions and the 'centre of gravity of livestock production' has shifted towards the Tropics and Sub-tropics (Steinfeld and Chilonda 2006, World Bank 2009). Indeed, trade is expanding much faster than production, and commodity chains for livestock products are rapidly transforming. The social and environmental costs of this growth and transformation are important since small scale livestock producers are increasingly marginalised and pressure on particularly fragile ecosystems such as drylands is increasing.[1]

In this context, Sub-Saharan Africa reveals itself as particularly vulnerable. On the one hand, demand for meat and milk is growing faster than production, making the continent an increasing net-importer of livestock products (Steinfeld and Chilonda 2006). On the other hand, drylands cover nearly 50 per cent of the land area of Sub-Saharan Africa. Therefore, pastoralism[2] is of particular importance for livestock production, accounting for over 33 per cent of livestock production in western and eastern Africa (including Chad); the pastoral population is estimated at 50 million people for Sub-Saharan Africa alone (Rass 2006). However, pastoral development has had little advocacy until today (Dobie 2001, Herrero et al. 2010, Pratt et al. 1997), although increasing scientific evidence shows that traditional pastoral production systems are actually adapted to the ecological non-equilibrium conditions of drylands: The opportunistic mobility for tracking variable opportunities and for avoiding risks, together with social flexibility, are conceived as important adaptive 'tools of sustainable production'. Traditional knowledge and

---

1   Jutzi (2006), Safriel and Adeel (2005), Suttie et al. (2005), Thornton and Gerber (2010).

2   Most commonly, 'pastoralism' is understood as a way of production in drylands in which at least 50 per cent of the income and the commodities created by a social unit come from the breeding of livestock. Taking into account social and cultural, as well as economic aspects (Khazanov 1994, Salzman 1995), the author refers to pastoralism as livelihoods in drylands based on livestock-breeding.

communication systems, as well as the social structures of pastoral communities – allowing for flexible and large-scale decision-making and mobility in conjunction with rapidly changing patterns of opportunities and risks – enable sustainable production in the particularly fragile ecosystems of Sub-Saharan Africa.[3]

Despite this potential, pastoral people in the drylands of Sub-Saharan Africa are, however, largely excluded from any investment in economic, human and social development at the present time (for example primary health care, basic education, and more recently, veterinary services).[4] Moreover, pastoralists are commonly confronted with legal frameworks and popular myths which give way to paternalist, if not hostile, attitudes of state authorities towards pastoral societies and their cultural legacy.[5]

Consequently, it is the people who are sustainable in their production who not only belong to the poorest communities worldwide – the incidence of extreme poverty ranges from 25 to 55 per cent among African pastoralists (Rass 2006: 1) – but who also belong to communities that benefit the least from any social infrastructure.[6] Under these circumstances, it is an open question how pastoral societies in Sub-Saharan Africa could actively contribute to solving problems and creating opportunities within a globalising economy.

Pastoral Livelihoods in Chad

Chad is a landlocked country in the intersection of West, East and Central Africa. With a population estimated at 10.9 million inhabitants in 2008 (OECD 2009: 149, UNDP 2009: 193), the country covers a surface of 1.28 million km² from the central Saharan mountains of Tibesti to the southern fringes of the Sudanese zone in the eastern Chad Basin (Fig. 4.1). The northern two thirds of the country are Saharan desert or Sahelian savannah where rural livelihoods are primarily based on nomadic pastoralism associated with the cultivation of date palms and cereals in the oasis, wadis and interdunal depressions (Arditi 1995, Chapelle 1982, Le Rouvreur 1989). The southern areas of Chad are situated in the Sudanese eco-climatic zone with rural livelihoods based on cotton cash-crop, subsistence agriculture of grain, manioc and peanuts, and horticulture (Magrin 2001).

Until the nineteenth century the Chad Basin was governed by three kingdoms – Baguirmi, Kanem-Bornou and Ouaddaï – that dominated the trans-Saharan

---

3   For example Bourgeot (1999), Niamir-Fuller and Turner (1999), Safriel and Adeel (2005), Scoones (1995), UNDP (2003).

4   Bonfiglioli and Watson (1992), Dobie (2001), Pratt et al. (1997), Rass (2006).

5   For example Azarya (1996), Scholz (1996), Swift (1996), Thebaud and Batterbury (2001).

6   See Aliou (1995), Foggin et al. (1997, 2006), Haraldson (1979), Hill (1985), Imperato (1974), Loutan (1989), Omar (1992), Sandford (1979), Schelling et al. (2008), Sheik-Mohamed and Velema (1999), Swift et al. (1990).

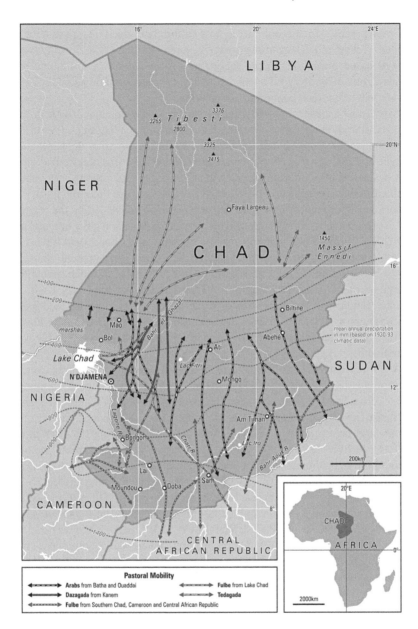

Figure 4.1   Pastoralists in Chad
Source: Author's fieldwork.

trade from central Africa to the Mediterranean coast and Arabian countries for one
millennium (Wright 1989, Zeltner 1980). However, with European colonialism in
Africa, Chad became a 'Cinderella' colony (Buitjenhuijs 1989: 54) – an appendix

of French central Africa economically and culturally divided into a southern zone labelled 'Tchad utile', according to the colonial politics of a 'mise-en-valeur' strategy based on cotton, forced labour and socio-cultural assimilation; and a northern zone of military maintenance with pastoral communities under indirect rule leading to its socio-economic dissociation.[7]

After independence in 1960, fratricidal civil wars coincided with droughts and agricultural devastation caused by locusts and cattle plagues (Azevedo 1998, Buitjenhuijs 1987, Yacoub and Gatta 2005). At the paroxysm of the pan-Sahelian drought in 1984, roughly 90 per cent of the country's infrastructure had already been destroyed by warfare especially in its northern and central parts, and the qualified elite was either killed or forced to flee (for example Foltz and Foltz 1991). A short period of reconstruction and of war with Libya was led by the regime of former president Hisseine Habré, who based internal power on governmental terror (ICG 2006: 30-31). Following the military coup by today's president Idriss Deby, taking power in end-1989, perspectives for democratic development opened up. However, the period of progress was short while political and human development became retrograde (May and Massay 2003, OECD 2009). Civil crises inside Chad are interrelated with the epicentres of some major crises in central and western Africa (Dickow 2005, ICG 2006): the Darfur humanitarian crisis to the East; the Nigerian oil-fuelled crisis to the West; and the chronic instability of an absent state to the South.

At present, Chad is considered one of the least developed and most impoverished countries in the world with its HDI (Human Development Index) ranking on place 175 among 182 countries and a Human Poverty Index ranking 132 among 135 Least Developed Countries (UNDP 2009: 173, 178.). Indeed, the oil-related revenue and investments, which had contributed to bolstering the GDP in early 2000 (2,090 PPP US$ in 2004 versus 871 PPP US$ in 2000), mask the actual deterioration of the human development situation in Chad during the same period: for example life expectancy (43.7 years in 2004) and adult literacy rates (25.7 per cent in 2004) declined (compare UNDP 2006: 286 with UNDP 2002: 152). Until present, the situation has not significantly improved and even degraded, such as GDP (1,477 PPP US$ in 2007: UNDP 2009: 173; OECD 2009: 151-152) and overall HDI (UNDP 2009: 169). No other Least Developed Country has a similar incongruity between GDP and HDI-ranking as that observed for Chad since the start of oil-production in 2003 (UNDP 2006, 2009). Despite – or precisely because of – the chronic and complex crisis situation, pastoral livestock production is at present time a very important pillar of the national economy in Chad.

Chad's national economy is based on subsistence agriculture, livestock-breeding, the export of raw-products (crude oil, cotton, livestock on hoofs, non-processed gum Arabic), and different forms of development aid. Apart from the national production of some consumer goods – beer, cigarettes and sugar – Chad imports nearly all industrial goods (OECD 2009).

---

7   Arditi (1994), Chapelle (1986), Lanne (1998), Magrin (2001), Stürzinger (1980).

The production of crude oil started in July 2003, exporting via the Doba-Kribi pipeline 172,600 barrels of low-quality crude oil per day. The direct contribution of the oil production accounted for 33.4 per cent of the GDP in 2004 (1.16 out of 3.47 billion Euro). It provided 163 million Euro in national revenues, covering 32.1 per cent of the national budget in 2004 (106[8] out of 331 million Euro). However, only 3,600 Chadian citizens were directly employed on the oil fields with six people placed in positions of higher responsibility (ICG 2006: 8, IMF 2007: 15). Indeed, contrary to the growing GDP, which includes values repatriated by foreign petrol companies, the gross national product, GNP, has actually decreased by about 2 per cent with the start of petrol exploitation in 2003. Consequently, the direct benefits of the oil-economy are rather imperceptible for most of the Chadian population which depends essentially on the redistribution of petrol-dollars via government spending (Dickow 2005, ICG 2006, IMF 2007). Such redistribution mechanisms are actually weak, fragile and biased towards elites and the urban centre of N'Djamena. The present political-institutional crisis of the country is particularly unfavourable to any effective redistribution of the national oil-revenues into social welfare and poverty reduction (Djindil et al. 2007, IMF 2007, OECD 2009).

In contrast to the oil-economy, agriculture and livestock breeding constitute the livelihood for 80 per cent of the population, accounting for 24 per cent of GDP in 2004. Livestock breeding alone produces 38 per cent of the agricultural GDP[9] (i.e. 320 million Euro) and supports the subsistence of 40 per cent of the active rural population in Chad.[10]

Throughout the last millennium, the Chad Basin[11] was not only home to long-lasting kingdoms, a centre of trans-Saharan trade, and a crossing of transcontinental migration and pilgrimage routes, but also, due to its particular pastoral potentials,

---

8   Economic data published in 2006/07 (for example IMF 2007, OECD 2006, UNDP 2006) apply for the year 2004. Conform to the legal framework fixing the basic rules for spending the national oil revenues of Chad, 57 million Euro of the total oil revenues (= 163 million Euro) did not directly contribute to the national budget in 2004, but alimented special funds, including a 'fund for future generations'. The modalities of distribution were changed by Chadian government in November 2005 (ICG 2006).

9   In contrast to the comparatively prospering livestock-production, the cotton-sector faces a major crisis (Magrin 2001, OECD op.cit.). Since the most rigorous and detailed review of evidence with regards to livestock production in Chad has been carried out in 2004 (Bonnet et al. 2004), the author will focus on this period more in detail. There is no indication that the general proportion and trends presented in this section have undergone major changes until today (compare recent data in OECD 2009 and UNDP 2009 with data provided for 2004 in OECD 2006 and UNDP 2006).

10   Data for 2004 from Bonnet et al. (2004: 5-10), OECD (2006: 525-526), UNDP (2006: 334). As outlined previously, this section examines in detail the period between 2000 and 2004 due to the exceptional availability of empirical data on livestock demography and pastoral development for this period. Recent data published by UNDP (2009) and OECD (2009) indicate no significant changes in the overall social and economic conditions in Chad.

11   For detailed description see Wiese (2001).

a pole of attraction for diverse pastoral societies throughout Africa (Wright 1989, Zeltner 1969, 1980). Some unique pastoral livelihood systems have developed inside the Basin (for example *Boudouma, Kouri* inside Lake Chad), in its southern periphery (for example *Moundang, Toupouri, Massa*), and in proximity to the Tibesti mountains ('Toubou': *Tedagada, Dazagada; Bidéyat-Zaghawa*).[12] Other societies have immigrated from West-Africa (*Fulbe*)[13] or from the Arabian Peninsula (*Arab Juhayna*).[14]

Pastoral livelihoods in the Chad Basin[15] are based on the nomadic breeding of dromedaries in the Saharan regions and on the breeding of cattle and other ruminants in the Sahelian and Sudanese climatic zones.[16] Complementary agricultural activities of the pastoral communities are the cultivation of dates in the Saharan oasis, of millet in the Sahel and of Sorghum cereals in the Sudanese zone. Wild cereals contribute significantly to food security (for example Tubiana and Tubiana 1977). According to differences in the management of water- and feed-resources three major types of pastoral livelihood systems are distinguished (Fig. 4.1): (a) The use of dry pastures and exploitation of groundwater beyond 10 meters depth (traditional mastery of well-construction even beyond 40 meters): for example *Dazagada* breeders of cattle. (b) The dependence on green feed and surface water: for example *Fulbe, Boudouma, Kouri*. (c) Mixed systems: use of dry/green feed with limited capacities to access ground water (traditional pastoral wells are simple and not deeper than 10 meters): for example *Juhayna* pastoralists.

Although only 5.7 per cent of the population in Chad were officially counted as 'nomadic' during the last national census in 1993 (BCR 1994), it is commonly estimated that 32 per cent of the rural Chadian population are pastoralists (Ministère de l'Elevage du Tchad 1998, OECD 2006, Toutain et al. 2000). Extensive and mobile pastoral systems manage at least 80 per cent of all livestock in Chad, estimated at 16 million Tropical Livestock Units (TLU), composed of 8.8 million cattle, 3.5 million dromedaries, and 7.5 million small ruminants, representing a standing capital of 2-3 billion Euro in 2004 (data from Bonnet et al. 2004: 5-10).[17] In general, little data exists on the structure of livestock ownership.

---

12   For details on 'Toubou' see Baroin (op.div.), Chapelle (1982), Clanet (op.div.), Fuchs (1979), Le Coeur (1950), Yosko (op.div.), Tubiana and Tubiana (1977).

13   For details on Fulbe in the Chad Basin see Boutrais (1995), Dupire (1962), Seignobos (2000).

14   For details on the history of the Arab migration Braukämper (1992: 107-16, 1994), Zeltner (1969, 1980: 85-103). Cf. Cunnison (1966), Fuchs (1979) and Owens (1994).

15   For typologies currently applied in Chad see Bonnet (2004: 11-15) and Reounodji et al. (2005): For example eco-climatic zoning, type of livestock and mobility, dependence on pastoral income etc.

16   For detailed overview see Bonfiglioli (1992, 1990), Bonnet et al. (2004), CIRAD & CTA (1996), CIRAD & IEMVT (1989), Clanet (1994), Fuchs (1979), Reounodji et al. (2005), le Rouvreur (1989), Yosko (1994).

17   The authors critically discuss the current estimations of livestock populations in Chad, which are commonly based on simple extrapolations of data from the last national

However, evidence for western Chad indicates trends of pauperisation among cattle-breeding pastoral communities, and a transfer of livestock-capital towards urban elites, who hire pastoral workforce with significant hazards to the ecological and social sustainability of livestock breeding (Wiese 2004).

National tax-income from livestock-exports officially increased to 99 million Euro in 2004, covering 29.9 per cent of the national budget (OECD 2006). It is estimated however, that only 35 per cent of cattle exports are registered (Bonnet et al. 2004, Duteurtre et al. 2002). Consequently, official accountancy might not only underestimate the export value of cattle by two-thirds, but the export of dromedaries is entirely informal and, hence, not legally taxed at all. The potential revenues from livestock export and the added value from pastoral livestock production may significantly exceed the direct contributions of oil-production to the national budget and the gross national product of Chad. However, numerous epizootic diseases are endemic in Chad (for example anthrax, brucellosis, tuberculosis, foot-and-mouth disease, and so on) which are prohibitive to the trading of Chadian livestock products on global markets (Diguimbaye et al. 2004, OIE 2004, 2009, Schelling et al. 2004). Consequently, Chadian livestock is only traded on a regional scale.

Cross-Border Cattle Trade[18]

Exported cattle livestock, estimated at 520,000 heads annually, consists mostly of animals 'on the hoof', literally walking from inner Chad to Cameroon, most of them heading for the important markets for beef in northern and central Nigeria (Bonnet et al. 2004, Duteurtre et al. 2002, Reounodji et al. 2005). The export facilities in Chad are simple: Cattle are driven by hired herders from the areas of production in inner Chad towards the national borders, most commonly walking several hundred kilometres. Livestock export from Chad consists, therefore, almost exclusively of animals raised in extensive, mobile pastoral systems.

The export of cattle has been organised by an oligopoly of Chadian families, descending from a pre-colonial, urban class of long-distance Muslim traders (Arditi 2003) affiliated to the former realms of Bornou, Baguirmi and Ouaddaï. Fluent in Arabic and experienced in the trans-Saharan trade of slaves and ivory towards North Africa and the Arabian world, their commercial activities were reoriented during colonial times (for example during the first half of the twentieth century) to the export of livestock and the trading of cereals, oriented towards the emerging economic and demographic centres South and West of Chad.

In connection with such long-distance trade in pre-colonial times, the pastoral societies of the Chad Basin played an indirect role offering transport services,

---

livestock census in 1976. Comparing such estimations with empirical evidence, the authors conclude that national (and consequently international) livestock statistics seriously underestimate present livestock populations in Chad.

18   This section does not account for the poorly known trade with dromedaries, exported to the northern and eastern neighbouring countries.

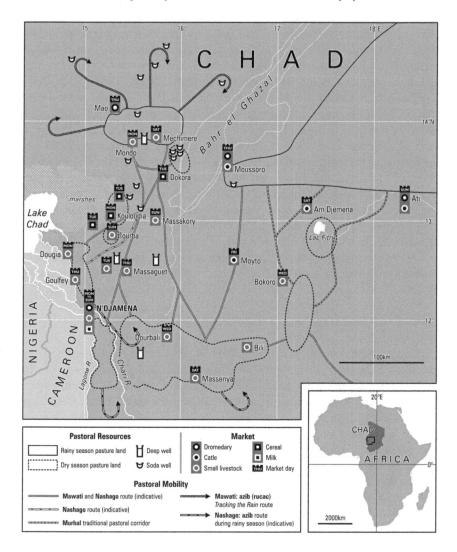

Figure 4.2   Pastoral livelihood system of the Arab Dazagada in western Chad
Source: Author's fieldwork.

security-guarantees, cavalry, and conducting raids. Until the twentieth century, the subsistence oriented exchange of livestock for cereals was largely non-monetarised and the lack of mastery in Arabic excluded major pastoral communities (for example *Toubou*) from trading opportunities (Arditi 2003). During colonial times monetarised taxation was introduced, forcing pastoral communities to commercialise livestock within an increasingly institutionalised setting (Bourgeot 1999).

	Dazagada of *Bahr el Ghazal*		Arab Juhayna of *Batha*	
	Kréda (Kéré)	*Daza*	Migrated into Kanem	Remained in Batha
Origin	Migration from meridional fringes of the Saharan desert *(Kiri)* southwards into Kanem, Bahr-el-Ghazal: breeders of cattle during 17th/18th century, breeders of dromedaries during 18th/19th century		Origins in Yemen arrived in Chad Basin in the course of 15th century; installed in Central Chad 18th/19th century; evoke a legacy of common descent to sedentary *'Arab Baggara'* in Chari-Baguirmi (Fig 1)	
Principal livestock	Cattle, sheep	Dromedaries, goats	Dromedaries, goats	Dromedaries, goats, cattle, sheep
Source of water & feed	Type A: dry pastures, groundwater >40m depth by traditional wells		Type C: dry/green pastures, surface- and groundwater (not deeper than 10 m)	
N-S extend of annual pastoral mobility	150-300km; cross central Chari-Baguirmi as level of groundwater beneath 50m		200-400km	250-600km
Changes in pastoral mobility (last 30 years)	Extension approximately 4 km/year southward in a 'saltatory' way; new routes westward to Lake Chad		Refugees from civil war and drought, took refuge 1985 in Kanem and Chari-Baguirmi	Shift and extension southward

Figure 4.3   Dazagada and Arab pastoral livelihood systems in western Chad
Source: Author's fieldwork.

Since the early twentieth century, an oligopoly of livestock traders managed the transfer of livestock from the markets in inner Chad towards the relay markets for regional livestock trading, which are presently located in Cameroon and northern Nigeria. The revenues from livestock-selling on these regional markets are locally reinvested into consumer goods to be imported into Chad. As this coupled system of export-import is particularly lucrative whenever legal taxation can be avoided, new actors have emerged since the 1980s, belonging to political elites who benefit from clientelism and nepotism at the highest levels in Chad (Arditi 2000, 2003).

Indeed, the risks connected with livestock export are high and diverse, so that investments into export facilities remain marginal. Theft and robbery, diseases, illegal taxes (accounting for even 25 per cent of the final commercial value of cattle), and the fluctuating value of the Nigerian currency, are the main hazards faced by livestock export from Chad (Reounodji et al. 2005). The role of livestock producers in livestock trading has therefore remained marginal until today and is limited to the selling of animals on the local livestock markets inside Chad (Fig. 4.2). Their share of benefit remains significantly under 50 per cent of the final value of cattle sold on the markets of Nigeria or Cameroon (Reounodji et al. 2005).

The case study is based on research conducted by the author between 1996 and 2004 among two pastoral communities in the region of Kanem and Chari-Baguirmi in western Chad:[19] the *Dazagada* from *Bahr-el-Ghazal*, who are nomadic breeders of cattle, and the *Arab Juhayna* of – supposedly – Yemenite origin, who

---

19 Participatory observation, mental mapping, focused group discussions and formalised interviews were combined with questionnaires in a random sample. Research was conducted on 'camel-back' covering two annual cycles (Wiese 2004, Wiese, Yosko and Donnat 2008 for details).

	Dazagada of *Bahr el Ghazal*	Arab Juhayna in *Kanem*
Physical, natural and human assets: Access to ecologically conditioned pastoral resources	**Cattle**: particularly susceptible to the deprivation of water, most critical situation in transitional period from dry to rainy season *(harshar)*; best pastures in Kanem; low demand for soda, moreover main soda sources owned by Dazagada. Each lineage possesses cattle, dromedaries and ouadis (i.e. wells & pastures) throughout Kanem. Particular capacity to construct wells of up to 60 metres depth and up to 3 times a year; rarely pay for access to village wells; strictly avoid any open source of water except ponds in ouadis during rainy season: • use most peripheral pastures especially during dry-season; • particular degree of dispersion (whenever considered to be safe) in areas linked by opportunist trajectories individually for each camp; • increased possibility of conflicts during dry season due to access to key-resources (pastures, wood for construction of wells) in regions with weak social ties or relationships not yet established (Chari-Baguirmi); particularly serious on the way to dry-seasonal pastures (access to water along route).	**Dromedaries**: resist drought, but particularly vulnerable to humid conditions; pastures are herbal as well as tree layers, therefore low competition with cattle in case that trees and bushes are present; low demand for water; but high demand for soda throughout the year; as 'refugees' do not own ouadis. Construct wells only if ground water less than 10 metres deep; also use open water of rivers and ponds if accessible: • depend on access to wells of sedentary people or to open water throughout year; • separation of livestock from camps allows concentration of nomadic communities close to urban centres for commercial activities (milk); • prone to conflicts during rainy season in Kanem: passage northwards to escape humidity under time-pressure, access to soda for dromedaries and to water for human consumption, but very limited and mostly conflictive social relationships.
Economic assets: Opportunities for income, commercial exchange and information	A general tendency among pastoralists and former agro-pastoralists towards diversified pastoral livelihood with opportunist cultivation in the southern pasture lands and with little land ownership; recent trend to diverse, secondary commercial activities including temporary working migration to Libya and Nigeria.	Specialised in pastoral production, Agriculture rare. Commercialisation of milk products in urban centres of Chari-Baguirmi became one pillar of Arab economy; breeding of dromedaries for export to Libya, Egypt, Sudan (and Saudi Arabia?); transport services.
Social & human assets: Language and education	Illiteracy in French language approximately 100%; attendance rate of primary school <3%	
	• More frequent attendance of advanced Arab school; traditional schooling outside own camp in mobile schools or village; considerable proportion of alphabetisation in Arab language	• Low formal education or attendance of traditional schools except for Koranic school in own camp; probably very low proportion of alphabetisation in Arabic
Experience outside the pastoral sector	• > 25% of male adults had once worked outside pastoral sector and visited a foreign country	• Rare

Figure 4.4   Dazagada and Arab pastoralists' endowment with assets
Source: Author's fieldwork.

are nomadic breeders of dromedaries (Fig. 4.3 and 4.4). The principal patterns of pastoral mobility are similar for Dazagada and Arab nomadic communities. They are basically north-south oriented with principal pasture-lands at the extremes: a meridian pole of dry-seasonal pasturelands in the wooded savannahs on the clay-plains of Chari Baguirmi and a northern pole of rainy-seasonal pasturelands in the Sahel scrublands covering the Pleistocene sand dunes (Old Erg) of Kanem.

	Dazagada breeders of cattle (%)	Arab breeders of dromedaries (%)
Based on sample size (household heads)	*N=129*	*N=137*
Livestock of family comprises several species (cattle, sheep, goats and dromedaries)	72.9	9.6
Women sell milk products	9.3	33.6
Engaged in agriculture during last year period since beginning of cultivation *(years)*	79.1 *7.8±8.7 (m=5)*	21.0 *1.5±1.2 (m=1)*
Family's demand in cereals covered by their own harvest *(months)*	*4.0±3.4 (m=3)*	*3.9±3.6 (m=3.5)*
Owner of livestock	85.3	92.0
Herder	14.7	8.0

Figure 4.5   Physical and economic assets among adult male Dazagada and Arab Juhayna

Note: Household heads only.

Source: Author's fieldwork.

Both systems have undergone important changes during the last three decades (before 2004), triggered by droughts and warfare. From an historical point of view, the southward extension of Dazagada pastoralism and the westward shift of Arab pastoralism from central into western Chad constitute continuity rather than an exceptional event.

Pastoral livelihoods are not only determined by the distribution of ecologically conditioned pastoral resources, but also by the availability of pastoral key-assets[20] the capitals and capabilities necessary to gain access to pastoral key-resources (for example water, pastures, and increasingly markets); to avoid conflicts and hazards; and to profit from opportunities for non-pastoral income; exchange and information. Each livelihood system is characterised by a specific mix of human, physical, social and financial assets (see Bohle, this volume; Fig. 4.4-4.6). However, the quality of social relationships is the single-most important determinant of pastoral spaces in addition to the presence of ecologically conditioned resources (Wiese et al. 2004).

On average, one male Dazagada income-earner is responsible for the well-being of nine other people ($\mu=9\pm9$ median=7), whereas one male Arab income-earner is responsible for the well-being of seven other persons ($\mu=7\pm4$ median=6; for details see Fig. 4.7). This responsibility includes covering expenditure (food and especially cereals, health, education, social events, etc.) as well as decision-making related to livelihood security. The difference observed for dependency ratios[21] is partly explained by child fostering and selective sedentarisation among

---

20   Asset = A wide range of tangible and intangible stores of value, investments or claims to assistance that can be mobilised to secure livelihood (based on Swift 1989: 11). Following current livelihood frameworks (based on Carney 1998), 'tangible assets' refers to 'capitals' and 'intangible assets' to 'capabilities' and 'opportunities'.

21   Dependence ratio = ratio between economically dependant persons and income-earners. Among the pastoral societies investigated by the author, income earners are male adult

	Dazagada breeders of cattle (%)	Arab breeders of dromedaries (%)
Based on sample size (male adults)	*N=160*	*N=164*
Attended Koranic school	93.8	74.4
in mobile school outside camp	46.7	3.3
in a village/town	41.3	9.8
Attended higher Arab school *(Madrasa)*	8.1	2.4
Once attended a public primary school	3.1	1.2
Able to read classic Arabic (as Koran)	28.3	4.3
Able to communicate in at least 2 languages	63.8	0.6
Speak some French (official language in Chad)	2.5	0.0
Once worked outside pastoral sector	26.9	7.9
Working-experience outside Chad	28.1	7.3

Figure 4.6   Social assets among adult male Dazagada and Arab Juhayna
Source: Author's fieldwork.

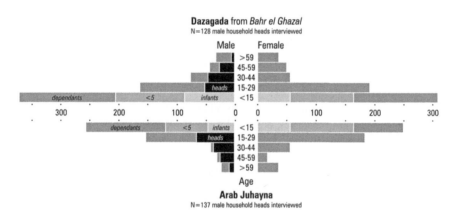

Figure 4.7   Dependency ratio in a random sampling among male Dazagada and Arab household heads
Source: Author's fieldwork.

the Dazagada communities: Better-off family members take care of children of closely related, poorer families who might sedentarise. Well-off pastoral families can thereby increase available workforce, whereas the children of poor families have the opportunity to acquire the capacities and knowledge of a successful pastoralist. The dependency ratio, which is an operational proxy-indicator of

household heads. Arab women, who sell milk, use their revenues for personal needs (except food and major medical expenditures), for buying the ingredients for sauces and for basic child-care. The paragraph is based on the results of a random sample among adult (age>15) male Arab (N=137) and Dazagada (N=128) household heads at the soda-wells of Am Harba (26 October to 22 November 1999). For methodology and statistics see Wiese (2004).

social-economic status in these pastorals societies, was statistically related to the pastoralists' endowment with social key-assets: a higher level of traditional education and/or a particular capacity; exclusive knowledge and experience among the Dazagada; and a position of power among the Arab.

Beyond the household, the lineage or the clan is a flexible genealogical 'unit' which is mobilised for providing social support during life-events and security in uncertain situations, which exceeds the families' or households' capacities to cope (for example Baroin 1985). Consequently, these units are best conceived as flexible and problem-oriented groups who temporarily collaborate to cope with specific events and to manage complex or difficult tasks including the selling of livestock on market sites and other major economic transactions (especially, if connected with social obligations, for example marriage, *diye*[22]).

Livelihood and Markets

The commercial activities of Arab and Dazagada pastoral communities in western Chad are related to the marketing of livestock, of dairy products, and of some services (for example herding; transport of livestock, cereals, dates and so on) in the immediate proximity to their customary pasturelands. Moreover, they might engage in limited speculative trading of livestock (for example buying exhausted or young animals to be sold after recovery or other increase in value) and of further traditional commodities (dates, cereals, soda).[23] Working migration and temporal employment outside the pastoral sector are complementary strategies used to build up and to secure pastoral livelihoods (Fig. 4.6; Clanet 1981).

In this context, markets are composed of different institutions. At one level government institutions regulate transactions and at another level a person (warrantor) from the genealogical unit of the lineage or clan embodies the longstanding obligations that result from market exchanges.

Cantons, basic units of central administration in Chad, have been superimposed on the genealogical matrix of pastoral societies since colonial time, intervening mainly at the level of a clan. As a result, a limited mandate, authority and salary have been attributed to selected local leaders, who were willing to cooperate with the central authorities. Even today they are expected to serve as an interface between pastoral communities and a central administration. On market sites and in administrative centres, the pastoral cantons are represented by a deputy of the cantonal chief, the *aliifa*, serving as the local relay on behalf of the central

---

22   *Diye* = 'blood price', i.e. a traditional compensation paid for killing a person or causing his/her accidental death.

23   Reliable detailed data on commercial activities among pastoralists are not known for Western Chad. A study conducted by CIRAD & VSF (2002) remained inconclusive since a straight-forward approach for the assessment of pastoral household economies failed (at least regarding Dazagada and Arab pastoral communities; Donnat 2006).

administration towards the pastoral communities. The *aliifa*, on behalf of the cantonal chief, usually receives fees whenever a member of his canton sells an animal on the market site or uses other services.

From the clan perspective, a warrantor (Arab *damiin*, Dazaga *dawré*, *küré*) who is a member of the pastoral community is expected to safeguard the quality and the legitimacy of each animal sold or bought on the market by a member of his own lineage or clan. In exchange he receives a fee for each transaction, which has to cover his obligations of hospitality and support. The warrantor also intervenes in the general relationship between the pastoral and other communities. His legitimacy is based on trust; his moral authority is rooted in traditional ties of kinship, solidarity, and family obligations.

Commercial activities – as they materialise at the market site – are organised at different levels: within local socio-economic frames, structuring day-to-day interaction (household, camp, chain of camps); according to genealogical patterns, structuring situational solidarity and major economic transactions (kinship: family, lineage, clan); and within political-administrative settings, structuring the interface towards state authority (canton, prefecture). The benefits from each commercial transaction are widely re-invested in terms of monetary and social assets throughout these structures. The ability to mobilise such assets when needed is a particularly important determinant of future livelihood security. Consequently, commercial activities and livelihood security of the Arab and Dazagada pastoral societies in western Chad are inseparably related by the same organisational framework.

Role of Livestock Commercialisation

Selling or less importantly, barter trade of livestock play different roles in the livelihood security strategies for the Dazagada and Arab breeders (Fig. 4.5). The role of livestock commercialisation is particularly critical for the cattle breeders specialised on the use of dry pastures, such as the Dazagada. Milk production is limited and rarely commercialised: only 9.6 per cent of interviewed household heads reported such activities. Consequently, the selling of livestock constitutes the almost sole source of monetary revenues to secure the pastoral livelihoods. Only well-off households may afford to keep some dromedaries, though for eco-climatic reasons, these cannot remain with the cattle. As for most pastoral communities, agriculture plays only a marginal role, which however is gaining importance in the case of the southern Dazagada: 79.1 per cent of the Kréda households interviewed had cultivated cereals over the last season, covering some subsistence requirements equivalent of 4 months of cereals for average family (Fig. 4.5). The rational is to reduce expenditures on cereals, since cattle products have lost purchase/ barter power against cereals during the last decades (loss of 50 per cent over 3 decades; see Fig. 4.8). To compensate for this high and, under the present market conditions, increasingly unfavourable degree of specialisation, families tend to diversify livestock (72.9 per cent of interviewed households keep several species).

CONTEXT	• Short- and medium-term climatic trends and general southward shift of pastoral systems; • Counterproductive legal frameworks as colonial heritage (e.g. *Loi cadre No.4*); • Weak democracy, bad governance, and heritage of warfare (erosion of conflict-management capacities); • Illiteracy of pastoralists.

PRESSURES	**Insecurity as experienced by pastoralists:**	
	Dazagada: Prolongation of annual routes into new pasture lands with generally poor pastoral resources. Increasing physical efforts to feed, water and guard livestock under growing time-pressure, with few social relationships to facilitate access to key-resources; consequently increased insecurity due to conflicts, abuse and further hazards; monetarisation of social relationships and increasing costs of conflicts whereas pastoral terms of trade decrease for cattle. Consider themselves as losers of recent trends.	***Arab Juhayna***: Emigrated from Central Chad in the mid-1980s to flee civil war; managed to open up new markets (e.g. milk in N'Djamena, local transport services, small livestock in urban centres,...) and pasture lands in Western Chad based on a legacy of common prestigious Yemenite descent with some local communities; experience an increasingly conflictive situation in their main pastoral corridor in Western Kanem due to competition with local owners of livestock. As breeders of dromedaries they are less affected by ecological degradation of Chari-Baguirmi. Consider themselves rather as winners of recent trends if compared to former situation encountered in Central Chad.
	Increased needs in financial assets resulting from:	
	• Demographic growth and expansion of agricultural cash-crops lead to degradation of already limited pastoral resources in Chari-Baguirmi; • Monetarisation of social relationships and access to pastoral resources; • Decentralisation in a context of weak democracy and bad governance: conflict management as source of income for local authorities; • Break-down of veterinary services after reform and privatisation; • Decreasing pastoral terms of trade for cattle: 50% in 30 years.	

Figure 4.8   Causes and consequences of pressured assets

Source: Author's fieldwork.

Additional income can be generated by temporary migratory or seasonal work, which helps constituting a herd and dowry to build a household. Temporary work outside the pastoral sector was quoted by 26.9 per cent of interviewed household heads, most of them went outside Chad – to seek opportunities but also to keep discretion with regards to work perceived as eventually disgraceful. As outlined in the next section, various factors increased the demand for monetary income for pastoralists in the central Chad Basin over the last decades, whereas the terms of trade have become increasingly unfavourable especially for cattle products.

In contrast to these highly specialised Dazagada cattle producers, the Arab Juhayna breeders of dromedaries currently find more favourable opportunities to sustain livelihoods. Market mechanisms and prices, though poorly understood by researchers in the context of Chad, seem favourable to sustain and even expand the breeding of dromedaries to an extent that official figures for livestock populations may underestimate the actual number of dromedaries in Chad by an order of magnitude (Bonnet et al. 2004). In addition to this, households breeding dromedaries can generate significant revenues by offering transport services and by selling milk. Since rearing of dromedaries requires specific skills and high degree of mobility, these pastoralists tend to specialise: 90.4 per cent of interviewed households kept dromedaries only. Agriculture is carried out less frequently: 22.1 per cent of households reported such activity for the last season (compared to 79.1 among cattle breeding Dazagada), covering approximately 4 months of household demand in cereals (Fig. 4.5). The market opportunities open to the breeders of dromedaries may, under favourable circumstances, determine the current mobility patterns of these communities (Wiese 2004). For example, the mid-80s migration of Arab Juhayna pastoralists into western Chad during civil war times, opened more recently new market opportunities especially for milk products in the urban centres and especially of the capital N'Djaména. 33.6 per cent of all Arab household heads interviewed in a cross-sectional survey by the author (Fig. 4.5 for details) reported that women members of their household usually sell dairy products (compared to 9.6 per cent among cattle breeders). Markets for milk products and opportunities for transport services have become so favourable to the Arab breeders of dromedaries, that these benefits outbalance the eco-climatically disadvantageous conditions of the southern Sahel. The pastoral communities have therefore been innovative in their social organisation, gender division of work, and patterns of mobility over the last three decades.

This brief overview implies, of course, schematisations. The next sections will hence examine in detail, how livelihood security and marketing of livestock are interrelated within the complex realities of two different pastoral livelihood systems.

Recent Pressures on Assets

Three recent developments impact on pastoral livelihoods: (a) Local civil and military authorities, which have multiplied as a result of a 'decentralisation'-policy engaged since 1996, have increasingly used conflicts as a source of income. Legally they are covered by a colonial heritage, including a pastoral code known as *Loi cadre No. 4* dating from 1959. This legal framework combines myths, prejudice and paternalism and directly criminalises the key-strategies of pastoral livelihood security: mobility; opportunist decision making; and social flexibility (Yosko 1999, 2000). (b) Because Dazagada and Arab pastoralists have almost no access to formal schooling (Fig. 4.6), they have particularly weak capacities in reading or writing and, consequently, in defending their rights and preventing abuse by authorities. Consequently, local authorities consider livestock as a source of wealth on four legs, easy to confiscate or to tax with little risk. (c) The national veterinary system has virtually broken down as a consequence of privatisation policies since 1996. From 50 private veterinaries working between 1996 and 1999, only three were still operational in 2005 (Reounodji et al. 2005: 61). Veterinary drugs and treatments are mainly acquired by pastoralists from non-licensed drug sellers on market sites, who are basically motivated by commercial aspects, with little investment and almost no qualifications (Reounodji et al. 2005).

Arab and Dazagada nomadic pastoralists in Chad were, however, among the most successful pastoral communities throughout the Sahel in coping with the droughts in the 1970s and 1980s: Their losses in livestock barely reached 30 per cent compared with average losses exceeding 80-90 per cent among most other livestock breeders in African Sahel (Clanet 1994: 329-40, 1977). This observation is even more remarkable since civil war and general insecurity devastated the country during the same period. Indeed, the mastery of long-distances, and the existence of strong and flexible networks proved to be key-assets for Dazagada and Arab pastoralists. In their words 'for nomadic people droughts do not exist' (Wiese 2004: 256).

Drought and civil war have had long-lasting impacts on the pastoral mobility in western Chad. Pastoral mobility has substantially increased among the Dazagada and some Arab communities have entirely shifted their pastoral systems into western Chad. While these traditional coping-mechanisms were highly effective during the crisis, the context of pastoral livelihoods has recently changed, and new pressures connected with today's multiple crises situation in Chad undermine their livelihood security. Their key-assets are increasingly monetarised (Fig. 4.8). Moreover due to short- and medium-term climatic trends, Saharan pastoral systems increasingly extend southward into the Sahelian pasture lands of Kanem. As a consequence, the Dazagada of Kanem are obliged to shift their dry-seasonal pastures further south into the clay-plains of Chari-Baguirmi with unfavourable conditions: a deep level of groundwater; poor pastures; and new hazards to the health of their livestock (anthrax, for example, is endemic in Chari-Baguirmi). Consequently, the pastoralists' workload, time-pressure and the dispersion of the family unit have considerably increased. High population growth in Chari-Baguirmi and an expansive cash-oriented production of counter-seasonal sorghum (*berberé* = Sorghum bicolour), of rice, horticulture, gum Arabic, and charcoal are leading to a serious degradation of pastoral key-resources (Acacia woodlands, seasonal ponds, pasture) in Chari-Baguirmi and the border-regions of Lake Chad (Ogier et al. 1998). Dazagada and Arab pastoralists increasingly search for new pasturelands. In the absence of social relationships in these new areas, the access to pastoral key-resources is increasingly restricted. Competition for key-resources in these areas triggers increasingly frequent violent clashes (for Dazagada in Chari-Baguirmi and for Arabs in Kanem close to Lake Chad), as traditional mechanisms of conflict-resolution were eroded during the civil war.

To sum up: The need for cash-income has increased and the pastoral terms of trade for cattle (i.e. the exchange value of cattle versus cereals) have decreased by 50 per cent on the local markets in western Chad during the past 30 years (Reounodji et al. 2005). Consequently, Dazagada breeders of cattle judge the recent changes as negative. In contrast, the Arab breeders of dromedaries have benefited from increasing terms of trade for dromedaries in Chad. Moreover, their animals are well adapted to the ecologically degraded conditions of Chari-Baguirmi allowing for new markets to sell milk and offer transport services. Consequently these Arab communities judge their overall situation to have improved since leaving central Chad during civil war.

Conclusion

The global livestock sector is undergoing profound changes in response to the globalisation of markets and the rapid transformation of commodity chains. The social and environmental consequences are important since small-scale livestock producers, especially in Sub-Saharan Africa, are increasingly marginalised. However, pastoral livelihood systems offer sustainable production in dryland areas. In Chad, pastoral livestock production remains an important pillar of the national economy even after the start of oil exportation in 2003. Livestock breeders manage 80 per cent of all livestock estimated at 16 million TLU, securing the livelihood of 40 per cent of the rural population, and contributing one third of the national budget through livestock export. This situation is particularly surprising since the facilities for livestock export from Chad are rudimentary. Although pastoral societies, like the Dazadaga and Juhayna, are proof of a remarkable resilience during the latest pan-Sahelian droughts and civil war in Chad, they face an increased encroachment upon day-to-day livelihood security. However, differences in livelihood securities between cattle and dromedaries breeders are remarkable.

References

Aliou, S. 1995. People on the move. *World Health*, 48(6), 26-27.

Arditi, C. 1994. Le commerce, l'islam et l'état au Tchad (1900-1990), in *L'identité tchadienne: L'apport des peuples et les apports extérieurs*, edited by J. Tubiana, C. Arditi, C. Pairault. Paris: L'Harmattan, 311-354.

Arditi, C. 1995. Le commerce des dattes du Borkou (Tchad). *Cahiers des Sciences humaines (ORSTOM)*, 31(4), 849-882.

Arditi, C. 2000. Du "prix de la kola" au détournement de l'aide internationale: Clientélisme et corruption au Tchad (1900-1998), in *Monnayer les pouvoirs. Espaces, mécanismes et représentations de la corruption*, edited by G. Blundo. Paris: IUED/PUF, 249-269.

Arditi, C. 2003. Le Tchad et le monde arabe: Essai d'analyse des relations commerciales de la période précoloniale à aujourd'hui. *Afrique contemporaine*, Automne 2003, 185-198.

Azarya, V. 1996. Pastoralism and the state in Africa: Marginality or incorporation? *Nomadic Peoples*, 38, 11-36.

Azevedo, M.J. and Nnadozie, E.U. 1988. *Chad: A Nation in Search of Its Future*. Boulder, Oxford: Westview Press.

Baroin, C. 1985. *Anarchie et cohésion sociale chez les Toubou: Les Daza Késerda (Niger)*. Cambridge: Cambridge University Press.

Baroin, C. 1986. Organisation sociale et prestations matrimoniales chez les Toubous. *Journal des Africanistes*, Paris: CNRS, 7-27.

Baroin, C. 1987. The position of Tubu women in pastoral production. Daza Kesherda, Republic of Niger. *Ethnos*, 52(1-2), 137-155.

Baroin, C. 1988. *Gens du roc et du sable. Les Toubou. Hommage à Charles et Marguerite Le Coeur*. Paris: Editions du CNRS.

Baroin, C. 2003. *Les Toubou du Sahara Central*. Paris: Editions Vents de Sable.

BCR. 1994. *Récensement géneral de la population et de l'habitat 1993*. N'Djamena: Ministère du Plan et de Cooperation & Ministère de l'Intérieur et de la sécurité.

Bonfiglioli, A.M. 1990. *Eleveurs du Tchad Oriental. Repères socio-économiques sur l'élevage du Ouaddai et du Biltine*. N'Djaména: GTZ–PEA.

Bonfiglioli, A.M. 1992. *L'agro-pastoralisme au Tchad comme stratégie de survie: Essai sur la relation entre l'anthropologie et la statistique*. Washington: World Bank.

Bonfiglioli, A.M. and Watson, C. 1992. *Sociétés pastorales à la croisée des chemins: Survie et développement du pastoralisme africain*. New York: UNICEF/UNSO, Project for Nomadic Pastoralists in Africa (NOPA).

Bonnet B., Banzhaf, M., Giraud, P.-N. and Issa, M. 2004. *Analyse des impacts économiques, sociaux et environnementaux des projets d'hydraulique pastorale finances par l'AFD au Tchad*. (Rapport provisoire). Montpellier-Paris: IRAM.

Bourgeot, A. 1999. *Horizons nomades en Afrique sahélienne*. Paris: Karthala.

Boutrais, J. 1995. *Hautes terres d'élevage au Cameroun*. Paris: ORSTOM.

Braukämper, U. 1992. *Migration und ethnischer Wandel. Untersuchungen aus der östlichen Sudanzone*. (Studien zur Kulturkunde 103). Stuttgart: Franz Steiner Verlag.

Braukämper, U. 1994. Notes on the origin of Baggara Arab culture with special reference to the Shuwa, in *Arabs and Arabic in the Lake Chad Region*, edited by J. Owens. Köln: Rüdiger Köppe Verlag.

Buijtenhuijs, R. 1987. *Le Frolinat et les guerres civiles du Tchad (1977-1984). La révolution introuvable*. Paris: Karthala.

Buijtenhuijs, R. 1989. Chad: The narrow escape of an African State, 1965-1987, in *Contemporary West African States*, edited by D.B.C. O'Brien, J. Dunn and R. Rathborn. Cambridge: Cambridge University Press, 49-58.

Carney, D. 1998. Implementing the sustainable rural livelihoods approach, in *Sustainable Rural Livelihoods. What Contribution Can We Make?*, edited by D. Carney. London: DFID, 3-23.

Chapelle, J. 1982. *Nomades noirs du Sahara*. Paris: Plon (rééd. of 1957), Paris: L'Harmattan.8

Chapelle, J. 1986. *Le peuple tchadien. Ses racines et sa vie quotidienne*. Paris: L'Harmattan.

CIRAD and CTA. 1996. *Atlas d'élevage du Bassin du Lac Tchad*. Wageningen, Pays-Bas: CTA.

CIRAD and IEMVT. 1989. *Elévage et potentialités pastorales sahéliennes. Synthése cartographique Tchad*. Montpellier: IEMVT.

CIRAD and VSF. 2002. *Etude sur les sociétés pastorales au Tchad, rapport de synthèse*. N'Djaména: Ministère de l'Elevage au Tchad, Programme de Sécurisation des systèmes pastoraux.

Clanet, J.C. 1977. Les conséquences des années sèches 1969-1973 sur la mobilité des éleveurs du Kanem. *Travail et documents de géographie tropicale*, 30, Bordeaux: CEGET.

Clanet, J.C. 1981. L'émigration temporaire des Toubou du Kanem vers la Libye. *Cahiers géographiques de Rouen*, 15, 17-33.

Clanet, J.C. 1994. *Géographie pastorale au Sahel central*. PhD Thesis, Paris IV Sorbonne, Paris.

Clanet, J.C. 1999. Stabilité du peuplement nomade au Sahel central. *Sécheresse*, 10(2), 93-103.

Cœur, C. le. 1950. *Dictionnaire ethnographique téda*. Paris: Larose.

Cunnison, I. 1966. *Baggara Arabs: Power and the Lineage in a Sudanese Nomad Tribe*. Oxford: Clarendon Press.

Dickow, H. 2005. Democrats without democracy? Attitudes and opinions on society, religion and politics in Chad. *Letters from Byblos*, 11, Byblos: UNESCO, International Centre for Human Sciences.

Diguimbaye, C., Schelling, E., Pfyffer, G.E. et al. 2004. Séroprévalence des maladies zoonotiques chez les pasteurs nomades et leurs animaux dans le Chari-Baguirmi du Tchad. *Médecine Tropicale*, 64, 482-485.

Djindil, N., Ndang, T.S. and Toina, M.A. 2007. *Who Benefits from Social Expenditures in Chad? An Incidence Analysis Using Survey Data*. (PMMA Working Paper 2007-11). PEP & IDRC.

Dobie, P. 2001. *The Global Drylands Initiative, 2001-09 – Challenge Paper: Poverty and the Drylands*. [Online]. Available at: http://www.undp.org/drylands/docs/cpapers/Poverty%20and%20the%20Drylands.doc [accessed: 3 October 2010].

Donnat, M. 2006. *Espaces pastoral, médical et sanitaire: Le recours aux soins en zone sahélienne. Le cas des communautés arabes Juhayna et Dazagara du Bahr-el-Ghazal au Tchad*. PhD Thesis, Université Paul Valéry de Montpellier III.

Dupire, M. 1962. Peuls nomades. *Traveaux et Mémoires de l'Institut d'Ethnologie*, 64, Université de Paris.

Duteurtre, G., Koussou, M.O., Essang, T. and Kadekoy-Tiguague, D. 2002. *Le commerce de bétail dans les savanes d'Afrique centrale: Réalités et perspectives*. (Actes du colloque: Savanes africaines: Des espaces en mutation, des acteurs face à de nouveaux défis, Garoua, Cameroun). [Online]. Available at: http://epe.cirad.fr/fr/doc/kousdut.pdf [accessed: 3 October 2010].

Foggin, P.M., Farkas, O., Shiirev-Adiya, S. and Chinbat, B. 1997. Health status and risk factors of seminomadic pastoralists in Mongolia: A geographical approach. *Social Science & Medicine*, 44(11), 1623-1647.

Foggin, P.M., Torrance, M.E., Dorje, D. et al. 2006. Assessment of the health status and risk factors of Kham Tibetan pastoralists in the alpine grasslands of the Tibetan plateau. *Social Science & Medicine*, 63, 2512-2532.

Foltz, A.-M. and Foltz, W.J. 1991. The politics of health reform in Chad, in *Reforming Economic Systems in Developing Countries*, edited by D.H. Perkins and M. Roemer. Boston: Harvard Institute for International Development, 137-157.

Fuchs, P. 1979. Nordost-Sudan, in *Die Völker Afrikas und ihre traditionellen Kulturen. Teil 2: Ost-, West-, und Nordafrika*, edited by H. Baumann. Wiesbaden: Steiner Verlag, 189-228.

Haraldson, S.R.S. 1979. Socio-medical problems of nomad peoples, in *The Theory and Practice of Public Health*, edited by W. Hobson. Oxford: Oxford University Press, 613-625.

Herrero, M., Thornton, P.K., Notenbaert, A.M. et al. 2010. Smart investments in sustainable food production: Revisiting mixed crop-livestock systems. *Science*, 327(5967), 822-825.

Hill, A.G. 1985. *Population, Health and Nutrition in the Sahel: Issues in the Welfare of Selected West African Communities*. London: Routledge and Kegan Paul.

ICG. 2006. *Tchad: Vers le retour de la guerre?* (Rapport Afrique, 111). Nairobi/ Bruxelles: International Crisis Group.

IMF. 2007. *Chad: Selected Issues and Statistical Appendix*. (IMF Country Report No. 07/28). Washington DC.

Imperato, P.J. 1974. Nomads of the West African Sahel and the delivery of health services to them. *Soc Sci Med*, 8, 443-457.

Jutzi, S. 2006. Foreword, in *FAO Livestock Report 2006*, edited by FAO. Rome: FAO, iii.

Khazanov, A.M. 1994. *Nomads and the Outside World*. 2nd Edition. Cambridge: Cambridge University Press.

Lanne, B. 1998. *Histoire politique du Tchad de 1945 à 1958*. Paris: Karthala.

Loutan, L. 1989. Les problemes de santé dans les zones nomades, in *Planifier, gérer, évaluer la santé en pays tropicaux*, edited by A. Rougemont and J. Brunet-Jailly. Paris: Doin Editeurs, 219-253.

Magrin, G. 2001. *Le Sud du Tchad en mutation: Des champs de coton aux sirènes de l'or noir*. Sépia.

May, R. and Massay, S. 2003. Presidential and legislative elections in Chad, 2001-2002. *Electoral Studies*, 22, 765-807.

Ministère de l'Elevage du Tchad. 1998. *Réflexion prospective sur l'élevage au Tchad.* (Rapport Principal). N'Djaména.

Niamir-Fuller, M. and Turner, M.D. 1999. A review of recent literature on pastoralism and transhumance in Africa, in *Managing Mobility in African Rangelands. The Legitimization of Transhumance*, edited by M. Niamir-Fuller. London: FAO, Beijer International Institute of Ecological Economics.

OECD. 2006. *Perspectives économiques en Afrique 2005-2006: Tchad*. [Online]. Available at: http://www.oecd.org/dataoecd/17/29/36803227.pdf [accessed: 3 October 2010].

OECD. 2009. *African Economic Outlook. Country Notes. Chad*, 149-163. [Online]. Available at: http://browse.oecdbookshop.org/oecd/pdfs/browseit/4109051E. PDF [accessed: 3 October 2010].

Ogier, J., Planel, S. and Magrin, G. 1998. Dynamiques d'un espace entre le lac Tchad et le Chari et relations agriculture-élevage. *Revue Scientifique du Tchad*, 5(2), 9-13.

OIE. 2004. *World Animal Health, Chad.* [Online]. Available at: ftp://ftp.oie.int/SAM/2004/TCD_F.pdf [accessed: 3 October 2010].

Omar, M.A. 1992. Health care for nomads too, please. *World Health Forum,* 13(4), 307-310.

Owens, J. 1994. *Arabs and Arabic in the Lake Chad Region.* Köln: Rüdiger Köppe Verlag.

Pratt, D., Le Gall, F. and Haan, C. de. 1997. *Investing in Pastoralism: Sustainable Natural Resource Use in Arid Africa and the Middle East.* (World Bank Technical Paper No. 365). Washington DC.: The World Bank.

Rass, N. 2006. *Policies and Strategies to Address the Vulnerability of Pastoralists in Sub-Saharan Africa.* (PPLPI Working Papers 37). Rome: FAO.

Reounodji F., Tchaouna, W. and Banzhaf, M. 2005. *Vers la sécurisation des systèmes pastoraux au Tchad: Enjeux et elements de réponse.* (PSSP Axe 1). N'Djaména: DDPAP and DED, Ambassade de France & Ministère de l'Elevage au Tchad.

Rouvreur, A. le. 1989. *Sahariens et Sahéliens du Tchad.* 2nd Edition. Paris: L'Harmattan.

Safriel, U. and Adeel, Z. 2005. Dryland systems, in *Millenium Ecosystem Assessment: Current State & Trends Assessment,* edited by R. Hassan and R. Scholes. Nairobi: UNEP.

Salzman, P.C. 1995. Studying nomads: An autobiographical reflection. *Nomadic Peoples,* 36/37, 157-166.

Sandford, S. 1978. Welfare and wanderers: The organisation of social services for pastoralists. *ODI Review,* 1, 70-87.

Schelling, E., Diguimbaye, C., Daoud, S. et al. 2004. Séroprévalence des maladies zoonotiques chez les pasteurs nomades et leurs animaux dans le Chari-Baguirmi du Tchad. *Médecine Tropicale,* 64, 474-477.

Schelling, E., Weibel, D. and Bonfoh, B. 2008. *Learning from the Delivery of Social Services to Pastoralists: Elements of Good Practice.* WISP/UNDP/IUCN. [Online]. Available at: http://cmsdata.iucn.org/downloads/social_services_to_pastoralists_english_2.pdf [accessed: 3 October 2010].

Scholz, F. 1996. *Nomadismus. Theorie und Wandel einer sozio-ökologischen Kulturweise.* (Erdkundliches Wissen, 118). Stuttgart: Franz Steiner Verlag.

Scoones, I. 1995. *New Directions in Pastoral Development in Africa.* London: International Institute for Environment and Development.

Seignobos, C. 2000. Les Fulbé, in *Atlas de la province Extrême-Nord Cameroun,* edited by C. Seignobos and O. Iyébi-Mandjek. Paris/Yaoundé: IRD, MRST, INC, 52-56.

Sheik-Mohamed, A., Velema, J.P. 1999. Where health care has no access: The nomadic populations of Sub-Saharan Africa. *Tropical Medicine and International Health,* 4(10), 695-707.

Steinfeld, H. and Chilonda, J. 2006. *Old Players, New Players.* (FAO Livestock Report 2006). Rome: FAO, 3-14.

Stürzinger, U. 1980. *Der Baumwollanbau im Tschad. Zur Problematik landwirtschaftlicher Exportproduktion in der Dritten Welt*. Freiburg, Zürich: Atlantis Verlag.

Suttie, J.M., Reynolds, S.G. and Batello, C. 2005. *Grasslands of the World*. (Plant Production and Protection Series 34). Rome: FAO.

Swift, J. 1989. Why are rural people vulnerable to famine? *IDS Bulletin*, 20(2), 8-15.

Swift, J. 1996. Desertification. Narratives, winners & losers, in *The Lie of the Land. Challenging Received Wisdom on the African Environment, African Issues*, edited by M. Leach and R. Mearns. London: The International African Institute, 73-90.

Swift, J., Toulmin, C. and Chatting, S. 1990. *Providing Services for Nomadic People. A Review of the Literature and Annotated Bibliography*. New York: UNICEF.

Thebaud, B. and Batterbury, S. 2001. Sahel pastoralists: Opportunism, struggle, conflict and negotiation. A case study from Eastern Niger. *Global Environmental Change*, 11, 69-78.

Thornton, P.K. and Gerber, P.J. 2010. Climate change and the growth of the livestock sector in developing countries. *Mitigation and Adaption Strategies for Global Change*, 15, 169-184.

Toutain, B., Touré, O. and Réounodji, F. 2000. *Etude prospective de la stratégie nationale de gestion des ressources pastorales au Tchad*. (Rapport 00-04). N'Djaména: CIRAD-EMVT.

Tubiana, M.J. and Tubiana, J. 1977. *An Ecological Perspective: Foodgathering, the Pastoral System, Tradition and Development of the Zaghawa of the Sudan and the Tchad*. Rotterdam: Balkema.

UNDP. 2002. *Rapport mondial sur le développement humain 2002*. Bruxelles/ New York: Éditions De Boeck Université.

UNDP. 2003. *The Global Drylands Imperative. Pastoralism and Mobility in the Drylands*. (UNDP Challenge Paper Series). [Online]. Available at: www.undp. org/drylands/docs/cpapers/PASTORALISM PAPER FINAL.doc [accessed: 3 October 2010].

UNDP. 2006. *Human Development Report 2006*. New York: Macmillan.

UNDP. 2009. *Human Development Report 2009*. New York: Macmillan.

Wiese, M. 2001. *The Pastoral Resource-Niches of Central Chad Basin*. (APT Reports, 12). Freiburg: University of Freiburg, Department of Physical Geography, 62-99.

Wiese, M. 2004. *Health Vulnerability in a Complex Crisis Situation. Implications for Providing Health Care to Nomadic People in Chad*. Saarbrücken: Verlag für Entwicklungspolitik.

Wiese, M., Yosko, I. and Donnat, M. 2004. La cartographie participative en milieu nomade – un outil d'aide à la décision en santé publique. *Médecine Tropicale*, 64, 452-463.

Wiese, M., Yosko, I. and Donnat, M. 2008. Contribution à une approche intégrée du pastoralisme Pauvreté, vulnérabilité et déséquilibre écologique. *Sécheresse,* 19(4), 237-243.

World Bank. 2009. *Minding the Stock. Bringing Public Policy to Bear on Livestock Sector Development.* (Report No. 44010–GLB). Washington DC.: The World Bank, Agriculture and Rural Development Department.

Wright, J. 1989. *Libya, Chad and the Central Sahara.* London: Hurst & Company.

Yacoub, M.S. and Gatta, G.N. 2005. *Tchad. Frolinat. Chronique d'une déchirure.* N'Djaména: Editions al-Mouna.

Yosko, I. 1993. Ethnoterritorialité Kreda et dynamique socio-spatiale nationale actuelle. *Revue Scientifique du Tchad,* 3(1), 28-31.

Yosko, I. 1994. *Les systèmes pastoraux du Tchad. Etat actuel des connaissances sur les systèmes traditionnels de gestion de ressources et leurs perspectives.* Abéché: GTZ.

Yosko, I. 1996. Le Bahr el Ghazal (Tchad): Occupation humaine et exploitation traditionnelle des ressources, in *Atlas d'élevage du Bassin du Lac Tchad,* edited by CIRAD and CTA. Wageningen, Netherlands: CTA, 66-70.

Yosko, I. 1999. *Législation foncière et pastoralisme au Tchad. Une nécessaire adaptation des textes.* Abéché: GTZ.

Yosko, I. 2000. *Pour un code pastoral au Tchad.* Actes des IIIe Journées Agro-Sylvo-Pastorales. N'Djaména: LRVZ, 111-115.

Zeltner, J.C. 1969. *Les Arabes de la région du Lac Tchad.* Sahr: CEL.

Zeltner, J.C. 1980. *Pages d'histoire du Kanem, pays tchadien.* Paris: L'Harmattan.

Chapter 5

Pastoral-Sedentary Market Relations in a War Situation: The Baqqāra-Nuba Case (Sudan)

Guma Kunda Komey

Introduction

The Baqqāra[1] of South Kordofan/Nuba Mountains in the Sudan, are a typical example of the *pastoral* and *nomadic*[2] groups that roam across the African arid and semi-arid lands. In the process of their well defined and rhythmic spatial mobility, as integral part of their life forms, they adopt constantly changing types of strategies and coping mechanisms to survive in ecological and human changing situations.[3]

Most of the pastoral and nomadic groups, including the Baqqāra, are neither proper traders nor have they themselves founded markets. Their economy does not allow them to be self-sufficient. Therefore, they are obliged to rely upon the resources of the local sedentary communities in order to supply their basic subsistence needs, primarily through regular local market exchanges. 'That is why, in all the areas where these cattle people live, one finds socio-economic symbioses between these two types of society' (Dupire 1962: 335).

This implies that the spatial mobility pattern of the pastoralists plays a prime role in stimulating the establishment and the continuation of the local markets in the regions they roam. This is in connection with the sedentary traditional economy in particular, and with the general economic system of the country with

---

1 The term *Baqqāra* (plural), which means cattlemen, applies to 'an Arab who has been forced by circumstances to live in a country which will support the cow but not the camel. ... The physical conditions upon which his existence depends are a dry district for grazing and cultivation in the rainy season connected by a series of waterholes with a river system where grass and water are available during the summer months' (Henderson 1939: 49).

2 Despite their distinct differences, of which I am aware, for the purpose of this chapter 'pastoral' and 'nomadic' terms are used here interchangeably to mean one broad category of mobility as opposed to a proper sedentary category. For some discussion on the distinction between 'pastoralism' and 'nomadism', see Azarya (1996: 3, footnote 1). See also Gertel (2007: 28, note 2) for some distinctions on 'nomadism' itself.

3 Abdel-Hamid (1986), Azarya (1996), Bovin and Manger (1990), Cunnison (1966), Gertel (2007), Haraldson (1982), Henderson (1939), Michael (1987a, 1998), Mohamed Salih et al. (2001).

all its regional and global dimensions. In this context, the complementarities and the constant interactions between the pastoral Baqqāra and the sedentary Nuba of the Nuba Mountains in Sudan are the predominant social and spatial features in the region. At the heart of these interactions a set of local market places act as a multifaceted intermediary spaces (Komey 2008a).

However, it is worth noting that the Baqqāra-Nuba contemporary 'symbiotic' relationship has been shaped and reshaped by cumulative and interwoven ecological, socio-economic and political dynamics. These dynamics can be traced throughout the pre-colonial, colonial, and post-colonial stages in the process of Sudanese state formation. The substantial and various influences of these cumulative dynamics in shaping the nomadic-sedentary symbiotic relationship have widely been dealt with in a range of burgeoning literature.[4]

What is fresh, however, is the new and far-reaching impact of the Sudan's recent civil war (1985-2005) on the long standing history of the relationship between the nomadic Baqqāra and the sedentary Nuba in the Nuba Mountains. Taking a 'bird's-eye-view' of the entire literature, it is evident that little attention has been paid to the impact of war on the cultural practises and the survival livelihoods of pastoralist and sedentary communities in the region. Nevertheless, some clues are observable in a few works that appeared during and after the civil war.[5] Thus, the novel point here lies in examining war-imposed transformations of the nomadic-sedentary relations, with the focus on the adopted responses, not only in their relentless efforts to maintain their economic and cultural practises, but in their strive for survival in the midst of the severity of war.

During the war, the severe fight between the Government of Sudan and the Nuba-led SPLM/A (Sudan People's Liberation Movement/Army) dictated its own logic and dynamics in the region. Shortly after its extension from the southern Sudan into the Nuba Mountains region in 1985, the war took on an ethnic dimension. The majority of the Nuba supported and, therefore, were supported by the SPLM/A against the central government. Simultaneously, the Baqqāra supported and, reciprocally, were supported by the government forces. As a result, the previously coexisting two groups were progressively divided into two heavily militarised politico-administrative zones along ethno-political lines (Manger 2003, 2006, Komey 2008a, 2008b, 2009a, 2009b, 2010a, 2010b).

Despite this war-imposed antagonism along the Nuba-Baqqāra divides, new and sporadic form of market places and exchanges emerged. Some key actors from both sides were able to 'strategically essentialise' themselves and managed

---

4   Abdalla (1981), Adams (1982), Ahmed (1976), Battahani (1980, 1986, 1998), Håland (1980), Henderson (1939), Ibrahim (1988, 1998, 2001), Komey (2008a, 2008b), Lloyd (1908), MacMichael (1912 (1967), 1922 (1967)), Manger (2004, 2008a), Mohamed Salih (1990, 1995a), Suliman (1999), Trimingham (1949 (1983)).

5   African Rights (1995), Africa Watch (1991, 1992a, 1992b), Elsayed (2005), Ibrahim (1998, 2001), Johnson (2006), Komey (2008a, 2008b), Manger (2004, 2006, 2007, 2008a, 2008b), Mohamed Salih (1995a, 1995b, 1999), Pantuliano et al. (2007), Suliman (1999).

to develop a new pattern of changeable market places across the war frontiers. The key question is, therefore, how and why the two groups, who were forced to fight each other, were simultaneously able *to collaborate* along economic-driven activities, centred on new forms of market places, spaces and exchanges in a situation of high insecurity and mounting risk?

Against this reasoning, the chapter intents to analytically describe the initiating driving forces, the function and the spatial pattern of the war-born markets in the context of the pastoral-sedentary relation in the Nuba Mountains, Sudan. The chapter pays attention to some adopted survival strategies and coping mechanisms deployed by the involved parties.

The chapter's main line of argument is empirically grounded on field-note-centred ethnographic material, obtained through a series of fieldwork trips, of a total duration of sixteen months, between 2005 and 2008. Theoretically, it is informed by 'strategic essentialism' concept (Spivak 1988), and, in a general way, by a range of conceptual and methodological insights centred on interconnected notions of 'social world' (Strauss 1978), 'social fields' (Grønhaug 1978) and 'social space' (Bourdieu 1985).

After this introduction, part two focuses on the local market institution as place, space and exchange in the context of the pastoral Baqqāra and the sedentary Nuba relations. It starts, however, with a brief description of the Nuba Mountains region as a spatial and social setting, while it ends with some notes on pastoral production and its local-national-global link. Part three focuses on the emergence of new forms of local markets as part of limited choices of survival strategies in response to the war situation. The Baqqāra-Nuba market-centred and locally brokered series of peace agreements are discussed in part four. In conclusion, some key findings are provided in the light of the overall analysis.

Market Institution as Place, Space and Exchange in the Baqqāra-Nuba Relations

The Nuba Mountains: Spatial and Social Settings

The Nuba Mountains region is located in the geographical centre of Sudan and covers an area of approximately 30,000 square miles in the South Kordofan State with Kadugli as its capital (Fig. 5.1). It is inhabited chiefly by a cluster of Nuba peoples of African origin, self-identified as indigenous to the region. They are predominantly sedentary groups that practice traditional rain-fed agriculture as their main means of livelihood. Writers like Lloyd (1908), MacMichael (1912 (1967)), Nadel (1947), Pallme (1844), Sagar (1922), Seligmann (1910, 1913, 1932 (1965)), Stevenson (1965), and Trimingham (1949 (1983)), among others, agree that the Nuba peoples were the first to settle in the area more than 500 years before other groups came in. Despite their statistical majority, 'they constitute a political minority due to their social and economic marginalization' (Mohamed Salih 1999: 1).

Figure 5.1   South Kordofan state/Nuba Mountains region
Source: Komey (2005).

The Baqqāra Arabs, who arrived in the region over 200 years ago as pastoral nomadic peoples, represent the second major ethnic group in the region.[6] They move seasonally southwards through the hilly Nuba areas towards the traditional homelands of the peoples of South Sudan during the dry season, and then back northwards during the rainy season. In recent years and due to some ecological and human-made changes, some of these nomads have gradually been transformed into sedentary groups with significant engagement in trading, and in traditional and mechanised rain-fed farming in the region.[7]

---

6   Cunnison (1966), Henderson (1939), Lloyd (1908), MacMichael (1912 (1967), 1922 (1967), Suliman (1999), Trimingham (1949 (1983)).

7   Abdel-Hamid (1986), Adams (1982), Henderson (1939), Ibrahim (1988), Mohamed Salih (1990).

In addition, there is a sizable number of Fellāta, West Africans, who migrated to the Nuba Mountains in search of work as agricultural labourers in the cotton fields in the 1920s. Moreover, there are small, but extremely influential, groups of the Jellāba of northern and central Sudan who migrated in several waves to the Nuba Mountains for slave raids and trade beginning at the turn of the seventeenth century (Manger 1984, 1988, Spaulding 1982, 1987, Trimingham 1949 (1983)).

The Jellāba merchants' extremely influential role stems from the fact that they control the economic and trade institutions in the region through well coordinated market chains and institutionalised trade networks at local, national, and international levels. Moreover, through their relations to the government, the Jellāba merchants have strong political influence and control over various levels of the state institutions as sources of power and wealth.[8]

Apart from the clearly defined territorial zones of urban settings and mechanised rain-fed farming schemes, the region's dominant land use pattern is characterised by the co-existence of two symbiotic sub-systems of subsistence: rain-fed cultivation and pastoral nomadism. In this context, the pastoral Baqqāra and the sedentary Nuba interact complementarily in farming, grazing, and market activities associated with subsequent far-reaching economic, cultural, and political consequences.

Local Market as a Multi-Faceted Intermediary Space

Prior to the civil war, there were three types of markets in the region: permanent small shops owned by the local Nuba and Baqqāra alike; permanent but usually larger shops owned by the Jellāba traders from the northern Sudan; and weekly markets usually located in villages that function as central places to the surroundings (see Vang and Granville, 2003). Unlike the two types of permanent shops, the weekly markets, which act as connecting points between the sedentary and nomadic groups, tend to flourish during the dry season and cease during the rainy season. The Baqqāra's spatial mobility patterns are compatible, in terms of timing and routing, with the times and the relative locations of the respective weekly markets. Therefore, the existence of the weekly local markets is an inevitable response to the sedentary-nomadic complementarities and symbiotic relationships.

Accordingly, the seasonal migratory routes of the Baqqāra's livestock tend to follow three separate, yet closely interlinked landscape features, namely: (a) the human-made or natural water sources, (b) the sedentary Nuba settlements and (c) the major weekly local market places. Also, the trade routes of the Jellāba merchants from and back to the urban centres in central parts of the Sudan follow a similar spatial pattern (see Fig. 5.2).

The clue here is that the water points and the weekly local market places are usually found within the vicinity of the sedentary Nuba settlements. Acting as

---

8   Battahani (1986), Ibrahim (1988), Komey (2009a, 2009b), Manger (1984, 1988), Mohamed Salih (1984).

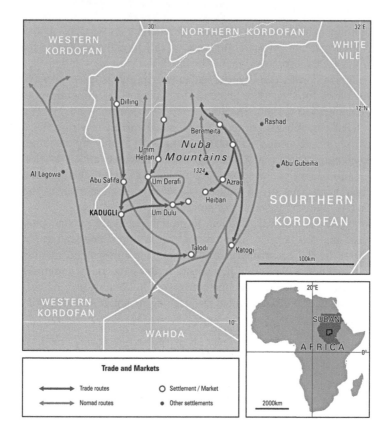

Figure 5.2   Relationships between Nuba settlements, Baqqāra mobility and trade routes during the pre-war situation

Source: Author's fieldwork.

intermediary spaces, these weekly markets bring together the farmers with their agricultural produce, the pastorals with their animal resources and the related products, and the Jellāba traders with their urban goods from the major towns of the northern and central parts of Sudan.

My participatory observations in a number of local markets in selected field sites reveal that local markets are not mere places for commodity exchanges. Rather, they are multifaceted intermediary spaces, full of interwoven social fields with significant implications for the involved actors. The weekly market place of the Keiga Tummero village, about 42 kilometres north of Kadugli, is a case in point. As a market place, Keiga Tummero acts as a set of socio-cultural, political and economic fields in the course of sedentary-nomads interactions (see Komey 2008a). The Friday weekly market brings together different societal actors with their respective functions and interests. It, thus, manifests the interdependent relationship between the pastoral and the sedentary groups in the area. Taking a critical look at the various forms of the transactions and interactions during that market day, it was palpable that the Keiga Tummero local market functions as:

a. A centre for economic and commercial transactions and exchanges among the involved local communities with their different ethnic, political and economic affiliations. The economic complementarities between the pastoral and the sedentary produces, on the one hand, and the Jellāba merchants' dominant role, on the other, are strongly felt during the market transactions. During the market exchanges, economic-driven interests tend to supersede all other political or ethnic-based interests or considerations.

b. A meeting point for networking and information exchanges between different actors. For example, information related to lost animals is usually found in the market where nomads from different farīq (livestock camps) meet not only for market transactions, but also for exchanging relevant information, views and news about their possible schedule of the migratory movements, potential gazing zones, available water sources, and other issues of common interests.

c. A forum for political campaigning and mobilisation by different political actors. For the government and non-governmental institutions, the market place remains the most effective institution in a rural setting for disseminating the relevant information. Most importantly, it also provides less expensive conditions for the state institutions to perform certain functions such as immunisation, agricultural extension, veterinary services, and tax estimates and collections.

d. A meeting point for negotiation, mediation, reconciliation, and conflict settlement including the payments of fines incurred as a result of court verdicts or gentlemen's agreements. Most conflict cases are mediated by the elders or native leaders during the market session because everybody can easily be voluntarily found or, otherwise, caught there. Equally, the weekly market and its ability to bring together people of different ethnic groups can also be a place that frequently triggers collective or individual conflicts, and the subsequent tendencies for retaliation.

e. An appropriate medium for developing social ties and acculturation among different socio-economic and ethnic actors. The selection of Friday as a market day in Keiga Tummero has a religious dimension as well. Its main mosque, located at the centre of the marketplace, represents one of the distinctive cultural landscape features. All Muslims of different ethnic backgrounds come together to perform the Friday communal prayers in that mosque. Furthermore, some friendships and personal relationships between people of different affiliations are stimulated and strengthened through market interaction. For instance, this can be observed in a gathering around a woman serving tea to customers from different ethnic backgrounds. By its very nature, the market imposes certain conditions of physical proximity to different people, to the extent that some warring parties may find themselves forced to peacefully face each other, simply because a third party has brought them face to face without prior arrangement.

The flow of the Baqqāra from their nearby *farīq*, and their effective participation in the market exchanges is a decisive factor that determines the success or failure of the market function. Indeed, the weekly markets in the Nuba Mountains tend to flourish during the dry seasons characterised by an intensive presence of the pastoral Baqqāra. As the rainy season approaches, the Baqqāra start their rhythmic movement northwards while the Nuba engage in cultivation. At this time, the flourishing markets start to shrink gradually to the point of complete but temporal disappearance awaiting the return of the Baqqāra in the following dry season to flourish again.

Pastoral Local Production and Its Link to National and Global Markets

While local market places act as intermediary spaces, where the pastoral Baqqāra and the sedentary Nuba interact, the Jellāba traders act as controlling agents of market exchanges. They have a long history of exercising effective control over most of the traded commodities at different levels of the market chain. In doing so, the Jellāba supply urban manufactured goods to the local peoples at expensive prices; while siphoning off cheaply local agricultural produce and animal resources.[9]

In this way, the local market operates as a starting point in a wider set of trade chains that systematically subjugate local economies to national and global markets. This implies that the multifaceted dynamics taking place in local markets across the different social fields are not merely local dynamics. Rather, they are, in most cases, local manifestations of national and globalisation processes with all their social, political and economic dimensions (Battahani 1986, Johnson 2006, Manger 1984, 1988, 2001, 2008b).

Taking pastoral production as an example it is not difficult to trace its local-national-global link. Local markets are collection and assemblage points for pastoral production, supplied by different households before they are transported for consumption at national and/or global markets. Two examples of this are: a certain type of traditionally processed white cheese, made primarily for national consumption; and the livestock trade transactions for export purposes.

During the dry season, some pastoral households, as an economic unit, make a prior deal with merchants' agents to locate their livestock farīq near a local market, a rural centre, or along a main road leading to some major urban centres. They then engage in the production of a set of dairy products, highly demanded by urban consumers, including the white cheese, *al-Jibna al-Bayda*. During my fieldwork in Keiga Tummero village, I came across numerous clusters of cheese

---

9   Battahani (1986), Ibrahim (1998), Komey (2005), Manger (1984, 1988), Mohamed Salih (1984, 1992).

cottage industry. They are not a separate but, rather, an integral part of the pastoral Baqqāra households, as economic units, in their respective camps nearby the Keiga Tummero village, as it is also a local market place.

The processing of cheese is, therefore, a joint venture between pastoral households, who provide fresh milk as the main raw material, and merchants' agents, who provide all other necessary inputs, particularly salt, packing materials and transport. The involved merchants, residing in major towns, either appear on a regular basis or send their own transport means to collect and transport the accumulated produce to the urban markets for final consumption. In this way, the highly localised economic units of the pastoral households are integrated systematically into wider national market and labour systems coupled with some form of the 'sedentarised' pattern (Michael 1987a, 1987b, 1991, 1998). This pattern of survival economies was reinforced during the war situation in response to the excessive and mounting insecurity and risk that restricted the pastoral spatial mobility.

Livestock trade transactions for export, that start at the local market level, are another conspicuous example of the national and global dimensions of the local pastoral production. Indeed, a close monitoring of the chains of the livestock trade transactions demonstrates further that pastoral households, as economic units, are not only linked to the national economic and market systems, but to the global market and labour systems at large (Michael 1987a, 1987b, 1991).

During the weekly local markets, there is always an intensive presence of local agents representing big merchants residing in the main cities. These merchants are connected to the global market system through a set of local production and national/ global-consumption chains. The sole role of their local agents is to buy most of the livestock supplied in the weekly local markets across the region. At some stage, the merchants arrive and pull together all the livestock collected by their agents, from the different local markets, and transport or trek them to the capital, Khartoum. In Khartoum, they are slaughtered and processed as fresh meat for national as well as global consumption. Daily cargo flights of the exported fresh meat from Khartoum and other regional airports to the global market, mostly to the oil-rich Gulf States, attests to this assertion (see El Dirani et al. 2009).

However, the crux of the matter here is that, in most cases, the real beneficiaries of these local-national-global market transactions are the Jellāba merchants residing in the main urban centres of the northern and central parts of Sudan. The local producers receive petty returns. What has been depicted in pastoral production and its national-global link is, to a large degree, applicable to the agricultural produce of traditional farming households as economic units in general.

The cumulative results of these local-national-global market links have been a long standing history of systematic and multiple exploitations and, therefore, marginalisation of the local pastoral and peasant communities. This market-linked economic marginalisation was a contributing factor that stirred up the political grievances among the local communities, particularly the sedentary Nuba. The

Nuba engagement in violent conflict in the context of the Sudan's civil war was partly due to this economic marginalisation.[10]

The Civil War and the Emergence of New Forms of Local Markets

The Nuba were forced to resort to an armed struggle when they joined the Sudan People's Liberation Movement/Army (SPLM/A) in 1985, during the course of the Second Civil War in Sudan (1983-2005). The war's impact on the entire region as a social world was, and still is, enormous.

The Impact of the Civil War Extension into the Region

The extension of the war into the Nuba Mountains region progressively introduced a new dynamic that has had significant repercussions on the historical, political, economic, and territorial relations between the state and the local community, on the one hand, and between the various ethnic groups, particularly the Nuba and the Baqqāra, on the other. The previously shared territory of the Nuba Mountains, was gradually divided up into two heavily militarised politico-administrative zones along ethnic lines: into areas controlled and administered by the government, with the Baqqāra maintaining the upper hand in the political space and in public affairs; and into areas held and administered by the Nuba-led SPLM/A, in which the Baqqāra nomads had no access to their traditional seasonal grazing lands in this part of the region (see Komey 2008a, Manger 2006, 2008a):

> During the war years, large tracts of the region, particularly at the foot of hills or in-between mountains ranges became off-limits to pastoralists who feared the SPLA. Pastoralists became less present and the interaction between the Nuba and the Baqqāra decreased. Traditional migration and transhumance routes were disturbed. Reciprocal agreements, both those rooted in tradition and those that were court-brokered, that had governed the passage of herds over agricultural land, fell into disuse. In other areas, farcically displaced Nuba no longer interacted with nomads (Manger 2006: 2).

Consequently, the previous co-existence of, and the cooperation between, the sedentary Nuba and the nomadic Baqqāra in the region ceased to exist. Shortly, there was a complete break-down of the relationships, including their normal and regular interaction in market places (Ibrahim 1998, 2001, Komey 2008a, 2008b, Manger 2006, 2008a).

---

10   African Rights (1995), Battahani (1998), Johnson (2006), Kadouf (2001), Komey (2004, 2005, 2008a, 2008b, 2010a, 2010b), Manger (2006, 2007, 2008a, 2008b), Mohamed Salih (1995a, 1995b, 1999), Rahhal (2001), Suliman (1999).

Moreover, with increased war intensity, the Jellāba traders of northern Sudan discontinued their business, following the cutting off of the trade routes between the region and the north.[11] Eventually, the Nuba in the areas under the control of the SPLM/A were completely cut off from the supplies of all urban basic commodities and services that used to flow from the major towns in northern and central parts of Sudan. They faced extraordinary hardship and suffering, in addition to the direct impact of the fighting.[12]

Simultaneously, the Baqqāra faced a similar but relatively slighter hardship when their livestock were squeezed into limited and poor grazing zones due to insecurity southwards. In addition, given the situation of insecurity along the war frontiers, there were repeated incidents of livestock looting resulting in enormous loss of animal wealth for the Baqqāra. Most importantly, the Baqqāra's own supply of the basic subsistence needs, particularly grains, which used to flow from the nearby Nuba hill communities, ceased completely.

In short, both communities were subjected to a set of externally induced, powerful, and unfavorable forces, brought about by the war dynamics. Hence, their livelihood and life forms were not only radically transformed, but their very survival was greatly endangered (see Komey 2009a, 2009b, 2010a, 2010b). It is within this context that new and sporadic forms of economic-driven transactions emerged. These were part of an extremely limited choice of coping mechanisms and survival strategies pursued by the two groups during the war.

These emerging working relations manifest themselves territorially in sporadic market exchange places, known as *aswāq as-salām* (singular: *sūq as-salām*) (peace markets). They were practiced and continued later in another modified version of cross-line transactions known locally as *aswsāq as-sumbuk* (singular: *sūq as-sumbuk*) (smuggling markets). Concurrently, these economically-driven and market-centered transactions were reinforced by a series of locally brokered peace deals, between some Baqqāra and Nuba groups in their respective localities. These new types of 'strategic', 'essential', and 'situational' cooperation emerged as an alternative to the pre-war normal symbiotic relations in order to counteract the severity of the formal war between the SPLM/A and the government forces.

The emergence of the aswāq as-salām and the subsequent version of aswsāq as-sumbuk, as new forms of market places, spaces and exchanges, signifies the necessity and inevitability of the Nuba-Baqqāra interdependency, particularly in the economic field. As described before, with the intensification of the prolonged war, the Baqqāra were in need of grain from their traditional suppliers, the Nuba. At the same time, the Nuba in the SPLM/A-controlled areas were indispensably in need of the essential urban goods and services particularly salt, soap, clothes, shoes, medicine, and agricultural tools.

---

11 African Rights (1995), Manger (2006, 2008a), Suliman (1999, 2002), Vang and Granville (2003).

12 African Rights (1995), Africa Watch (1992a, 1992b), Komey (2009b, 2010a, 2010b), Manger (2001, 2003), Rahhal (2001).

The conditions of the Nuba in the SPLM/A-controlled areas worsened further due to two main adverse political circumstances that reinforced each other: for one they were subjected to total isolation from the outside world by the central government policy that amounted to genocidal intent; and secondly, the flow of military supplies to the Nuba SPLA fighters was cut-off completely from the main SPLM/A in South by the then defected forces of Riek Machar in 1991.[13] Obviously, the Nuba in the SPLM/A-controlled areas were not only in critical need of the basic subsistence supplies, but they were also desperately in need of an alternative way of getting ammunition supply, in order to be able to continue fighting the government forces.

Aswāq as-Salām (Peace Markets)

In a taped interview with my key informant, Simon Kalo, in Kadugli, 9 January 2007, it was clear that peace markets, and the later version of sumbuk markets, played a vital role in sustaining the Nuba in the SPLM/A-held areas during the critical war time. Kalo was the second SPLA Commander in the Buram area in the Nuba Mountains during the early 1990s. He was in charge of looking for an alternative form of market exchanges with pastoral Baqqāra in the government controlled areas, at the time the government had effectively sealed off the Nuba Mountains region from any outside world contact.

By October 1991, the government had applied a tight network of security checkpoints along the frontiers, prohibiting any travel or flow of people or goods from and to the SPLM/A-controlled areas. By 1992, the SPLM/A controlled areas in the Nuba Mountains were subjected to a complete isolation from any national, regional or international transactions. The government not only sealed off the region, but it also prevented the extension of the UN-led Operations Lifeline Sudan (OLS), to the Nuba Mountains, although the OLS was covering the entire war-torn southern Sudan.[14]

The aggregate result was the death of thousands due to hunger, disease and fighting. At the same time, there was enormous suffering among Nuba communities whose survival was endangered, in the absence of basic supplies. Their military capabilities were weakened drastically, to the extent that they were unable to counteract the central government's systematic campaigns of genocides and ethnocide in the context of the waged war (African Rights 1995, Africa Watch 1991, 1992a, 1992b, Meyer 2005).

In this critical context, the aswāq as-salām (peace markets) emerged in the rural centre of Buram under SPLM/A control in the early 1990s. However, these peace markets did not last for long. They were soon discovered, and therefore,

---

13   African Rights (1995), Africa Watch (1991, 1992a, 1992b), Johnson (2006), Meyer (2005), de Waal (2006).

14   African Rights (1995), Africa Watch (1991, 1992a), Johnson (2006), Meyer (2005), de Waal (2006).

targeted by the government through a series of offensive military interventions. My informant, Simon Kalo detailed the initiation, the operation and the demise of the Buram peace market, and the emergence of the aswāq as-sumbuk afterwards. The peace market as a practice started initially in the northern part of Bahr al-Ghazāl in southern Sudan, between the pastoral Baqqāra of Meseīrīya and the SPLM/A forces. In this regard, Ibrahim reports that:

> In 1993, the Meseīrīya Arab groups ... and the Ngok Dinka of Abyei area have sat together and through the revitalization of their historic tribal alliances and arrangements came to peaceful co-existence pact. According to the pact, they organized their grazing and farming rights on each other's Dār, i. e., tribal homeland. They as well, have institutionalized their inter-tribal episodic conflicts over these natural resources and to be resolved through these recently revitalized arrangements. Moreover, the pact has granted the trading rights of all parties through what are called peace market. ... These markets are established inside the SPLM/A held areas in Bahr al-Ghazāl. It enabled them to exchange goods and services (i.e., mostly consumer goods such as salt, tea, coffee, clothes and shoes and perhaps, sometimes ammunition according to the official government allegations against and in condemnations of these markets. To the both parties, these peace markets are but a symbolic gesture, for an overall peace in all aspects of the two parties' life (Ibrahim 2001: 47, footnote 25).

What is obvious here is that the pastoral Meseīrīya were not interested in trade relations as an end in itself, but as a means to secure access for their traditional grazing land and water inside the SPLM/A-controlled areas during the dry season. This implies that the extension of market-centred relations to the Nuba Mountains was a travelling model. It was initiated by the pastoral Baqqāra of the Meseīrīya Humr, who had secret on-going trade relations with the SPLM/A along the north-south border line during the civil war. The initiative was enthusiastically supported by the SPLM/A leaders in the Nuba Mountains. Subsequently, a certain Meseīrīya SPLM/A officer acted as an intermediary between the SPLM/A, and his Meseīrīya people in the government controlled areas. The plan was to agree and coordinate efforts aiming at introducing some sort of trading transactions led by the Meseīrīya in selected local markets·in the SPLM/A controlled areas. The initiative was received by the Meseīrīya leaders and merchants with a high level of interest.

Towards that end, some pastoral Baqqāra of the Meseīrīya sent a delegation, composed of some merchant representatives and native leaders to the SPLM/A-controlled areas. The aim was to negotiate the possibility of establishing some form of market exchange relations with the SPLM/A leaders. The Meseīrīya delegation, which travelled secretly without government detection, managed to arrive safely at Buram in the SPLM/A-controlled areas in a situation characterised by high insecurity. Upon their arrival in Buram, they were well received by the Nuba people and their leaders. In a short space of time, the two parties were able to successfully negotiate and conclude an economic-driven and market-centred peace agreement, known as the Buram agreement in February 1993. In the words of my key informant, Simon Kalo:

> We held a joint meeting in Buram under the leadership of Telephūn Koko, the then Commander and myself as second Commander in the Buram area. After we discussed and agreed on the practicality of implementing the initiative, a committee was formed under my chairmanship and entrusted with the task to supervise and ensure safe implementation of the transaction. The required security arrangement was the main concern. This is because of the high risk involved in the expected journeys between the SPLM/A-controlled areas in the central and southern parts of the Nuba Mountains and the Meseīrīya areas in the western part of South Kordofan. The challenges were twofold: first, how to provide effective security measures to these commercial convoys during their journeys from, and to, the SPLM/A-controlled areas; second, how to make sure that the SPLM/A's security and military systems and affairs are not penetrated by some possible pro-government elements within the convoys (Author's interview with Simon Kalo: Kadugli, 9 January 2007).

Following their return home, the Meseīrīya sent a convoy of over 20 merchants and numerous camels carrying various basic commodities needed desperately by the Nuba people. The day of the convoy arrival at Buram signified the beginning of the agreement implementation. In a short period, a big market in Buram was in operation under the name of 'peace market'. My informant Kalo continued his testimony and described how the Nuba were relieved by the convoy's arrival:

> The first Meseīrīya convoy, composed of more than twenty camels, arrived and supplied the Buram market with commodities from the government controlled areas. It was an historic day for the local people! Not only for the people in and around the Buram area but all over the SPLM/A-controlled areas in the Nuba Mountains. They rushed into Buram market from all directions to obtain some basic commodities like salt, soap, sugar, clothes, medicine and shoes, after several years of deprivation. Most importantly, the Meseīrīya also supplied us with some ammunition at the time our military supplies from the South were cut-off. In return, the Meseīrīya went back with a sizable number of livestock. Initially, the exchange system was almost in the form of barter trade because there was no money in the SPLM/A-controlled areas. Since then, the commodity supplies continued to flow regularly through the Meseīrīya convoys (Author's interview with Simon Kalo: Kadugli, 9 January 2007).

In the process, new local markets were opened up in Ekurchi in Moro and other places in Kawalib and Shwai. The Buram market was acting as a central place, diffusing the supplied commodities into the different parts in the SPLA-controlled areas. At a certain stage, the government security agents discovered the on-going transactions and started tracing, arresting and eventually killing some of the involved parties inside the government controlled areas. Despite this mounting threat from the government, the market continued to flourish because the involved Meseīrīya in the government controlled areas and the SPLM/A were both determined to sustain it. It was a profitable trade business for the pastoral Meseīrīya while it was a survival strategy for the Nuba in the SPLM/A-controlled areas.

Due to frequent ambushes by government forces and the militia, a new pattern of mobility for trade convoys was invented. The Meseīrīya convoys started to move in the form of small but well armed groups to protect themselves and their property. These pastoral trade convoys continued to frequently encounter and, therefore, engage in fighting with government forces and militia. Despite this, they continued to pursue their economic gains in collaboration with the Nuba in the SPLM/A-controlled areas. In fact, according to my key informant, Kalo:

> The Meseīrīya joined their hands with us in fighting back against the government forces in many instances. For example, they sided with us during the big battle of Buram in December 1993 when the government forces launched an offensive attack during a busy market day. The Meseīrīya merchants who were present in the market had to join with the SPLM/A in fighting back against the government forces. During the attack the government forces regained Buram and the surrounding plain areas while the SPLM retreated to the nearby hills. The SPLM/A, however, managed to protect and eventually evacuate all the Meseīrīya out of the area with no fatalities among them. In this way, these Baqqāra Arabs who were perceived as the enemy of the SPLM/A, proved to be not only supportive, but strategic and trustworthy partners during a very critical period in the history of the Nuba struggle (Author's interview with Simon Kalo: Kadugli, 9 January 2007).

Obviously, this economically-driven and market-centred cooperation contributed, to a large degree, to the survival of the sedentary Nuba as it supplied them with essential goods after several years of isolation. It also reinforced the military ability of the SPLM/A forces through ammunition supply, at a time when their military supplies, from the main SPLM/A in the southern Sudan, had long been discontinued. The demise of the peace market due to drastic war developments on the ground, led to its rebirth in another form, which came to widely be known as sūq as-sumbuk.

Aswāq as-Sumbuk (Smuggling Markets)

Historically, the word 'sumbuk' refers to a risky but lucrative trade business of 'slave smuggling' along the Red Sea, during the early days of Anglo-Egyptian rule. In his 'Report on Slavery and Pilgrimage, 1926', C.A. Willis, Assistant Director of Sudan Intelligence (1915-1926), defined sumbuk or sombuk as an act of 'smuggling or illegal [*sic*] transporting across the Red Sea to Saudi Arabia by cheap but poor quality, and therefore, risky boats, or Dhows' (Sudan Archive

212/2/1-94: 21). The involved risk was partly due to the fact that the British Government in the Sudan was very harsh with slave dealers if caught red handed; because slave trade was already banned internationally. It seems that sumbuk as concept and practice was somehow revisited, travelled and re-utilised in similar profitable business, stimulated, this time, by the extremely risky and insecure situation of large-scale civil war.

The recapture of the Buram area by government forces led to an immediate termination of the Meseīrīya-SPLM/A market exchanges and the related connections. In fact, it put an end to all peace markets in the SPLM/A-controlled area. However, the closure of the peace markets led, as mentioned, to the emergence of aswāq as-sumbuk. The main distinction between these two market exchange patterns is that the aswāq as-salām or peace markets operated inside the SPLM/A controlled areas with formal back up by the SPLM/A officials. Unlike the aswāq as-salām, the aswāq as-sumbuk or smuggling markets were pure community–community transactions with no involvements of the SPLM/A or the government authorities. In other words, sūq as-sumbuk was practiced by some community members from both sides, behind the back of the government and the SPLM/A military and security authorities, and their respective politico-administrative institutions. It was a sporadic and highly mobile market exchange, operating in a number of strictly confidential and strategic sites along the transitional war zones.

After the collapse of the peace market and the withdrawal of the Meseīrīya dealers, sūq as-sumbuk soon became the trading pattern prevailing along the war frontiers, particularly around al Hamra, Umm Serdeba, and Umm Derafi. It was practiced by some local Baqqāra of Hawāzma Rawāwqa, in government-controlled areas, and some Nuba in SPLM/A-controlled areas. The pre-war social ties along the Baqqāra-Nuba line were very instrumental in this new form of market exchanges. Sūq as-sumbuk was very lucrative but an extremely risky trade business undertaken by some Baqqāra dealers. For the Nuba in the SPLM/A-controlled areas, the activity was an alternative survival strategy, or coping mechanism, in the context of the war situation.

In sūq as-sumbuk, a group of people from both sides of the antagonistic parties, loaded with commodities potentially needed by the other party, meet in a specific, hidden, but agreed upon, strategic site to quickly exchange their commodities before they depart. The next site and time for a similar market exchange is determined depending on the war dynamics on the ground. Thus, the mobility pattern of each group from the Baqqāra and the Nuba sides, with their traded goods, is adjusted constantly in a situation of high insecurity and risk. It is, therefore, secret in timing and location, exclusive in membership, and mobile in nature.

Several testimonies and narratives of involved actors from both sides of the conflict attest to this claim. The practice reflects that the complementarities between the coexisting sedentary Nuba and nomadic Baqqāra groups are both necessary and inevitable for their livelihoods. Responding to my question as to whether the Baqqāra Hawāzma participated in sūq as-sumbuk, one of my key Baqqāra informants, who preferred to withhold his name, affirms that:

Yes, we participated effectively in sūq as-sumbuk dealings. Some of our people used to smuggle some essential commodities like clothes, sugar, salt and oil, from Kadugli to the SPLM/A controlled areas especially to Buram, Shat, Kololo and Saraf al-Jamūs markets. They use bicycles, donkeys and camels to carry the smuggled commodities in small groups. The goods are usually either exchanged for US$ from the international NGOs operating in the SPLM/A-controlled, livestock from the Nuba people, or some automatic weapons from the SPLA forces. To avert risk, we exchanged the US$ money for local currency in al-Obayed or Khartoum but not in Kadugli.

Sūq as-sumbuk was a dangerous and a risky trading and exchange business; but it was worth it, for it was a lucrative business. Estimates of eight to twelve of our young people were shot dead by the government army and security agents during their sumbuk transactions along the frontiers. In fact, during the smuggling journeys, it was safer to encounter the SPLM/A troops than that of the government. On several occasions, the sumbuk participants were safely escorted close to the government controlled areas by the SPLM/A army after finishing the transaction (Author's taped interview with an anonymous informant: Reikha, 17 February 2007).

Apart from market-driven relations, there were other forms of economically or socially driven interests and cooperation between individuals or groups on antagonistic sides of the war. For instance, some maintained ties motivated by the fixed economic assets left behind in the SPLA-controlled areas by Baqqāra and Jellāba traders. At the same time, many Nuba communities, who had joined the SPLA, left behind their fixed assets, farming lands, and well established villages in the government-controlled areas. In many cases, these economic assets seem to have facilitated the continuation of some family-to-family ties along the Nuba-Baqqāra divisions during the war and thereafter.

In this regard, my anonymous informant cited above revealed that the Awlād Tayna of Delamīya of Rawāwqa-Hawāzma in Reikha were forced to relocate themselves to Kadugli during the war. Many of them left behind most of their immovable assets like shops, mills, stores and houses. After the signing of the peace agreement, they were eager to return to the area and restart their local trading business. Unlike other Baqqāra who were unable to return to their pre-war areas, Awlād Tayna of Delamīya's return to Reikha was very smooth. They ascribe this to their ability to reposition themselves in such a way that allowed them to maintain their economic interests and personal ties in the area during the war. While staying in the government-controlled areas, they managed to uphold their old ties with the Nuba community leaders, including some of the SPLA commanders. This enabled them to cut across the war zones and practice aswāq as-sumbuk at the frontiers.

Hāmid Sattār, the Shaykh of the Baqqāra of 'Yātqa, is another key informant who testified that when the war caught them by surprise, in al-Azraq village in the eastern part of Heiban, they were forced to retreat with their livestock northwards into the government controlled areas in Umm Berembeita and Kortala within South Kordofan, and 'Alūba in North Kordofan. Several of their settled families in al-Azraq left behind most of their fixed assets. 'My own family', he lamented, 'left behind in al-Azraq several orchard farms with a total of 224 mango trees, in addition to two well built shops and houses. At the same time, we took along with us some of the Nuba cattle which used to be under our care, together with some of their boys, when the insecurity forced us to move out of the al-Azraq area' (Author's interview with Hāmid Setār: Khartoum, 10 June 2006). It is these cross-cutting social ties that maintained some sort of group or individual links across the dividing lines during the war and thereafter.

These narratives, among so many others, indicate that the pre-war Nuba-Baqqāra ties, in terms of neighbourhood, social and economic relationships were very instrumental in facilitating the initiation and the maintenance of the sūq as-sumbuk. In this way, Sūq as-sumbuk was an opportunity for some individuals, who had strong social ties or economic assets across official conflict lines, to maintain positive connections throughout the civil war. The ultimate objective was to secure a smooth return to their pre-war situation after the accomplishment of peace.

The nature of both aswāq as-salām and aswāq as-sumbuk, and the way they were initiated and managed, demonstrates the dominance of the economic-based interests. These interests are rooted in the socioeconomic complementarities between the Nuba and the Baqqāra. These various forms of cooperation, particularly peace markets and/or sūq as-sumbuk became an entry point for subsequent, locally brokered peace initiatives in the region between some Baqqāra and Nuba communities. These peace agreements strengthened further the Baqqāra-Nuba market relations, despite the war's intensity and the related high insecurity and risk along the frontier zones.

Some Market-Motivated and Community-Based Agreements[15]

In the context of their market relations during the war, different groups of the Nuba and the Baqqāra concluded several community-based deals: the most noticeable ones were the Buram, the Regifi and the Kain agreements. They were brokered by some pastoral Baqqāra communities in the government-controlled areas and their respective sedentary Nuba counterparts in the SPLM/A-controlled areas in 1993, 1995 and 1996 respectively.

There were several reasons these community-based peace agreements were established. For example, the Baqqāra lamented that they had been misinformed by the government and made to fight against the Nuba, despite both being marginalised by the same government. They lost many people and sizable animal wealth and most households were displaced. They expressed their strong desire to trade with the Nuba. Simultaneously, the Nuba emphasised their intention to fight the government and not the Baqqāra; and expressed interest in trading with

---

15   For detailed discussion on these agreements, see Suliman (1999: 215-18).

the Baqqāra, in terms of exchanging their crops and animals for the urban goods namely, clothes, shoes, salt, soap and medicine, from the government areas.

In February 1993, the first community-based peace agreement took place in Buram. Most of the principles enshrined here were echoed in similar subsequent agreements. Both sides agreed to abide by the following (see Suliman 1999: 215-16):

1. To immediately stop all military actions against each other, and exchange relevant military and security information.
2. To safeguard free movement of people and goods to and from both sides, and to assist them, if necessary, to reach final destinations safely.
3. To settle disputes, or cases of peace violation, through a joint committee.
4. To collaborate in putting an end to the recurring incidents of animal looting across the frontier zones.
5. To safeguard the trade exchanges between people of the two territories.

The direct result of this agreement was the opening up of a trade route into the Buram peace market. This market flourished until the end of 1993, when the government troops recaptured the Buram area and put an end to the market exchanges.

On 15 November 1995, the *Regifi* agreement was reached. It reiterated the principles enshrined in the previous agreement. Both sides acknowledged that a brokered community-based peace agreement was crucial for their co-existence during the conflict. The government did all it could to sabotage the agreement:

> It targeted the leaders of the Baqqāra who signed it: 'Abdalla, the Meseīrīya leader at the negotiations, was shot dead; others were assassinated or imprisoned; a few were bribed and skillfully used by the government to undermine the spirit of trust and cooperation between the Baqqāra and the Nuba which had begun to spread in the region (Suliman 1999: 217).

The crux of the matter is that the forces that continued to distance the Baqqāra from the Nuba through persistent policies of 'divide and rule,' were always generated at the centre and beyond the reach of both. In June 1996, the Nuba began another initiative towards peaceful cooperation with the Baqqāra of Rawāwqa-Hawāzma. A five-person delegation sought the Rawāwqa on neutral ground in Zangura, west of Tima, in the Lagawa area. They invited the Baqqāra to move their market close to the 'liberated areas', for example areas under SPLM/A-control for better mutual cooperation along trade and market exchanges. The Baqqāra accepted the initiative, and in 1996 negotiated what became known as the Kain peace agreement. It was almost identical to the previous two agreements. However, a special trade committee was established this time to supervise the implementation of trade and market exchanges. What was remarkable in the agreement was that:

1. The Rawāwqa were so confident in the stability of the agreement that they began to bring in ammunition and army uniforms to sell to the Nuba;
2. The Baqqāra traders began to come unarmed to the markets and were gradually accompanied by women and children; and
3. The first test of the agreement came shortly after signing it, when an Arab attacked a Nuba, took his weapon, and left him [nearly] dead: the Baqqāra brought the weapon back, paid for the treatment of the victim, and promised to deliver the attacker to the Nuba authority (Suliman 1999: 217-18).

Shortly after the agreement began and the market started, government security forces began to clandestinely attend marketplaces in civilian clothes. The attempt was to sabotage the agreement through murder, imprisonment, and bribery. The Nuba leadership became alarmed and ordered the market closed. Yet, the government continued the policy of targeting the leaders on both sides of the divide with murder, imprisonment and bribery: 'In one known case, government officials offered a would-be assassin four million Sudanese pounds and a license for a mill in return for killing a leading Nuba signatory to the agreement (in 1999, 2576 Sudanese pounds [SDP] = 1 United States dollar)' (Suliman 1999: 218).

These community-based peace agreements were driven by mutual interests of both parties. This implies that the macro agenda of the central government proved to be neither appealing nor viable for substantial portion of the common people involved (Ibrahim 2001). But it is equally true that not all Baqqāra or Nuba recognised, or adhered to these peace agreements, as many also sided with the government army and/or the Peoples Defense Forces (PDFs) in the context of the civil war.

Moreover, there were frequent clashes between some Baqqāra armed groups and the SPLA along the frontiers because both were unaware of the peace agreements. Another challenge was that some Baqqāra traders were playing double roles for their own economic interests. 'On the one hand, they traded with Nuba and even sold them ammunition; on the other, they supplied the government with information about rebel troops' (Suliman 1999: 218). Despite these enormous challenges and insecurity, the two groups managed to engage in a constantly changing form of market relations during the entire war period.

Nuba-Baqqāra relations continue this changing after the signing of the Nuba Mountains Ceasefire Agreement in 2002 and the conclusion of the Comprehensive Peace Agreement in 2005, which allowed for free movement of people and commodities in the Nuba Mountains. However, the impact of the post-war dynamics on the Baqqāra-Nuba relations, which are still in the making, is beyond the scope of this chapter, and, indeed, deserves a separate scholarly undertaking.

Conclusion

The chapter recognises that the current relation between the pastoral Baqqāra and the sedentary Nuba, of the Nuba Mountain region in Sudan, has been shaped and reshaped by various and cumulative historical and contemporary forces. However, the chapter's main point of departure is that Sudan's recent civil war is repressive, with far-reaching and multiple ramifications, not only for the Baqqāra-Nuba coexistence, but also for their respective livelihoods and survival. Focusing on the role of market institutions in pastoral-sedentary relations in a war situation, the chapter analyses the forces, functions and spatial patterns of the war-born economic-driven cooperation among the two groups. Through the narratives of key informants, the study reveals some of the survival strategies and coping mechanisms adopted during the war.

The analysis has generated empirically grounded and theoretically guided insights that may further deepen our understanding of the complex nature of a number of interconnected issues. Firstly, the empirical dimension of this study not only confirms the inevitability of the pastoral-sedentary complementarities but also their reinforcement even during a war situation. The Sudan's prolonged and severe civil war imposed its own logic and dynamics on the livelihoods of both communities in the region. As a result, their shared landscape was progressively divided into two heavily militarised politico-administrative zones along ethno-political lines. Moreover, they were made to fight a proxy war that came to have destructive consequences on their mutual cooperation along their interlinked socioeconomic practices.

Secondly, while the two groups were forced to fight each other, some were simultaneously able to strategically essentialise themselves, and developed a new pattern of cooperation during the war. The cooperation was driven chiefly by the 'economies of survival' (Gertel 2007: 18), resulting in the emergence of a new pattern of local market exchanges along the war frontier zones. It was pursued jointly as part of extremely limited survival choices. Aswāq as-salām (peace markets), aswāq as-sumbuk (smuggling markets), and a number of related community-based and market-motivated peace deals, were nothing but manifestations of survival strategies by the two endangered communities for the sake of counteracting the adverse ramifications of war.

Thirdly, the continuation of the Baqqāra-Nuba relations during the war, though reduced in intensity and sporadic in pattern, indicates that the driving motives that forced them to engage in fighting against each other, in the context of the larger civil war, were not internally generated. Rather, they were externally driven and state induced factors. Some of these factors are empirically traceable through a set of local-national-global chains that tend to constantly push the local communities into unfavourable economic, social and political conditions. In this respect, the partial analysis of the pastoral production of the Baqqāra and their involvement in market business during the war concurs with a number of similar recent studies in the Nuba Mountains[16] and elsewhere in Africa (see Azarya 1996, Bovin and Manger 1990, Mohamed Salih et al. 2001). Specifically, the study substantiates the assertion that 'a pastoralist can be involved in several economic activities at once, and revenues from different family members are pooled in a joint household. Here the risks of social reproduction are redistributed among the generations' (Gertel 2007: 18).

Finally, like other pastoral groups in Africa, the Baqqāra are neither proper traders nor have they themselves founded markets. Despite this, they play a key role in the evolution and the continuation of the local market as place, and space of exchanges. The empirical case of the weekly local market of Keiga Tummero reveals that the flow of the pastoral Baqqāra from the nearby livestock camps, and their effective participation in the market exchanges, is a decisive factor that largely determines the success or failure of such market institutions. In this way, the Baqqāra-Nuba market relations, and their joint unrelenting strive to maintain them, during the war severity, demonstrate beyond doubt the intrinsic and inevitable socio-economic complementarities between the pastoral nomadic and the sedentary communities in the region and elsewhere.

Acknowledgments

The fieldwork ethnographic material used in this chapter is part of an ending research project titled 'Contested autochthony: Land and water rights in the relation of nomadic Baqqāra and sedentary Nuba people of South Kordofan/ Nuba Mountains, Sudan'. The project is part of Collaborative Research Centre: 'Difference and Integration' of the Universities of Halle and Leipzig (http://www. nomadsed.de). It is funded by the German Research Foundation (DFG) for the period 2004-2012. The wider context of this research was published in 2010 by the author under the title: *Land, Governance, Conflict and the Nuba of Sudan.* Woodbridge: James Currey, Eastern African Series.

References

Abdalla, A.J. 1981. Sudan pastoral-nomads in transition: Problems of evolution and adjustment, in *Die Nomaden in Geschichte und Gegenwart.* (Beiträge zu einem internationalen Nomadismus-Symposium am 11. und 12. Dezember 1975 im Museum für Völkerkunde Leipzig), edited by R. Krusche. Berlin: Akademie-Verlag, 203-211.
Abdel-Hamid, M.O. 1986. *The Hawazma Baqqāra: Some Issues and Problems in Pastoral Adaptations.* M.A Thesis, University of Bergen.

---

16   Ibrahim 1998, 2001, Michael 1991, 1998, Mohamed Salih 1990, Suliman 1999.

Adams, M. 1982. The Baggara problem: Attempts at modern change in southern Darfur and southern Kordofan (Sudan). *Development and Change*, 13(2), 259-289.

African Rights. 1995. *Facing Genocide: The Nuba of Sudan.* London: African Rights.

Africa Watch. 1991. *Sudan: Destroying Ethnic Identity. The Secret War Against the Nuba.* London and New York: Africa Watch.

Africa Watch. 1992a. *Sudan: Refugees in Their Own Country.* London/New York: Africa Watch.

Africa Watch. 1992b. *Sudan: Eradicating the Nuba.* London/New York: Africa Watch.

Ahmed, A.G.M. (ed.). 1976. *Some Aspects of Pastoral Nomadism in the Sudan.* Khartoum: Economic and Social Research Council.

Azarya, V. 1996. *Nomads and the State in Africa: The Political Roots of Marginality.* Aldershot: Ashgate Publishing Ltd.

Battahani, A.H. 1980. *The State and the Agrarian Question: A Case Study of South Kordofan 1971-1977.* Khartoum: University of Khartoum.

Battahani, A.H. 1986. *Nationalism and Peasant Politics in the Nuba Mountains Region of Sudan, 1924-1966.* PhD Thesis, University of Sussex.

Battahani, A.H. 1998. On the transformation of ethnic-national policies in the Sudan: The case of the Nuba People. *Sudan. Notes and Records*, 2, 99-116.

Bourdieu, P. 1985. The social space and the genesis of groups. *Theory and Society*, 14(6), 723-744.

Bovin, M. and Manger, L.O. (eds) 1990. *Adaptive Strategies in African Arid Lands.* Uppsala: Scandinavian Institute of African Studies (SIAS).

Cunnison, I. 1966. *Baqqāra Arabs: Power and Lineage in a Sudanese Nomad Tribe.* Oxford: Clarendon Press.

Dupire, M. 1962. Trade and markets in the economy of the nomadic Fulani of Niger (Bororo), in *Markets in Africa*, edited by P. Bohannan and G. Dalton. Evanston/Chicago: Northwestern University Press, 335-362.

El Dirani, O.H., Jabbar, M.A. and Babiker, I.B. 2009. *Constraints in the Market Chains for Export of Sudanese Sheep and Sheep Meat to the Middle East.* Nairobi: ILRI.

Elsayed, G.F. 2005. *The Politics of Difference and Boundary Making Among the Nuba and the Baggara of Southern Kordofan State, Sudan.* M. Phil Thesis, University of Bergen.

Gertel, J. 2007. Mobility and insecurity: The significance of resources, in *Pastoral Morocco: Globalizing Scapes of Mobility and Insecurity*, edited by J. Gertel and I. Breuer. Wiesbaden: Reichert, 11-30.

Grønhaug, R. 1978. Scale as a variable in analysis: Fields in social organization in Herat, Northwest Afghanistan, in *Scale and Social Organization*, edited by F. Barth. Oslo: Universitetsforlaget, 78-121.

Håland, G. (ed.). 1980. *Problems of Savannah Development: The Sudan Case.* Bergen: University of Bergen.

Haraldsson, I. 1982. *Nomadism and Agriculture in the Southern Kordofan Province of Sudan*. Uppsala: Sveriges lantbruksuniversitet.

Harragin, S. 2003. *Nuba Mountains Land and Natural Resources Study: Part 1 – Land Study*. Missouri: University of Missouri and USAID.

Henderson, K.D.D. 1939. A note on the migration of the Messiria tribe into South-West Kordofan. *Sudan Notes and Records*, 22(1), 49-74.

Ibrahim, H.B. 1988. *Agricultural Development Policy, Ethnicity and Socio-political Change in the Nuba Mountains, Sudan*. PhD Thesis, University of Connecticut.

Ibrahim, H.B. 1998. *Development Failure and Environmental Collapse: Re-Understanding the Background to the Present Civil War in the Nuba Mountains (1985-1998)*. Paper presented at the symposium: Perspectives on tribal conflicts in Sudan, 11-12 May 1998, Khartoum: University of Khartoum, IAAS and FES.

Ibrahim, H.B. 2001. *In Search of the Lost Wisdom: The Dynamics of War and Peace in the Nuba Mountains Region of Sudan*. Khartoum: Unpublished manuscript.

Johnson, D.H. 2006. *The Root Causes of Sudan's Civil Wars: Updated to Peace Agreement*. Oxford: James Currey.

Kadouf, H.A. 2001. Marginalization and resistance: The plight of the Nuba people. *New Political Science*, 23(1), 45-63.

Komey, G.K. 2004. *Regional Disparity in National Development of the Sudan and Its Impact on Nation-building: With Reference to the Peripheral Region of the Nuba Mountains*. PhD Thesis, University of Khartoum.

Komey, G.K. 2005. Dynamics of the marginalization process in Sudan: The Nuba Mountains case, in *Sudan: The Challenge of Peace and Redressing Marginalization*, edited by H.A. Abdel Ati. Khartoum: EDGE for Consultancy and Research and Heinrich Böll Foundation, 54-86.

Komey, G.K. 2008a. The autochthonous claim of land rights by the sedentary Nuba and its persistent contest by the nomadic Baqqāra of South Kordofan/ Nuba Mountains, Sudan, in *Nomadic-Sedentary Relations and Failing State Institutions in Dārfur and Kordofan, Sudan*. (Orientwissenschaftliche Hefte 26/2008, Mitteilungen des SFB 12), edited by R. Rottenburg. Halle: University of Halle, 101-127.

Komey, G.K. 2008b. The denied land rights of the indigenous peoples and their endangered livelihood and survival: The case of the Nuba of the Sudan. *Ethnic and Racial Studies*, 31(5), 991-1008.

Komey, G.K. 2009a. Autochthonous identity: Its territorial attachment and political expression in claiming communal land in the Nuba Mountains region, Sudan, in *Raum-Landschaft-Territorium. Zur Konstruktion physischer Räume als nomadische und sesshafte Lebensräume*. (Nomaden und Sesshafte 11), edited by R. Kath and A.-K. Rieger. Wiesbaden: Reichert, 205-228.

Komey, G.K. 2009b. Striving in the exclusionary state: Territory, identity and ethno-politics of the Nuba, Sudan. *Journal of International Politics and Development*, 7(2), 1-20.

Komey, G.K. 2010a. Land factor in war and conflicts in Africa: The case of the Nuba struggle in Sudan, in *Wars and Peace in Africa*, edited by T. Falola and R.C. Njoku. Durham NC: Carolina Academic Press, 351-381.

Komey, G.K. 2010b. *Land, Governance, Conflict and the Nuba of Sudan (Eastern Africa)*. Woodbridge: James Currey.

Lloyd, W. 1908. Appendix D: Report on Kordofan province. Sudan, in *Sudan Archives: 783/9/40-86*. Durham: University of Durham Library.

MacMichael, H.A. 1912 (1967). *The Tribes of Northern and Central Kordofan*. London: Frank Cass.

MacMichael, H.A.1922 (1967). *A History of the Arabs in the Sudan and some Account of the People Who Preceeded Them and of the Tribes Inhabiting Dār Fur, Vo. 1*. London: Frank Cass & Co. Ltd.

Manger, L.O. 1984. Traders and farmers in the Nuba Mountains: Jellāba family firms in the Liri area, in *Trade and Traders in the Sudan*, edited by L.O. Manger. Bergen: University of Bergen, Department of Social Anthropology, 213-242.

Manger, L.O. 1988. Traders, farmers and pastoralists: Economic adaptations and environmental problems in the southern Nuba Mountains of the Sudan, in *The Ecology of Survival: Case Studies from Northeast African History*, edited by D.H. Johnson and D.M. Anderson. Boulder: Westview Press, 155-172.

Manger, L.O. 2001. The Nuba Mountains: Battlegrounds of identities, cultural traditions and territories, in *Sudanese Society in the Context of War*, edited by M.-B. Johannsen and N. Kastfelt. Copenhagen: University of Copenhagen, 49-90.

Manger, L.O. 2003. *Civil War and the Politics of Subjectivity in the Nuba Mountains, Sudan.* Paper presented at a workshop of the XV ICAES Congress: 'Humankind/Nature Interaction: Past, Present and Future', 5-12 July 2003, Florence, Italy.

Manger, L.O. 2004. Reflections on war and state and the Sudan, in *Social Analysis, 48(1): The State, Sovereignty, War, and Civil Violence in Emerging Global Realities, Critical Interventions, 5*, edited by B. Kapferer. New York: Bergham Books, 75-88.

Manger, L.O. 2006. *Understanding the Ethnic Situation in the Nuba Mountains in the Sudan. How to Handle Processes of Groups-making, Meaning Production and Metaphorization in a Situation of Post-conflict Reconstruction.* Paper presented at the 1st International Colloquium of the Commission on Ethnic Relations (COER) for IUAES, 7-9 July 2006, Florence, Italy.

Manger, L.O. 2007. Ethnicity and post-conflict reconstruction in the Nuba Mountains of the Sudan: Processes of group-making, meaning production and metaphorization. *Ethnoculture*, 1, 72-84.

Manger, L.O. 2008a. Land territoriality and ethnic identities in the Nuba Mountains, in *Nomadic-sedentary Relations and Failing State Institutions in Dārfur and Kordofan, Sudan*. (Orientwissenschaftliche Hefte 26/2008, Mitteilungen des SFB 12), edited by R. Rottenburg. Halle: University of Halle, 71-99.

Manger, L.O. 2008b. Building peace in the Sudan: Reflections on local challenges, in *Between War and Peace in Sudan and Sri Lanka: Deprivation and Livelihood Revival*, edited by N. Shanmugaratnam. Oxford: James Currey, 27-40.

Meyer, G. 2005. *War and Faith in the Sudan*. London/Cambridge: William B. Eerdmans Publishing Company.

Michael, B.J. 1987a. *Cows, Bulls and Gender Roles: Pastoral Strategies for Survival and Continuity in Western Sudan*. Lawrence: University of Kansas.

Michael, B.J. 1987b. Milk production and sales by the Hawazma (Baggara) of Sudan: Implications for gender roles. *Research in Economic Anthropology*, 9, 105-141.

Michael, B.J. 1991. The impact of international wage labor on Hawazma (Baggara) pastoral nomadism. *Nomadic Peoples*, 28, 56-70.

Michael, B.J. 1998. Baggara women as market strategists, in *Middle Eastern Women and the Invisible Economy*, edited by R.A. Lobban. Gainesville: University Press of Florida, 60-73.

Mohamed Salih, M.A. 1984. Local markets in Moroland: The shifting strategies of the Jellāba merchants, in *Trade and Traders in the Sudan*, edited by L.O. Manger. Bergen: University of Bergen, 189-212.

Mohamed Salih, M.A. 1990. Agro-pastoralists response to agricultural policies: The predicament of the Baggara, western Sudan, in *Adaptive Strategies in African Arid Lands*, edited by M. Bovin and L.O. Manger. Uppsala: Scandinavian Institute of African Studies (SIAS), 59-75.

Mohamed Salih, M.A. 1992. Traditional markets and the shrinking role of the African state, in *Traditional Marketing Systems*. (Proceedings of an International Workshop, Feldafing, July 6th to 8th, 1992), edited by L.F. Cammann. Feldafing: German Foundation for International Development, 188-199.

Mohamed Salih, M.A. 1995a. Pastoralists and the war in southern Sudan: The Ngok Dinka/Humr conflict in South Kordofan, in *Conflict and the Decline of Pastoralism in the Horn of Africa*, edited by J. Markakis. Houndmills: Macmillan, 16-29.

Mohamed Salih, M.A. 1995b. Resistance and response: Ethnocide and genocide in the Nuba Mountains, Sudan. *Geojournal*, 36(1), 71-78.

Mohamed Salih, M.A. 1999. Land alienation and genocide in the Nuba Mountains. *Cultural Survival Quarterly*, 4, 36-38.

Mohamed Salih, M.A., Dietz, T. and Abdel Ghaffar, M.A. (eds). 2001. *African Pastoralism: Conflict, Institutions and Government*. London: Pluto Press.

Nadel, S.F. 1947. *The Nuba: An Anthropological Study of the Hill Tribes of Kordofan*. London: Oxford University Press.

Pallme, I. 1844. *Travels in Kordofan*. London: J. Madden and Co. Ltd.

Pantuliano, S., Buchanan-Simth, M. and Murphy, P. 2007. *The Long Road Home: Opportunities and Obstacles to the Reintegration of IDPs and Refugees Returning to Southern Sudan and the Three Areas: Report of Phase 1*. London: Overseas Development Institute.

Rahhal, S.M. 2001. *The Right to be Nuba. The Story of a Sudanese People's Struggle for Survival.* Lawrenceville: The Red Sea Press.

Sagar, J.W. 1922. Notes on the history, religion and customs of the Nuba. *Sudan Notes and Records*, 5, 137-156.

Seligmann, C.G. 1910. The physical characters of the Nuba of Kordofan. *The Journal of the Royal Anthropological Institute of Great Britain and Ireland*, 40, 505-524.

Seligmann, C.G. 1913. Some aspects of the Hamitic problem in the Anglo-Egyptian Sudan. *The Journal of the Royal Anthropological Institute of Great Britain and Ireland*, 43, 593-705.

Seligmann, C.G. and Seligmann, B. 1932 (1965). The Nuba, in *Pagan Tribes of the Nilotic Sudan*, edited by N.A. London: Routledge & Kegan Paul, 367-412.

Spaulding, J. 1982. Slavery, land tenure and social class in the Northern Turkish Sudan. *The International Journal of African Historical Studies*, 15(1), 1-20.

Spaulding, J. 1987. A premise for pre-colonial Nuba history. *History in Africa*, 14, 369-74.

Spivak, G.C. 1988. Introduction: Subaltern studies: Deconstructing historiography, in *Selected Subaltern Studies*, edited by R. Guha and G.C. Spivak. Oxford: Oxford University Press, 3-32.

Stevenson, R.C. 1965. *The Nuba People of Kordofan Province: An Ethnographic Survey.* (Graduate College Publications Monograph 7). Khartoum: University of Khartoum.

Strauss, A. 1978. A social world perspective. *Studies in Symbolic Interaction*, 1, 119-128.

Sudan Archive: SAD: 212/2/1-94. *Report on Slavery and Pilgrimage, 1926 by C. A. Willis, Assistant Director of Sudan Intelligence 1915-1926.* Durham: University of Durham Library.

Suliman, M. 1999. The Nuba Mountains of Sudan: Resource access, violent conflict, and identity, in *Cultivating Peace: Conflict and Collaboration in Natural Resource Management*, edited by D. Buckles. Ottawa/Washington D.C.: World Bank Institute, International Development Research Centre, 205-20.

Suliman, M. 2002. Resource access, identity, and armed conflict in the Nuba Mountains, southern Sudan, in *Transformation of Resource Conflicts: Approach and Instruments*, edited by G. Baechler, K.R. Spillmann and M. Suliman. Bern: Peter Lang, 163-183.

Trimingham, J.S. 1949 (1983). *Islam in the Sudan.* London: Frank Cass & Co. Ltd.

Vang, K. and Granville, J. 2003. *Aiding Trade: An Assessment of Trade in the Nuba Mountains.* Missouri: University of Missouri and USAID. [Online]. Available at: http://cafnr.missouri.edu/iap/sudan/doc/aiding-trade.pdf [accessed: 22 May 2008].

Waal, A. de. 2006. *Averting Genocide in the Nuba Mountains, Sudan.* New York: SSRC. [Online]. Available at: http://howgenocidesend.ssrc.org/de_Waal2/ [accessed: 5 October 2007].

Interview Partners

Simon Kalo, SPM/A Commander, Kadugli, 9 January 2007.
Anonymous Key Baqqāra Informant, Reikha, 17 February 2007.
Hāmid Setār, Skeikh of the 'Ayātqa Baqqāra, Khartoum, 10 June 2006.

Chapter 6

Pastoral Integration in East African Livestock Markets: Linkages to the Livestock Value Chain for Maasai Pastoral Subsistence and Accumulation

Fred Zaal

Introduction

There has been an upsurge in interest in dryland development and pastoral societies in recent years (McPeak and Little 2006). New developments in land tenure, land use, and changes in pastoral production strategies and tourism development have triggered new interest and new conceptual ways of looking at pastoralism. In the early 1980s, with most of the pastoral development projects in a state of crisis or abandonment, it became obvious that the assumptions on which these projects were based must be wrong. It took some years before the 'mainstream view', as Sandford (1983) called these earlier assumptions, became the 'old orthodoxy' (Lane and Swift 1989, Lane 1991) and new ways of understanding the pastoral world developed under what was obviously called the 'new thinking' (Scoones 1995). This shift took place when it was realised that instead of focusing on *pastoral* development, the old orthodoxy projects had been focusing on *livestock* development and commercial *meat* production for the urban market rather than the support of milk production for home consumption. There was renewed interest in the logic of pastoral production. The new research themes and policy buzz words were flexibility in the face of variability, institutional arrangements to accommodate change and indigenous management of scarce resources.

However, the main focus was on the pastoral system itself and not on the linkages of pastoral production with the wider economy. The implicit presumption was that these pastoral systems were autonomous or geared to self-sufficiency. However, external relationships have always existed in pastoral societies. Only a limited number of studies have focused exclusively on these relationships, though recently there is an increased interest.[1] Studying this outward-looking aspect is important to our analysis, as it means we must look at evidence of integration,

---

1 Aklilu and Catley (2010), Bekure et al. (1991), Dietz (1993), Ensminger (1996), Karim (1991), Kariuki and Kaitho (2009), Kerven (1992), Komen et al. (2009), Little

analyse what differentiation this has caused in pastoral society and economy, and assess the impact of this on pastoral systems in the future. Who has benefited, and who hasn't? And for whom therefore is market development beneficial? Is this approach the way to pro-poor development?

The present chapter develops the view that marketing relationships are shaped not only by the pastoral system itself, but also by the economic world around it, though the 'rationalities' of pastoral systems and the outside economic world differ (Behnke 1983). Consequently, pastoral society changes profoundly when these relationships intensify. However, this does not mean that typical 'pastoral' aspects of livestock production cease to exist. Rather, income opportunities and benefits for pastoralists provided by the market are always compared to the risks of dependency and loss of control, and also to the benefits for third parties such as traders. In this chapter I present a case study on the Maasai from the northern part of Tanzania and southern part of Kenya that reveals how pastoral economies and societies are changed by inclusion in the meat value chain.

Marketing in East African Pastoral Systems

In a groundbreaking study, Kerven (1992) highlighted the existence of long distance livestock trade flows in early colonial Sub-Saharan Africa. Many historians seem to have neglected this trade, perhaps because they were more interested in (pre) colonial export goods such as ivory and slaves. Most livestock was traded in networks whose core business was oriented towards luxury products such as gold or textiles rather than livestock. Livestock was a 'return product'. Nevertheless, livestock and livestock products, such as skins and hides and leather, featured in these trading caravans too.

In eastern Africa, long distance trade developed from the early nineteenth century onwards and was mainly dominated by Swahili and Arab traders from the coast. Fage mentions gold, in addition to ivory, slaves, cloves and only occasionally skins, as the major trade products of this region. All of these were considered to be export products (Fage 2002). Due to the relative proximity of different ecological zones in East Africa (Blench 1999), the small exchange of food grains and livestock products that existed between farmers and pastoralists was usually short distance trade. Kerven (1992), however, taking a closer look at East African pastoral economies, not only presents evidence of local trade between pastoralists and farmers but of long distance trade as well. After the devastating pandemics of the late nineteenth century, Maasai demand for stock was such that present-day Kenya and Tanzania were included in long distance livestock trade from Somalia and southern Ethiopia (de Haan et al. 1999, Zaal 1998). Somali traders would barter camels for heifers and young bulls in the northern rangelands

(1987), Little et al. (2001), McPeak and Little (2006), Moritz (2009), Sutter (1982), Swift (1986), Teklewold et al. (2009), Zaal (1998), Zaal et al. (2006).

of Kenya. These animals would be taken to the Maasai area and exchanged for mature bulls, which were sold in the emerging consumer centres.

These patterns of cattle trade were disrupted with the introduction and penetration of colonial rule. Sales from African livestock producers in Kenya were generally hampered by veterinary quarantines. Trade between indigenous producers and white settlers or with the newly established consumer markets was only allowed when demand for local stock was (too) high, or with a drought crisis that forced local producers to sell. This was the case when settlers needed stock to start or restock their ranches, or when demand for meat accelerated such as during the First World War. During the droughts of 1918-21 and the late 1920s, increasing numbers of animals came onto the market, causing massive price slumps.

Illegal trade (or rather unofficial trade, as trade to more densely inhabited agricultural areas has always taken place) also took place between cattle producing areas and areas of emerging African small-scale agriculture and the developing urban centres. In the mid 1930s, double the number of animals were marketed in Kenya compared to the 1920s (Kerven 1992), as by then producers were encouraged to sell to emerging African consumer markets. Official figures indicated an increase from 8,000 to 24,000 head of cattle and from 14,000 to 48,000 head of small stock sold annually in the course of the 1930s. During the Second World War, in order to supply meat to the armed forces, Maasai producers sold 25,000 head of cattle per year (de Haan et al. 1999).

The pattern that developed in the 1950s was more complex. White settlers in Kenya marketed most of their livestock through the Kenya Meat Commission (KMC), established in 1950. KMC mostly served the settler population as a channel along which livestock could easily and cheaply be sold to the meat conservation industry or local urban centres, and it also arranged export to Europe. Considerable numbers of African livestock came from northern Kenya and Somalia to the coast, from the southern rangelands to both Nairobi and Tanzania, and from the western part of Kenya, the Rift Valley, to Nyanza Province and Uganda. A separate organisation, the African Livestock Marketing Organisation (ALMO) was set up to market these animals. In 1957-58, for example, official records suggest that between 50,000 and 63,000 head of cattle were marketed to Nyanza and Uganda from Rift Valley producers and Trans Mara Maasai (Republic of Kenya 1959). Some cattle came from southern Sudan to Uganda as well. Cattle would also come from the central parts of Tanzania to the coastal town of Dar es Salaam. However, in the late 1950s, meat prices in Kenya had gone up so much, that cattle started to come from Tanzania to the north into Nairobi (Mittendorf and Wilson 1961). Overseas exports were also important, mainly from Kenya and Tanzania to the UK, Germany, and Mediterranean countries. Low quality animals from African producers, processed by Liebig's Company under supervision from KMC, were exported as corned beef. In the late 1950s, this amounted to 13,000 tons from Kenya, and almost 10,000 tons from Tanzania (Mittendorf and Wilson 1961).

Geographical patterns in most East African countries did not change much in the first years after independence. Trade flows, being basically dependent on relatively slowly changing levels of demand and supply, continued to be directed from the drylands to the main consumer areas in the various countries. With Kenya as the main consumer market and Somalia and Tanzania as its main foreign suppliers, cross-border cattle trade was high during the drought periods of the early 1970s (around 70,000 head annually) and 1980s (150,000 head), but since then reduced to about 20,000 head in 1990. Finally, due to the civil war and a complete avoidance by Kenyan traders of Somalia and the neighbouring region of Kenya, direct cross-border trade from that area stopped completely in 1992, while supply from Tanzania increased (Zaal 1997, Zaal et al. 2006). More recently however, trade with the northern parts of Kenya has flared up again as Somali traders have switched direction taking animals from the Somali and southern Ethiopian rangelands through Kenya to Nairobi, avoiding the conflict-ridden southern part of Somalia. Prices are dependent on the animals of course, but also on the occasional high demand in Nairobi, and facilitate exports (Teklewold et al. 2009). 200,000 to 250,000 head of cattle were estimated to have been taken across the border (McPeak et al. 2006), sometimes to ranches around Voi first, and sometimes directly to Nairobi (see Mahmoud, this volume). Officially, cattle trade from Tanzania to Kenya has been limited since the early 1990s. However, fieldwork by Zaal (1997, Zaal et al. 2006) proved that thousands of heads of cattle cross the border unofficially, attracted by the high prices in nearby Nairobi. Based on national statistical data, we may assume that 1.2 million head of cattle are consumed within Kenya annually, but what share Nairobi takes is difficult to ascertain. We estimate that in addition to this, unofficially and unsurveyed, between 60,000 and 100,000 head of cattle come from Tanzania annually, and another 5,000 from Uganda (Zaal et al. 2006). Exports from Kenya are now almost absent after a slow decline between the 1980s and 1990s.

Post-colonial policies in East Africa were very much a continuation of colonial ones, as marketing boards continued to play a role. However, in the course of the 1970s and 1980s, quite a number of these boards ran into financial problems. In Kenya, KMC continued to buy cattle from ranches in limited numbers, while ALMO was responsible for purchasing cattle from pastoral areas. With the Kenya Livestock Development Project in 1968, a new department took over from ALMO, the Livestock Marketing Department (LMD). Providing holding grounds, stock routes, price information and various other services, it became a service oriented organisation, and no longer bought cattle from producers, a task which proved too costly. Like KMC, high running costs ultimately pushed ALMO out of the market. KMC, like other livestock related services, extended its clientele to include the African small scale producers (Heyer et al. 1976). However, as private slaughterhouses had slowly taken over the market in the late 1970s, the two enormous slaughterhouses owned by KMC deteriorated, and were finally closed in 1992, to open again and again in an unreliable sequence since 2005.

In Tanzania, a similar situation occurred. The *Ujamaa* philosophy (African Socialism) exerted an even greater pressure for government policies to control cattle trade. However, the livestock and meat marketing boards, due to their unfavourable fixed prices, did not adequately control the livestock sector. In addition, cattle producers made only a token contribution of animals to co-operative farms and to state controlled dairy units. As a consequence the cattle sector had a fairly high degree of competition between official and parallel markets. The Tanzania Livestock Marketing Company (TLMC) had a market share of 20-40 per cent in the 1970s and even less in the 1980s (de Wilde 1984). An International Livestock Centre for Africa (ILCA) study (based on extensive work done in the region on the role of marketing on livestock production) demonstrated that a small trade flow of animals had developed from Tanzania to Kenya in reaction partly to this unfavourable situation in Tanzania (ILCA 1983).

Kenya, primarily Nairobi and the densely populated highlands in the West of the country, has become the main consumer market of cross-border livestock trade in eastern Africa, with Somalia, Ethiopia, Uganda and Tanzania as its main foreign suppliers. As the huge market of Nairobi continues its rapid growth, the area from which it draws its food expands. The future of livestock marketing is painted in the most glowing colours by researchers who assume that population growth, growth in wealth, and related growth in meat consumption, both domestically and in export markets, will continue faster than levels of productivity in agriculture and in household consumption in general (Delgado et al. 1999, Rich 2009). If these projections are correct, meat production will be profitable for the next few decades to come. Producers in the areas relevant to this chapter (Maasai areas in Narok and Kajiado districts of Kenya and parts of northern Tanzania) seem to recognise this potential and have changed their production strategies towards meat production. This is most clearly seen in the shift from cattle towards smallstock in their herds, a strategy to increase meat rather than milk production. Therefore, there is a very good case to be made for increased attention to the livestock marketing system and the impact of this on the production system.

Livestock Marketing Systems in the Tanzania/Kenya Border Area

Flows of Animals

The Maasai of Kajiado and Narok districts in Kenya and the neighbouring area in Tanzania exemplify the predicament of many pastoralists (Zaal and Dietz 1995). This predicament is one of increasing human and fluctuating livestock populations (Zaal 1998), continued loss of land to conservation efforts (Homewood et al. 2001, Sindiga 1985) and urban and arable land use, land tenure changes and changing demand for their products. Despite these difficulties, they have adapted themselves by intensifying their use of increasingly limited resources and by commercialising their production strategies, partly based on traditional breeds. This has been facilitated by the relatively favourable trends in market prices due to high demand in the market, as predicted by Delgado et al. (1999).

Most animals from Tanzania and the Narok and Kajiado area are taken on the hoof towards Nairobi. In most instances animals were previously trekked and are

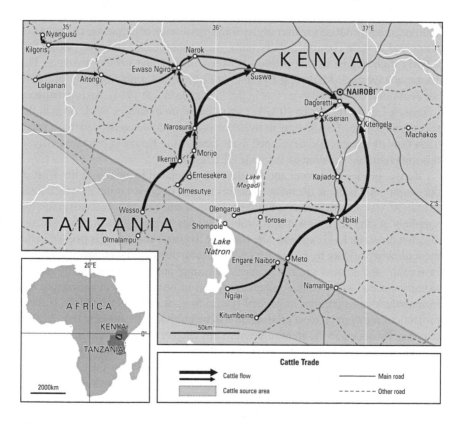

Figure 6.1   Flows of animals from Tanzania through Kajiado and Narok districts towards Nairobi (Dagoretti market), 2006

Source: Own survey.

now moved using trucking, except for the route from Suswa and Ewaso Njiro to Nairobi and its surrounding peri-urban area (see Fig. 6.1). Most of the Tanzanian and Trans Mara/Narok cattle are trekked along established routes towards either Narok Town and Suswa or directly towards Nairobi, with markets directly to the South increasingly taking over from the more westerly ones as entry points from Tanzania. The larger traders have started trucking their animals as a direct transport link between Narok and Suswa towards Nairobi, which necessitates holding pens and loading ramps. They do this to be able to react rapidly to higher prices in the end-markets. At present, an estimated 45,000 head of cattle is transported through Suswa market. In Kajiado, meat traders have a different solution to this need for rapid action to suddenly favourable prices. They use local slaughterhouses to slaughter the cattle and truck the carcasses to their customers within two hours. Both systems are recent innovations, not supported by Government or NGO programmes, but developed locally. Trucking and transporting meat also avoids the regular quarantines: The animals can't contract any diseases on the way nor infest other animals. That discussion, so intensely held in the colonial period and later independence years, has finally been laid to rest in this area.

Officially the number of cattle exported from Tanzania is reported as minimal. Unofficially however, a large number of animals are transferred from Tanzania to Kenya. In April–June 2006, around 10 per cent of cattle in Suswa market came from Tanzania. Higher percentages (around 70 per cent in 2006) were found in markets closer to Tanzania such as Ilkerin, while for smallstock these percentages were closer to 40 per cent. Data from a survey undertaken in 2004 suggests that imports from Tanzania were between 60,000 and 100,000 head of cattle that year. Data from our survey in 2006 indicates that, at least temporary, the imports from Tanzania surged to an estimated 120,000 head or higher.[2] Estimates on smallstock imports are harder to obtain. Percentages of Tanzanian smallstock on the southernmost Kenyan markets may be as high as 60 to 65 per cent, and for the markets in Narok district as a whole, they may be about 40 per cent.

Animals come from various directions within the production zone, but some major trade routes have developed where basic facilities are supplied by local people, mostly women, who offer a wide range of services, including the preparation of food, the provision of water and a safe place to sleep.

Prices of Animals

In general there has been a steady increase in prices for all markets in recent years. However, the drought of late 2005 and early 2006, caused a very steep decline and a slump in the market as producers dumped their animals to at least partially recoup costs. The following data[3] (Fig. 6.2) have been collected as part of an assignment for Heifer International Kenya (Zaal et al. 2007). This data applies to the Narok district in the period of April–June 2006. Prices generally increase from the Mara area and southern Narok district towards the north (the Mau area and the northeast).

This also applies to the Kajiado district, where prices increase from the border area towards the north (Fig. 6.2). The largest market influence in that direction is Nairobi, though the high population density in the peri-urban area around Nairobi is so vast that it generates separate marketing possibilities.

The price gradient in Figure 6.2 implies a rising price for animals going from the South (Olmalampu and Wasso) towards the North-east (Dagoretti, close to Nairobi). Due to the strong Kenyan Shilling in recent years, the price hikes

---

2   We refer to this development as temporary as it may have been caused by distress sales by producers due to the drought rather than for other reasons. The drought ended in March 2006.

3   These figures cover markets from Tanzania (Olmalampu and Wasso) to Nairobi (Dagoretti). When data were lacking, the prices were estimated on the basis of prices before and after the market in the marketing system for that particular type of animal. It concerns Olpusimoro, Oloolaimutia, Naikarra (Mature bull) and Olmesutie, Oloolaimutia and Naikarra (Mature Oxen). We gratefully acknowledge HIK for allowing us to use these data.

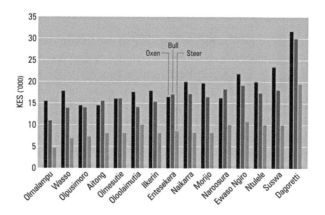

Figure 6.2   Prices of cattle between Tanzanian markets (left in the graph) towards Nairobi (right), in Kenyan Shilling, April–June 2006

Source: Own survey.

immediately on crossing the border into Kenya. The difference due to currency conversion is an additional profit to the international trader, an idea corroborated by discussions with traders in the field. The second sharp rise in cattle prices occurs between Suswa and Dagoretti. Can this be explained by transportation costs? Between Suswa and Dagoretti, transport costs for animals on the hoof are KES 200 per head, by truck KES 400 per head, while for the longer distance between Ewaso Ngiro and Dagoretti, it costs KES 600 for animals taken by truck. Transportation costs further increase for Naroosura to Dagoretti. For animals on the hoof they amount to KES 1,500, and per truck to KES 2,000. Therefore, transport costs alone cannot account for the sharp hike in prices, even taking into account additional indirect transport costs, such as police blocks, food and shelter for the trader, water and grazing on the way. Certainly when we see that the price per head of cattle rises only slowly from Tanzania to, for example, Suswa, we would expect a slower rise for the shorter distance between Suswa and Dagoretti, even though the availability of water and grass between those two places is less than between Tanzania and Suswa. The only possible conclusion is the margin made by traders between Suswa and Dagoretti. It is rather high, probably to the order of. KES 5,000-6,000 gross per head or more. This is partly because of a 'congestion rent' (as it may be called); the fact that there are so many customers on the Dagoretti market that sale is almost always assured. Private individuals, butchers and large institutions (the army, hospitals, schools and universities) generate an almost insatiable demand, hence prices are high. Also however, large-scale traders dominate this market, and they have colluded to organise a semi-monopoly where very few smaller traders can survive. Smallstock may be of increasing importance if present trends towards a focus on meat production persist. The following figure shows this data (Fig. 6.3).

As noted earlier, in the Maasai production areas, there has been a shift towards the keeping of smallstock with flock sizes of goats and sheep increasing and herd

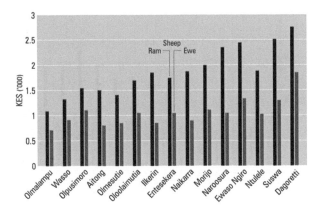

Figure 6.3   **Prices of sheep across Narok district between Tanzanian markets (left in the graph) towards Nairobi (right), in Kenyan Shilling, April–June 2006**

Source: Own survey.

sizes of cattle remaining relatively stable. A number of points can be made here. First of all, there is not such a dramatic price hike between Tanzania and Kenya markets in absolute terms as with cattle. However, even with low prices, the relative increase is still considerable. Certainly with mature rams, the more commercial type of animal, the gradient is steep and steady and the resulting margin by traders between Suswa and Dagoretti is therefore comparatively small, and certainly not larger than between other markets. For the less commercial mature ewe, the gradient is less steep and the price ultimately lower. Thirdly, there seem to be small dips in prices, but these may be caused by temporary dips when that particular market was visited rather than that the prices on that market are structurally lower. This may still however be the case when the market is small and caters for local needs only (as in the case of Ntulele). As with cattle, the price between the Kenyan border and Dagoretti market doubles as a consequence of transport costs and trader margins.

The same analysis was repeated for goats. The following figure presents the results (Fig. 6.4). This time we again see a steep and steady gradient from Tanzania to Kenya, especially so for the more commercial mature male goat. Here again, as with sheep, the prices rise steadily towards the final market of Dagoretti, though there is a slightly higher margin in the latter market than for sheep. And thirdly, the prices in Ewaso Ngiro seem relatively high, and in retrospect this is also seen with sheep, though less markedly. The reason for this is that Ewaso Ngiro is a loading point for smallstock traders, taking animals to the Dagoretti market in trucks and pickups. The prices for smallstock are therefore slightly closer to the prices of Dagoretti. Also, there is a slaughterhouse in Ewaso Ngiro that caters for the local population, and thus customers are numerous here and demand is high.

A summary of the findings has been used to generate the map (Fig. 6.5). For the markets mentioned, the prices of selected types of animals are presented. We

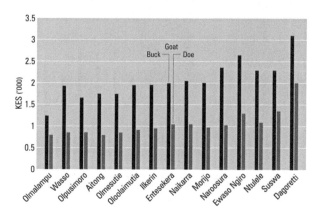

Figure 6.4   Prices of goats across Narok district between Tanzanian markets (left in the graph) towards Nairobi (right), in Kenyan Shilling, April–June 2006

Source: Own survey.

see the increases in price from the south-west to the north-east, and the relative coherence in the prices between cattle and smallstock. In monetary terms, there is a factor ten between smallstock prices and cattle prices.

Impact of Livestock Marketing in Pastoral Livelihood Systems

The livestock marketing system is increasingly important as the consumer market continues to grow both domestically and in export markets, and also because livestock producers (in Maasai areas in Narok and Kajiado districts and parts of northern Tanzania) recognise this potential and have changed their production strategies toward introducing improved breeds (Scarpa et al. 2001) and more smallstock. From the producer perspective there is a continuous need to improve the productivity of herds and flocks, in a sustainable way to increase off-take without doing damage to pastoral livestock production systems and the society that relies on it (Mwacharu and Drucker 2005). In that sense the introduction of new breeds, water and animal health facilities have been more successful than the full-scale introduction of the ranching system (Goldschmidt 1981, Homewood et al. 2005). The question is however: 'To what degree have pastoral production systems and society changed and adapted to this highly competitive marketing system? Can we detect this in their livelihood strategies?'

It is of course unlikely that all producers profit from these opportunities in a like manner or to the same degree. It is generally found that pastoral communities are highly skewed in ownership of livestock, and thus marketing returns may also be skewed. The possibility of participation in the pastoral community may likewise be

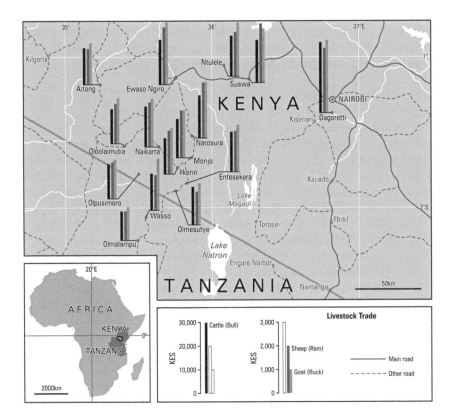

Figure 6.5   Livestock prices, Tanzania-Dagoretti marketing system, in Kenyan Shilling, 2006

Source: Own survey.

affected, as livestock increasingly translates into capital and a source of cash flow rather than into a carrier of social relationships, mutual obligations and status.

In the Kajiado district, an area with a long history of marketing, where many of the developments described above are found, we studied changes in livelihood systems. A set of households, visited in the mid 1990s (70 households were surveyed at the time) (Zaal 1998) was revisited in the early 2000s (37 were found and revisited, a further 30 households added randomly selected from the area). This was done to analyse the impact of commoditisation of livestock, land, labour and water. A principle component analysis was undertaken based on present and past production and livelihood characteristics, to see whether commercial rather than traditional strategies had led to a reduction of vulnerability (Seevinck 2004).[4] The results were rather remarkable, in the sense the most important livelihood

---

4   In the principal component analysis, 8 components were extracted with an eigenvalue higher than 1, explaining 73.9 per cent of the total variance. Almost 60 per

strategies in this area, close to Nairobi, were still of a more traditional pastoral character.[5] Most important for our argument however is the conclusion that only a small group of local pastoral producers really benefit from the markets directly.

Successful Traditional Pastoralists

The first and most important livelihood strategy can be called successful traditional pastoralists. These are pastoralists that enjoy a high household security. They own a large number of animals, of which a high percentage is of an improved breed. Often labour is hired to look after the herd. Inputs per animal are low. This group receives and gives out large numbers of animals and continues to maintain strong reciprocal ties with other members of the Maasai community. Because of the high number of non-commercial transactions, these households are not considered to be very commercially active. Interestingly, these successful traditional pastoralists actively took up crop cultivation as an innovation some years ago. Crop cultivation (especially for home consumption) is now a readily adopted activity and, because of the livestock wealth of the successful traditional pastoral households, investment in agriculture can be easily financed. The cultivation of crops is an accumulation strategy, whereby financial capital that was previously spent on food purchases can now be utilised for other purposes (e.g. animal purchases). Another important income generating activity for these households is the marketing of milk, in one sense an obvious activity given the large number of animals owned. However, these households only sell surplus milk after their own household needs have been met.

To position all this in a wider timeframe, variables concerning wealth and commercial behaviour in 1995 were also considered. The households that are now successful pastoralists were already wealthy in 1995 in terms of the number of animals owned. Somewhat surprisingly these households displayed a high level of commercial activity in 1995. We may conclude that a commercial strategy in time leads to greater security and wealth, and ultimately to a return to traditional pastoral production. This is similar to Little's findings among the Il Chamus, who used irrigated cultivation to generate money, buy animals, and return to pastoral life (Little 1992). It is also possible that in 1995 these households were still building their herds and that the importance of reciprocal relationships (social capital) and non-commercial transactions increased only when the households had matured. In that sense it is an important indication of the household life-cycle concept linked to Chayanovs work[6] (Bernstein 2009). The success of this group of pastoralists today

cent of the total variance is explained by the first five components. We distilled different household livelihood strategies from the components.

5   The detailed results can be obtained through the author. I present here the general conclusions only.

6   The life cycle here means that small families (between 5 and 12 people) start with a small herd and a number of dependent persons (children), towards an increasingly large household of a man with a number of wives and large resources (between 12 and 25

may be related to the abundance of livestock owned in 1995. Today they still own large herds from which they do sell numbers of male animals unused for breeding, generate income from selling surplus milk and actively pursue crop cultivation for household consumption purposes. These households have secured their leading position within the pastoral production system by means of a (temporarily high level of) involvement in the livestock market, and later by means of giving livestock to establish social relationships. Due to their number these pastoralists collectively are very important as suppliers to the livestock market.

Marginal Pastoralists

A second type of livelihood strategy involves households that are further on in the household life cycle, and are lead by an old household head. They own few animals and spend a relatively disproportionate amount of money on inputs, which is partly a result of their small herd size. There is very little commercial activity and hardly any livestock is sold or purchased. Herd size is virtually stagnant, but the high inputs per head of cattle, and the labour that is hired for livestock herding, indicate that the cultural importance of livestock has not diminished for these households. Lack of financial capital and the absence of household labour to manage the herd and flock properly prohibit them from expanding their herds. For these households, crop cultivation is not an accumulation strategy; it is adopted as a response to permanent financial difficulties. Although many of these households cultivate crops, it is in a more haphazard manner than the successful traditional pastoralists. When money or labour is lacking, fields are not prepared and seeds are not purchased. Their own labour is frequently exploited in order to generate income; accepting all sorts of formal, seasonal and odd jobs can thus provide them with the financial capital necessary to sustain their livelihood. Women are also often involved in charcoal burning.[7] These households are the future drop-outs of the pastoral system. The households own too few animals to sustain their livelihood strategy by exclusively relying on livestock. They cultivate crops (often very opportunistically) mostly for home consumption, and they sell their labour and burn charcoal to generate extra income. It can be expected that these households will continue to participate in the pastoral production system in view of the importance attached to the pastoral way of life, but as wage labourers for large herd owners instead of productive livestock keepers. They may be hired by the successful pastoralists to herd their cattle and to carry out other (agricultural) jobs. They do depend on the market, but on the labour market rather than the livestock market.

---

people), towards a fragmenting household where the herd is split and given to children, and the parents remain behind, or in the care of the children.

7   This corresponds with findings from Smith (in Little et al. 2001) on diversification behaviour of pastoralists in North-Kenya and South-Ethiopia, whereby poor pastoralists frequently mentioned charcoal burning as an income generating activity and rank it as a relatively important money-making option.

Labour Migrants

The third type of household livelihood strategy is characterised by a number of household members living in the original homestead but, at the same time, a relatively large proportion of household members that are working elsewhere. Those living away are still considered to be part of the household production and consumption unit by standards of the household head. The income generated by these absent household members contributes to household income through a process of diversification. Often on a regular basis remittances are received. This may significantly improve household security, especially when the absent household members hold a formal job, or are employed as non-farm wage labourers, in which case they are less affected by seasonality and environmental crises. These households still have some animals, and sometimes they can purchase additional animals with the proceeds of the salaries received. They use the livestock market for survival and for the growth of their herds.

Pastoral Quitters

The fourth type of pastoral livelihood strategy is very different. This type of household has opted to provide (most of) their children with at least a primary level education. Children sometimes live away from the homestead to receive (secondary) education. These households do not own a large number of animals, nor do they actively pursue crop cultivation, but they are often involved in specific diversification activities, such as selling of products in small shops and other activities outside the pastoral production system to generate income. For these households, the economic importance of livestock has diminished substantially, but animals are still kept as a kind of insurance against times of hardship, when they can be sold to acquire financial capital. In this sense access to livestock markets is of occasional importance. Furthermore, the households may keep some animals for social and cultural practices. The assumption that these households have exited the pastoral production system is underlined by the fact that they do not often purchase livestock to extend their herds. This may be related to the lack of financial capital obtained from their income generating activities. When the high number of children being educated is considered, it is likely that this group of households will ultimately adopt a permanent livelihood strategy that lies completely outside the pastoral sphere. They use the sales from their herds to change towards a non-pastoral livelihood, rather than vice versa.

Commercial Pastoralists

The fifth clearly recognisable pastoral livelihood strategy can be characterised by its highly commercial behaviour. These pastoralists practise a form of fenceless ranching, though low hedges made of branches and live plants may increasingly be found. Of the total number of transactions conducted by the household, a relatively large number are commercial transactions. Animals are bought to make the herd grow more quickly and to introduce new genes and improved breeds in the herd, and are sold to finance new livestock purchases and to pay for services and inputs. Interestingly, in 1995, these households were already exhibiting strong commercially orientated behaviour. Maintaining social networks by giving and receiving animals appeared, and appears, to be secondary to decisions concerning herd management for meat production and other related economic considerations. This group of households does not sustain their social ties by giving out large numbers of livestock to relatives and affiliates, and thus one may wonder whether they can still be considered pastoral in outlook. Their production strategies are increasingly focused on ranching for meat production. They have more male cattle and relatively more small animals in their herds and flocks. These households also frequently hire labour to work on their land. Labour is almost exclusively paid in cash, no longer in the form of animals. This again indicates an increasing level of participation of these households in the monetary economy. It is expected that these pastoralists continue to sustain their livelihood by commercially exploiting their herds; though a lack of animals and little income from other activities may prevent them from building herds sizeable enough for a reliable pastoral livelihood. The large number of labourers hired for crop cultivation may be an indication of a longer-term commitment to cultivation. Still, this type of household is the most commercially oriented, and the livestock trade relies heavily on them.

Remaining Less Dominant Types of Livelihood Strategies

The remaining types of livelihood strategies have less explanatory value for understanding the Maasai pastoral system. However, some interesting observations can be made. One type of livelihood can be defined as that of the pastoral 'starters'. The household heads are young and the households are generally of a small to average size (between 5 and 12 people). Although the number of livestock owned is low, crop cultivation and other activities are not pronounced, and the fact that they are not active in charcoal burning indicates that this group of young households is not immediately vulnerable. According to the households' own perception, they are able to obtain either a formal or informal loan or other support if needed. The last strategy belongs to households likely to adapt to either a changed role within the pastoral production system or to a livelihood outside the pastoral sphere. These households owned a fairly large number of animals in 1995, but in 2003 this number had declined markedly. Crop cultivation is seldom pursued, but the households are often involved in (wage) labour for others, which is their single most important income generating activity. Similar to the marginal pastoralists, these households may be hired by the wealthier pastoral households in the area to perform all sorts of temporary and (semi) permanent jobs related directly or indirectly to livestock keeping or crop cultivation. On the other hand, household members may find formal employment outside the pastoral production system.

These livelihood strategies reveal that an increased level of livestock production and marketing, in particular – in terms of absolute numbers of animals and their importance to the income of the households – is happening in a limited number of households only. The key actors are the commercial pastoralists and those who use the market to either survive crises or develop their herd to become pure pastoral producers (as far as they can). Other households focus on the sale of their labour, their grazing or the water facilities they have on their land. A type of specialisation has thus developed, that is based on there being numerous means of production in each household. The fact that they have access to these resources, and relatively secured access too, is caused by the recent privatisation of land that has taken place in Kajiado district (and Kenya in general). This allows the marginal households to still survive in a pastoral setting by leasing out their resources to neighbours. There is however also a steady outflow of households from the pastoral system as some households decide to focus on education, migration and non-agricultural livelihood strategies.

Conclusion

Inclusion of traditional pastoral livestock systems in national and international meat markets has progressed to the point that functional specialisation has developed within various pastoral production systems. Though successful traditional pastoral production systems still exist, commercial capital-based and labour intensive commercial systems have also developed, often in the same area and within one pastoral system. These (now varied) systems are linked to flexible and hierarchical, culture-based and commercial international value chains, but not similar for everyone.

The case of northern Tanzania and southern Kenya illustrates this development and demonstrates the impact of developing meat value chains on the transformation of pastoral economy and society. There is a diversified response to the opportunities provided by the market, with only certain types of households making use of the increasing number and benefits of the livestock value chains oriented towards Nairobi. Though traders take a major part of the profits, especially around Nairobi, certain categories of livestock producers obtain a large part of the ultimate consumer price. Most of these producers are either successful traditional pastoralists who sell their surplus male stock, or commercial producers who focus on selling high value heavy animals for meat. More so than in many other societies, pastoral societies are characterised by very skewed ownership of livestock, and it is only the large, wealthy and populous households (the households in the prime of their years when the market began to develop) that have had the opportunity to choose. Some remain traditional and sell surplus young male cattle, others have developed into commercial producers. A number of other households are involved and incorporated in the market too, but more indirectly, as they provide labour and other means of production. These are households that had herds that were too small for self-sustained growth. The fact that resource ownership has 'solidified' due to the land privatisation drive, that made even small households own land, created a class of pastoralists that maintains itself by leasing out resources (pasture and water), profiting indirectly from the market. And ultimately there is a group that may see this as a stagnating situation: They opt out and invest in education and other businesses.

Interestingly, a large number of households who were successful in 1995 as traditional and commercial producers were also successful in 2003, though there is evidence that some households have fallen out of the system (particularly as they go through the last stages of the household life cycle), while others get in as young families. Those 1995 households that could not be found may have been more marginal than average, while those selected later and added to the group may have been more successful than average. Our conclusions (based on a less marginal group selected than is found in reality) should therefore be seen as relatively optimistic. It seems therefore that inequality within pastoral society and economy is now quite permanent, dividing the Maasai into successful producers, directly linked to the markets, and unsuccessful producers, indirectly linked through the sale of their labour and other resources, or opting out.

Acknowledgements

I would like to acknowledge the work and cooperation of a large number of people who have contributed to this chapter. Morgan Ole Siloma, Kihiu Mwangi and Joel Leposo cooperated in a study of the marketing chain in Narok district, and I would like to thank Alex Kirui of Heifer International Kenya who allowed me to use the data collected. Julia Seevinck did a follow-up study in 2003 of a group of pastoral producers that I studied in Kajiado district in 1995, the results of which are also used in an aggregated way in this chapter. I would like to thank Meeli and Nkele for assisting in the original data collection in that district. I would like to thank Paul Quarles van Ufford and Leo de Haan for data on earlier trade flows in Africa. I therefore use the collective 'we' in the text to acknowledge all of these contributions. All errors remain my own.

References

Aklilu, Y., Catley, A. 2010. *Livestock Exports from Pastoralist Areas: An Analysis of Benefits by Wealth Group and Policy Implications*. (IGAD LPI Working papers 01-10). Tufts University: Feinstein International Centre.

Behnke, R.H. 1983. Production rationales, the commercialisation of subsistence pastoralism. *Nomadic Peoples*, 14, 3-34.

Bekure, S., de, Leeuw, P.N., Grandin, B.E. and Neate, P.J.H. 1991. *Maasai Herding. An Analysis of the Livestock Production System of Maasai Pastoralists in Eastern Kajiado District, Kenya*. (ILCA Systems Studies 4). Addis Ababa: ILCA.

Bernstein, H. 2009. V.I Lenin and A.V. Chayanov: Looking back, looking forward. *Journal of Peasant Studies*, 36(1), 55-81.

Blench, R. 1999. Why are there so many pastoral groups in East Africa?, in *Pastoralists Under Pressure?*, edited by V. Azarya, A. Breedveld, M. de Brujn and H. van Dijk. Leiden: Brill.

Delgado, C., Rosegrant, M., Steinfeld, H., Ehui, S. and Courbois, C. 1999. *Livestock to 2020. The Next Food Revolution. Food, Agriculture and the Environment.* (Discussion Paper 28). Washington DC/ Rome/ Nairobi: IFPRI/ FAO/ ILRI.

Dietz, T. 1993. The state, the market, and the decline of pastoralism, challenging some myths, with evidence from Western Pokot in Kenya/Uganda, in *Conflict and Decline of Pastoralism in the Horn of Africa*, edited by J. Markakis. London: Macmillan Press, The Hague, ISS, 83-99.

Ensminger, J. 1996. *Making a Market. The Institutional Transformation of an African Society. The Political Economy of Institutions and Decisions.* Cambridge: Cambridge University Press.

Fage, J.D. 2002. *A History of Africa.* London: Routledge.

Goldschmidt, W. 1981. The failure of pastoral economic development programs in Africa, in *The Future of Pastoral Peoples*, edited by J.G. Galaty, D. Aronson, P.C. Salzman and A. Chouinard. Ottawa: IDRC, 101-119.

Haan, L. de, Quarles van Ufford, P. and Zaal, F. 1999. Cross-border cattle marketing in Sub-Saharan Africa since 1900. Geographical patterns and government induced change, in *Agricultural Marketing in the Tropics. Contributions from the Netherlands.* (ASC research series 15), edited by L. van der Laan, T. Dijkstra and A. Van Tilburg. Aldershot: Ashgate.

Heyer, J., Maitha, J. and Senga, W. (eds). 1976. *Agricultural Development in Kenya: An Economic Assessment.* Nairobi: Oxford University Press.

Homewood, K., Lambin, E.F., Coast, E. et al. 2001. Long-term changes in Serengeti-Mara wildebeest and land cover: Pastoralism, population or policies. *PNAS*, 98, 12544-12549.

Homewood, K., Trench, P., Randall, S. et al. 2005. Livestock health and socio-economic impacts of a veterinary intervention in Maasailand: Infection-and-treatment vaccine against East Coast fever. *Agricultural Systems* (in press).

ILCA. 1983. *Pastoral Systems Research in Sub-Saharan Africa.* (Proceedings of the IDRC/ILCA workshop Addis Ababa, 21-24 March 1983). Addis Ababa: ILCA.

Karim, H.A.G. 1991. *Trade, Trade Links and Market Behaviour in the Red Sea Province (Sudan).* RESAP.

Kariuki, G. and Kaitho, R. 2009. *Deteriorating Terms of Trade and Food Security Among Pastoral Livestock Producers in Kenya.* (GL-CRSP Research Brief 09–03). University of California.

Kerven, C.K. 1992. *Customary Commerce: A Historical Reassessment of Pastoral Livestock Marketing in Africa.* (Agricultural Occasional Papers no. 15). London: ODI.

Komen, M., Kariuki, G. and Kaitho, R. 2009. *Factors Influencing Beef Cattel Marketing Behaviour in Pastoral Areas of Kenya: The Role of Livestock Market Information*. (GL-CRSP Research Brief 09–02). University of California.

Lane, C.R. and Swift, J. 1989. *East African Pastoralism: Common Land, Common Problems*. (Report on a Workshop on Pastoral Land Tenure, Arusha, 1-3 December 1988. Drylands Programme Paper No. 8). London: IIED.

Lane, C.R. 1991. *Alienation of Barabaig Pasture Land, Policy Implications for Pastoral Development in Tanzania*. PhD thesis at University of Sussex. Brighton: University of Sussex: IDS.

Little, P.D. 1987. Domestic production and regional markets in Northern Kenya. *American Ethnologist*, 14(2), 295-308.

Little, P.D. 1992. *The Elusive Granary. Herder, Farmer and State in Northern Kenya*. (African Studies Series no. 73). Cambridge: Cambridge University Press.

Little, P., Teka, T. and Azeze, A. 2001. *Cross-Border Trade and Food Security in the Horn of Africa*. (Research Report of the Project on Cross-border Trade and Food Security in the Horn of Africa). Binghamton/New York: Institute of Development Anthropology.

McPeak, J.G. and Little, P.D. 2006. *Pastoral Livestock Marketing in Eastern Africa. Research and Policy Challenges*. Rugby: ITPublications.

Mittendorf, H.J. and Wilson, S.G. 1961. *Livestock and Meat Marketing in Africa*. (Report of a Survey). Rome: FAO, Chad, Fort Lamy.

Moritz, M. 2009. Crop-livestock interactions in agricultural and pastoral systems in Africa. *Agriculture and Human Values*, 27(2), 119-128.

Mwacharo, J.M. and Drucker, A.G. 2005. Production objectives and management strategies of livestock keepers in South-East Kenya: Implications for a breeding programme. *Tropical Animal Health and Production*, 37, 635-652.

Republic of Kenya. 1959. *The Marketing of African Livestock. Report of an Enquiry Made by Mr. P.H. Jones into the Whole Problem of the Marketing of African Stock*. Nairobi: MoA, Animal Husbandry and Water Resources.

Rich, K.M. 2009. What can Africa contribute to global meat demand: Opportunities and constraints. *Outlook on Agriculture*, 38(3), 223-233.

Sandford, S. 1983. *Management of Pastoral Development in the Third World*. Chichester: Wilet & Sons, ODI.

Scarpa, R., Kristjanson, P., Drucker, A. et al. 2001. *Valuing Indigenous Cattle Breeds in Kenya: An Empirical Comparison of Stated and Revealed Preference Value Estimates*. Milano: Fondazione Eni Enrico Mattei.

Scoones, I. 1995. *Living with Uncertainty. New Directions in Pastoral Development in Africa*. London: Intermediate Technology Publications.

Seevinck, J. 2004. *Between Opportunity and Opportunism. Livelihood Strategies and Diversification Behaviour of Pastoral Maasai in Osilalei Group Ranch Area, Kenya*. MA Thesis, Amsterdam: University of Amsterdam.

Sindiga, I. 1984. Land and population problems in Kajiado and Narok, Kenya. *African Studies Review*, 27(1), 23-39.

Sutter, J.W. 1982. Commercial strategies, drought and monetary pressure: WoDaaBe nomads of Tanout Arrondissement, Niger. *Nomadic Peoples*, 11, 26-60.

Swift, J. 1986. The economics of production and exchange in West African pastoral societies, in *Pastoralists of the West African Savanna*. (Selected studies presented at the 15th IAS), edited by M. Adamu and A.H.M. Kirk-Greene. Manchester: Manchester University Press.

Teklewold, H., Legese, G., Alemu, D. and Negasa, A. 2009. *Determinants of Livestock Prices in Ethiopian Pastoral Markets: Implications for Pastoral Marketing Strategies*. (Paper to the Conference: International Association of Agricultural Economists Conference, Beijng, China, August 16-22, 2009). Addis Ababa: ILRI.

Wilde, J.C. de. 1984. *Agriculture, Marketing and Pricing in Sub-Saharan Africa*. Los Angeles: African Studies Centre.

Zaal, F. 1997. *Livestock and Cereals Markets in Kajiado District, Kenya*. (MDP Report No. 11, NIRP Project on Growing Market Dependence of Food Security Arrangements for Pastoralists in Southern and Eastern Kenya). Amsterdam: University of Amsterdam.

Zaal, F. 1998. *Pastoralism in a Global Age. Livestock Marketing and Pastoral Commercial Activities in Kenya and Burkina Faso*. Amsterdam: Thela Thesis.

Zaal, F. and Dietz, T. 1995. *Of Markets, Meat, Maize and Milk. Pastoral Commoditization As a Necessary but Risky Livelihood Strategy*. Paper to the workshop: Poverty and Prosperity of East African Pastoralists, Uppsala, 15-17 September 1995.

Zaal, F., Ole Siloma, M., Andiema, R. and Kotomei, A. 2006. The geography of integration. Cross-border livestock trade in East Africa, in *Pastoral Livestock Marketing in Eastern Africa. Research and Policy Challenges*, edited by J.G. McPeak and P.D. Little. Rugby: IT Publications.

Chapter 7

Livestock Marketing Chains in Northern Kenya: Re-Aligning Exchange Systems in Risky Environments

Hussein Mahmoud

Livestock traders in northern Kenya are faced with a myriad of risks and uncertainties and are forced to develop social relationships and diversification strategies to avert and cope with these constraints (see Little 1992a, Little et al. 2001, Smith et al. 2000). In the past, many pastoral communities in East Africa, and particularly those in Kenya, were restricted from active livestock trade. Colonial policies protected European livestock owners from competition with African traders and herders, constraining local patterns of trade.[1] Recently, livestock marketing has been the focus of several studies on East Africa highlighting new constraints, such as market uncertainty, macroeconomic decline, and political instability.[2] The social organisation and livelihood systems of pastoral communities of northern Kenya and southern Ethiopia have been well documented.[3] The current pastoral literature covers such aspects as the commodification of livestock and the process by which pastoral communities are gradually drawn into the larger economic system of the state and world system (Barrett et al. 1998, Dietz 1999, Fratkin et al. 1996). This explains the extent to which livestock marketing in arid areas is increasingly significant in the lives of local pastoralists. Local uses of livestock are rapidly being replaced by cash transactions in organised markets, which, in some cases, include auctions where livestock are openly bought and sold.

This chapter focuses on northern Kenyan livestock trading networks, which present an array of different risks and a multi-ethnic setting. The cattle trading chain in Kenya/Ethiopia borderland area follows an ascending sequence starting from smaller (primary) market places in southern Ethiopia through a medium size market (secondary) in Moyale, Kenya, to a large (terminal) market in Nairobi (Fig.

---

1   See Dahl (1979, 1981), Fratkin (1991, 1992), Kerven (1992), Kitching (1980), Little (1992a), Zaal, this volume.

2   See for example Behnke (2008), Chabari and Njiru (1991), Little (1996b), Little et al. (2001), McPeack and Little (2006), Nunow (2000), Teka et al. (1999), Zaal (1998).

3   See for example Almagor (1978), Behailu (2001), Coppock (1994), Dahl (1976), Fratkin (1998), Hjort (1979), Hogg (1986), Kassa (2002), Legesse (1973), Lusigi (1984), McPeak (1999), McPeack and Little (2006), O'Leary (1985), Smith (1997).

Figure 7.1   Cross-border cattle marketing chain
Source: Author's fieldnotes.

7.1).[4] Hence, southern Ethiopia and northern Kenya are the primary production areas, while Nairobi to the south is the major consumption centre (Fig. 7.2). However, considerable differences exist in the utilisation of trading networks, according to the ethnic composition of market actors, ethnic politics and relations.[5]

Cattle trading is a dynamic enterprise, which has led to the construction of organised market places to facilitate commercial exchanges. Specifically, the recognition of three sets of livestock market places (southern Ethiopia, Moyale, and Nairobi) are based on their unique characteristics (Fig. 7.2). Cattle markets in southern Ethiopia are smaller, scattered, and usually within comparatively shorter distances (less than 200 km) and are mostly held weekly, while the Moyale market is relatively larger, shows a higher ethnic diversity, and handles a large number of animals. The Njiru and Dagoreti centers in Nairobi are not livestock markets in the real sense but rather are market-cum-slaughterhouses. The location of the two

---

4   Livestock movements and livestock trading activities cover large geographical areas. To gain insights into the dynamics of regional cattle commerce, a spatial link must be established between southern Ethiopia (the major source of cattle) and northern Kenya on the one hand, and between northern Kenya and Nairobi (the major terminal market for northern cattle), on the other. The study region is expansive covering an area that traverses the Ethiopia/Kenyan international boundary, as well as regional and administrative boundaries on both sides of the border. It is defined by the main cattle producing areas in southern Ethiopia and northern Kenya and the cattle markets of southern Ethiopia, northern Kenya, and Nairobi.

5   Markets were perceived as unimportant in early studies in anthropology. Although indigenous marketing systems may not have contained the complexity of Western marketing structures, they were significant mechanisms of exchange conceded by the participants (see Plattner 1985).

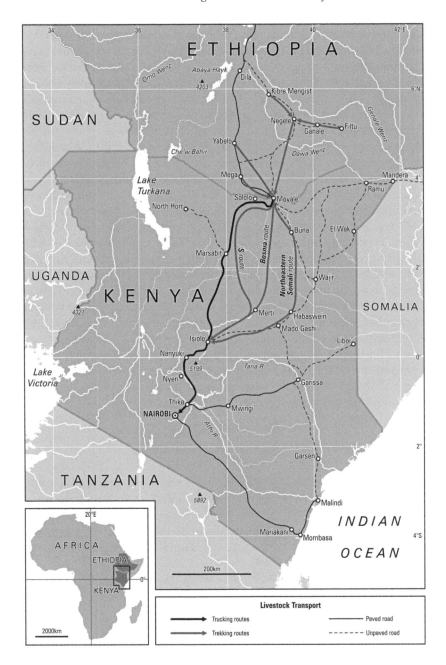

Figure 7.2   Livestock transport routes from southern Ethiopia to Nairobi, Kenya

Source: Author's fieldnotes.

major cattle markets in Nairobi close to these slaughterhouses is convenient for traders, wholesalers, and the slaughterhouse staff as well as for other market actors.

In this chapter I argue that social relations based on trust reduce market transaction costs. Cattle traders in northern Kenya find it increasingly necessary to forge trading partnerships and establish networks to avoid or reduce trading risks. My argument departs from previous studies by emphasising seller-seller relationships as opposed to buyer-seller relationships. While the latter relationships are important in the market, the former strategy is used to minimise risks among traders. Here ethnicity is not only important for the construction of trust in market relations it is also an important factor in assuring the legitimacy of business dealings in contexts of economic uncertainty, as has been documented both in Africa and elsewhere.[6] Several ethnic groups, including the Boran, Burji, Gabra, and Somali groups of Ajuran, Degodia, and Garre, make up the cultural mosaic in northern Kenya, often engaging with each other in various economic activities. Livestock trade is one of the major economic pursuits that draws people together in a complicated network of relationships. Different market actors build relationships that are based on such factors as kinship, ethnicity, patron-clientelism, and livestock loaning.[7] These social relations can influence how much one charges or pays for an animal, or whether to give a loan or ask for it. Above all, relationships based on kinship and ethnicity are critical in deciding whether to trust a business colleague or not, in the absence of cash advance payments or tangible collateral. The Boran are predominantly herders[8] and 'bush traders'[9] at the lower end of the market chain, while the Burji are urban-based livestock traders and livestock transporters at higher levels of the market.

Challenges of Livestock Marketing in Northern Kenya

The risks and constraints of cattle marketing are widely experienced in pastoral areas of Africa.[10] While most constraints are similar across the continent, some are

---

6   See Bian (1999), Cohen (1969), Dalleo (1975), Ensminger (1992), Landa (1994), Little (1992b), Little et al. (2001), Nunow (2000), Smart (1993), Teka et al. (1999), Trager (1981), van Ufford (1999), Zaal (1998).

7   See Ensminger (1992), Little (1996a), Trager (1981), van Ufford (1999), Zaal (1998).

8   In this context herders, also referred to as pastoralists, are the primary livestock producers residing in southern Ethiopia and northern Kenya.

9   Little (1992b) coined the term 'bush trader' to refer to small-scale livestock traders in southern Somalia. According to Little, bush traders are characterised by, firstly, their large number and secondly, their small-scale operations. Finally, they lack full-time commitment to livestock trading. The major characteristic of bush traders is their satellite operation of buying livestock from individual herders in remote locations and selling them to medium- and large-scale traders based in urban areas.

10   See Barrett et al. (2002), Bonfiglioli (1993), Holtzman and Kulibaba (1994), Kerven (1992), Little (1992b), Sandford (1983), van Ufford (1999).

fundamentally unique to specific areas and specific market actors. Cross-border stock thefts and other cross-border issues especially hamper cattle marketing in certain parts of Africa for example the Kenya/Tanzania border (Fleisher 2000); Algeria/Tunisia border (Homewood 1993); Lesotho/South Africa border (Kynoch and Ulicki 2000); Kenya/Somalia border (Little 2000, 2003); and Ghana/Burkina Faso border (Tonah 2000). Certain cross-border zones frequently incur ethnic conflicts, human and livestock movement controls, and border closures, which all can constrain marketing activities. In East Africa, the Ethiopia/ Kenyan borderland is crucial because it connects prime cattle producing areas of southern Ethiopia to the region's largest market, Nairobi.

Cross-border constraints may have slowed down the pace of cattle commerce in the area, but not as a permanent deterrent to the trade. Nevertheless, the severity of certain trading constraints during particular periods has forced several Ethiopian and Kenyan traders to abandon cross-border commerce.

Specific trading concerns that I discuss in this chapter include insecurity, high transport costs, loan defaults and a poor credit system in Nairobi. Other forms of trader risks include road infrastructure, communication systems, illiteracy, and intimidation by cattle wholesalers in Nairobi. Some risks are more common in certain markets than others (see Little 2000, cf. Mahmoud 2003). The eight most serious constraints, in order of importance, to the cattle business in northern Kenya include (a) insecurity (includes Moyale-Isiolo road robbery, stock theft in Moyale, and cash theft in Nairobi), (b) high transport costs, (c) loan defaults in Nairobi, (d) high Moyale County Council fees, (e) poorly maintained Moyale-Isiolo dirt road, (f) lack of and poor pasture in and around Moyale town, (g) scarcity of water in Moyale for transit cattle to Nairobi, and (h) stiff competition from an ever increasing number of cattle traders. The first three risks are discussed below.

Insecurity and Cattle Trading in Northern Kenya

Insecurity is not a new phenomenon in livestock producing and trading areas in East Africa. The problems of insecurity and raids in the livestock sector have been discussed in the literature on African pastoralism and particularly on pastoral livestock marketing (e.g. Fleisher 2000, Little 2000, Nunow 2000, van Ufford 1999). Insecurity in cattle marketing along the Kenya-Somali borderlands is rampant, but traders assert that the problem is more acute on the Kenyan side of the border than on the Somalian side. There, clan negotiations and pacts minimise the rate of attacks on livestock traders. Therefore, the establishment of a sound relationship or pact with clans on trade routes is a fundamental consideration in trekking livestock over long distances in Somalia (Little 2003).

In contrast, in Benin, West Africa risks increase not in the small rural markets but as traders approach the terminal markets. Common forms of insecurity in West Africa include cattle theft on treks and in urban areas, theft by herders, and cattle loss because of unfamiliarity with urban areas. Livestock terminal markets in

West African cities of Abidjan, Ibadan, and Lagos are famous for armed robberies against cattle traders. Ambushing of cash-laden cattle traders returning to the north is also common in parts of Nigeria (van Ufford 1999).

Insecurity in northern Kenya, in all forms, is a devastating constraint to traders, the business community, government employees, visitors, Non-Governmental Organisations, and to the public. It affects all facets of the social and economic life of residents and every sector of the local economy. For the cattle traders of northern Kenya, insecurity is disguised in numerous ways and manifests itself in several forms – ethnic conflicts, highway robbery (also called banditry or *shifta*), livestock theft in Moyale, and to a lesser extent livestock and cash theft in Nairobi.

Ethnic conflicts in northern Kenya can hinder the operations of cattle markets, but may not necessarily stop cattle from moving to Nairobi. Only a tiny fraction of traders in northern Kenya cited a low flow of cattle to markets as a constraint. It was observed that cattle supplies to Moyale dwindle a little during conflicts but this usually is only for a few days. Moreover, after ethnic feuds are over, traders return to trading routines because market actors recognise the centrality of cattle trading in their livelihood systems. Burji traders are good at maintaining extra stock of cattle so that livestock movement to Nairobi is adequately protected against conflict-instigated shocks. However, eruption of serious conflicts in the area can halt or seriously paralyse all forms of business. Because of widespread conflicts in the area, the government authorised the issuance of arms to individuals and called them police reserves. Some of these individuals come from wealthy sections of the community, who were able to convince the authorities that they are in imminent danger of attacks by bandits or rival ethnic groups.

Trekking and Trucking Livestock

In northern Kenya, both trekking and trucking are used to transport livestock to markets in other parts of the country. Three trekking routes exist from Moyale to Isiolo. The use and control of trekking routes are ethnic-based; that is, they are ethnically 'owned'. The use of the routes is also dependent on the weather conditions and the level of insecurity. Some trekking routes are used during the rainy season when surface water is abundant and are avoided during the dry season. Certain routes are not used during heightened levels of insecurity. There are three main trekking routes, and one trucking route, the main road from Moyale to Nairobi (see Fig. 7.2).

a. The 'S' route
    This route passes through the towns and centers of Moyale, Sololo, Turbi, Bubisa, Marsabit, Merti, and Isiolo. The route is 'owned' and used largely by the Boran because most of it traverses their territory. This route is mainly utilised when there are conflicts between the Boran and the northeastern Somali ethnic groups, such as the Degodia and Garre. The route has the

shape of the letter S (see Figure 7.2). The construction of the S shape in this route is both strategic and for convenience. The route almost follows the motor road until the town of Marsabit. This allows access to water. After Marsabit, the route heads towards the town of Merti then to Isiolo. This bend is created to avoid the Samburu territory part of which includes a section of the main Marsabit-Isiolo road. The section between Merti and Isiolo is Boran territory. Marsabit traders also use this route to trek livestock to Isiolo. This route is long although it has many water pans except between Turbi and Bubisa. Boreholes are used when water pans dry up.

b. The 'Bosnia' route

This route is popularly known as the 'Bosnia' route because of its hardships and innumerable problems. Although this route is the shortest one from Moyale to Isiolo, it is also the riskiest. It has a chronic scarcity of water, no communication, and no towns or trading centers along one long stretch from Moyale to Merti (a distance of over 200 km). Moreover, traders must hire armed security. This route is communal and can be used by any ethnic group. It is avoided, however, during serious conflicts and is used with great caution at other times. It is a no man's route and passes between Marsabit and Wajir heading south from Moyale to Merti and on to Isiolo.

c. The 'Northeastern/Somali' route

Route three is known as the northeastern route because the Somali ethic groups of the northeastern Province of Kenya, such as the Ajuran, Degodia, and Garre generally use it. Particularly the Degodia and Garre groups prefer to use this route when they are in conflict with the Boran. The route passes through Moyale, Buna, and Habaswein and on to Isiolo. The trek from northern Kenya ends at Isiolo from where animals are trucked to Nairobi or locally sold to Meru traders.

The significance of analysing trekking routes is to demonstrate how some aspects of cattle trading are ethnic-based. The Burji own the majority of the trucks in Moyale and they prefer to use their trucks to transport livestock to Nairobi. It generally can be noted that the Burji own trucks, and the other groups own the trekking routes.

In Kenya and other African countries (for example, Botswana and Somalia) livestock traders use trucks as a means of transporting livestock to markets at times when trekking is problematic (Sandford 1983). There are two motorised routes from Moyale in northern Kenya to Nairobi and other markets. The first one is the unpaved road that passes through Marsabit, Isiolo, Nanyuki, and to Nairobi. The section between Isiolo and Nairobi is paved, but with major stretches of poor surface. The road distance between Moyale and Nairobi is about 780 kilometers (488 miles). In many parts, this road is a security risk. In addition, during rainy seasons, some parts of the road are impassable. However, the road is more manageable during the wet season than the alternative route that traverses the northeastern Province. Vehicle wear and tear on the Moyale-Isiolo road is very high.

The second motorised route passes through the northeastern province towns of Buna, Wajir, and Garissa (see Fig. 7.2). The section between Moyale and Garissa is unpaved while the Garissa to Nairobi section is paved and has a good surface. This route is slightly longer than the first route by about 100 km (63 miles) and safer in terms of security. Since the road is mostly sandy, wear and tear on vehicles is also minimal. However, during rainy seasons, this road usually is closed.

Most of the trucks that operate in Moyale are involved in trucking livestock to Nairobi. For northern Kenyan traders, the choice between trekking and trucking is not easy. Certainly, trekking poses a serious security risk, while trucking is very costly.

Trucks are loaded early in the morning starting about 6am at the cattle market. They leave for Nairobi with the main convoy at 9am each morning.[11] They spend their first night in Marsabit town (about 250 km from Moyale) and take off from there at 6am on the next morning. They stop the second night at Makutano near the Embu junction (about 430 km from Marsabit). They arrive in Nairobi (about 100 km from Makutano) on the third day at about 9am in the morning and immediately head for Njiru market where they offload the cattle. This journey lasts approximately 48 hours. The size of the convoys varies daily with some days busier than others. Livestock trucks constitute a large proportion of the convoy. The largest number of livestock trucks depart on Tuesdays and the fewest on Wednesdays, other factors being constant. The largest number of consumer cargo trucks headed for northern Kenya leaves Nairobi on Saturdays because truck drivers resent spending Sundays in Nairobi (which implies idleness and unnecessary expenses, so they would rather spend this day on a desert road. Similarly, truck owners do not like leased trucks to remain idle). They spend Saturday night in Isiolo, Sunday night in Marsabit, and arrive in Moyale on a Monday afternoon. A similarly large convoy is ready to head back to Nairobi, this time with loads of livestock. Since most trucks depart from Nairobi on Saturdays, a few or no trucks depart on Sundays. The second reason for numerous departures on Saturdays is that most merchandise wholesalers in Nairobi are closed on Sundays. Absence of trucks departing on Sundays from Nairobi implies no arrivals in Moyale on Tuesdays, consequently few or no departures from Moyale on Wednesdays (this is the 'Sunday Effect'). While this is the dominant trend, exceptions to this pattern exist because of delays or other reasons.

---

11   This convoy of vehicles is escorted by the Kenya Police and all vehicles must leave no earlier or later than 9am each morning. Northern Kenya is and has been a 'security zone' since colonial days. Indeed the Ethiopian government does not allow its vehicles or any other vehicles bearing its registration to cross into northern Kenya. Despite the stigma of 'security zone', commercial trucks, private and government vehicles, eco-tourists and holiday revelers, and researchers like the author have traversed the area very frequently without ugly insecurity incidents. While vehicles using the Isiolo-Moyale road are susceptible to bandit attacks, so are Rumuruti-Maralal, Kitale-Lodwar and even the busy Nairobi-Mombasa roads. Yet, these roads have not been declared 'security roads' by the Kenyan government or the United Nations. Similarly, some pockets in southern and southeastern Ethiopia are equally unsafe for motorists.

The main cargo from Moyale to Nairobi is livestock and transporters back-haul foodstuffs, second-hand clothes, and a range of consumer items from Nairobi. The costs of transporting livestock are much higher than the cost of back-hauling goods from Nairobi. It is not uncommon for trucks to go back to Moyale empty counting on transporting livestock to Nairobi. The activity pays well enough that this option still allows a profit.

Transport Costs

Trucking can be a costly item in cattle trade. While the increased efficiency in the transportation sector in West African countries improved cattle marketing (van Ufford 1999), these types of infrastructure investments have not yet been achieved in cattle producing areas of the Horn of Africa. In the period 1996-1998, transport costs in northeastern Kenya cattle trade increased by 52 and 72 per cent along the major market routes of Garissa-Nairobi and Garissa-Mombasa, respectively (Little 2000: 25). Similarly, trucking costs in northern Kenya also have increased dramatically in the past few years. In the past, transport costs were based on the number of animals transported from northern Kenya to Nairobi. The current transport pricing system in Moyale is not based on unit price, but on a negotiable price. This has led to escalation in trucking costs in the area. Livestock transportation costs in northern Kenya are a function of several factors, including distance between markets, mode of transportation, and cattle supply. Trekking cattle to markets can be a cheaper way of transportation, but the decision to trek is influenced by yet a different set of factors: security, availability of water and pasture on the trekking route, and the kind of relationship between a trader and ethnic groups residing en route.

Security for in-transit animals is an important cost element in cattle trekking and its provision is always given consideration. Although the cost of security may be insignificant in some instances, it usually depends on two major factors. The first is the region through which livestock are trekked. If hostile groups and bandits are known to inhabit livestock transit areas, appropriate caution is taken by hiring adequate security men to accompany animals. Secondly, the volume of animals being trekked determines the type and size of security hired. If the volume of livestock trekked is high – for example, in excess of, say, 200 head of cattle – more security men are hired. Trekking a high volume of livestock is riskier even if the trekking route passes through lands occupied by friendly clans or ethnic groups.

Transport comprises a significant portion of the costs of cross-border trade along the Kenya-Somali borderlands. More than 95 per cent of cattle traders in northeastern Kenya, working in Garissa and across the border with Somalia, pay for transport-related expenses. Like the Ethio-Kenyan border, the mode of transportation in the cross-border cattle trade along the Kenya-Somalia borderlands involves both trekking and trucking.

Trekking is more convenient and cheaper on the Somalia side, while trucking is safer and faster on the Kenyan side. In northern Kenya, trucking is usually undertaken by traders who are well established and have a high turnover and, most importantly, can afford the costs of trucking. However, some cattle traders cannot afford the high trucking costs that sometimes account for as much as 33 to 40 per cent of the price of cattle. Transport cost in Moyale ranges from KES 1,500 to KES 3,500 (US$ 19.2 to US$ 44.9) per head of cattle. The main commodity transported from Moyale to Nairobi is livestock, while a variety of merchandise – used-clothes, construction materials, foodstuffs, timber, plastic ware, and household goods – are back-hauled from Nairobi to Moyale. The costs of hiring a truck from Nairobi to Moyale average about KES 20,000 (US$ 256.4), while the costs of transporting livestock from Moyale to Nairobi ranged between KES 30,000 and KES 70,000 (US$ 384.6 and US$ 897.4) per trip during March 2001 to February 2002. Confronted with such circumstances, transporters make up for losses by charging cattle traders exorbitantly high trucking costs.

Loan Defaults and the Culture of Cattle Credit in Nairobi

Loans are a notable feature of livestock marketing in Africa.[12] While butchers in Benin markets lure traders with higher prices to obtain cattle on credit, traders in Nairobi set higher than usual prices if they intend to sell their cattle on credit. In both scenarios, the main goal of traders is to complete sales of their cattle as soon as they can, so that they can begin arrangements to purchase another consignment of cattle.

Terminal markets for cattle are generally located in large cities where incomes and demand for meat are higher than in rural areas. In an effort to reduce living expenses and the risk of robbery, cattle traders are often under pressure to dispose of their cattle as soon as possible. Such circumstances not only compel businessmen to sell cattle on credit, but also deny them the opportunity to explore different options in the market. Cattle wholesalers in Nairobi are fully aware of the predicament of cattle traders. Consequently, they do not hesitate to exploit the situation. Cattle wholesalers in Nairobi customarily operate in networks and collude against traders.

In another context, merchants in northern Nigeria establish relationships of trust with their urban landlords when dealing with butchers in the south of the country. In such circumstances, landlords play a dual role. Firstly, they provide accommodation to traders and secondly, act as intermediaries to help traders sell their cattle. Indeed, landlords in southern Nigerian cities are the first contact for traders from northern Nigeria. Landlords appoint a commission agent for traders, whose duties are to facilitate the sale of animals and collect money. The commission agent charges a fee called *lada*. Cattle are sold on credit hence the duration of the money collection period can extend for a period of 2-4 weeks

---

12   See Cohen (1969), Holtzman and Kulibaba (1994), van Ufford (1999).

(Cohen 1969). In Benin, traders sell their cattle with the help or mediation of middlemen (van Ufford 1999). Likewise, merchants in Nairobi deal with cattle wholesalers through the mediation of brokers. The major role of cattle wholesalers in Nairobi is to slaughter purchased cattle and sell the meat to butchers in various parts of the city. Butchers in turn are the final outlet to consumers.

The Predicament of Northern Cattle Traders in Nairobi

Why do northern Kenya traders sell cattle on credit in Nairobi when most business enterprises operate on a cash basis, including cattle sales in the large Dagoreti market? What forms of collateral should be used to protect cattle traders against loan defaults? Would northern businessmen survive over the long-term in a market dominated by credit and loan defaults? These concerns do not have easy solutions as acknowledged by several traders. Northern Kenyan cattle dealers push the limits of trust too far by lending out large parts of their lifetime investments, in the form of cattle, to total strangers. Consequently, many traders have become victims of wholesalers abusing their trust. The stranger's identity and collateral are only his or her initials painted on the acquired cattle, initials that cannot even be verified. Under such frightening trading circumstances, the entire working capital of a northern businessman is put at risk. In addition, livelihood systems of traders' entire families are endangered. If these merchants do not receive their money or receive it in a reasonable time, they jeopardise their cattle trading businesses. Many cattle traders have gone out of business in this way and many more continue to do so. Thirty-four per cent of northern businessmen in Nairobi express deep concerns about the dangers of the cattle credit business in Nairobi.

Credit in Nairobi is a cattle traders' nightmare. They fear the consequences of loan defaults in Nairobi more than they fear bandits' bullets in the northern and northeastern rangelands of Kenya. Nevertheless, credit in cattle marketing in Nairobi's Njiru market is a reality that every northern dealer has to contend with on a daily basis. Credit is not common in the Dagoreti market, which is mainly occupied by traders from southern, western, and Rift Valley regions of Kenya (see Zaal, this volume). Northern and northeastern businessmen dominate Njiru market, where cattle are sold on credit. Based on trader surveys in northern Kenya, I found that 97 per cent of traders stated that their business is based solely on a credit system. Another three per cent said they do not provide credit under any circumstances because they had been victims of credit defaults in the past (N=58).

Several reasons compel Kenyan traders to sell cattle on credit in Nairobi. Fifty-nine per cent of the sample of traders said they provide credit because there are no cash buyers in the market. Second, selling cattle on credit in Nairobi's Njiru market has become the norm, hence, deviation from the 'custom' is perceived as an anomaly and traders may have trouble finding buyers. Many traders express concern about credit relations, but exclaim that 'everyone sells on credit'. There is a general lack of consensus among cattle traders in Nairobi's market regarding the elimination of credit. They would occasionally agree verbally to stop selling on credit. However, some businessmen secretly violate the verbal agreement by selling on credit. Under these circumstances, traders who truly tried to enforce the agreement feel deceived. Several businessmen are sceptical about stopping the credit culture at the market, because they say that it has become an institution, and sometimes state, 'how would you deny a credit to your customer of 10 years?' Some cattle wholesalers manipulate certain traders and lure them into selling their cattle on credit.

Risks are, however, perceived and buffered differently: If bandits robbed a cattle trader of all his/her cattle on the road, his clan members and trader associations will contribute some money to assist the victim to return to business as is illustrated in the following statement from an official of the Moyale Livestock Traders Association:

> The cattle traders' association in Moyale gave to one of our members KES 50,000 as a support – the money is not repayable. Another member received KES 200,000 from the association's accounts and a further KES 100,000 from clansmen. He is doing very well in business. Another member's cattle were raided in Butiye (a suburb of Moyale town) – out of the seven cattle that were stolen, four were recovered and three had been eaten. Association members contributed from their pockets a total of KES 30,000 to aid the victim.

If, on the other hand, a trader lost his/her working capital to a mischievous cattle wholesaler in Nairobi, he would attract very little sympathy from his clansmen because he would have been considered imprudent – this is the curse. To control the credit problem and loan defaults, northern traders have recently formed a trader association in Nairobi to counter credit defaults. The main objectives include minimisation or eradication of the unpopular cattle credit system.

Innovations in Livestock Trading in Northern Kenya

It was shown in the previous section that cattle trading in the northern Kenya-Nairobi corridor face a myriad of risks stemming from multiple sources. A meticulous assessment of the full extent of cattle business risks in the study area is certainly an overwhelming task. Certain forms of risk can be linked to institutional failures, market forces, and/or local and cross-border politics. Others are caused by weather-related infrastructural breakdowns and other factors. A combination of interrelated issues in the marketing chain generates constraints for cattle traders and other participants in the market. While improvement in security in the market and improvement and upgrading of roads and telecommunications are responsibilities of the state, there are certain constraints that cattle traders are capable of addressing.

In situations where market forces pose unfavourable circumstances, such as low demand and low prices of cattle in Nairobi, or lack of choice for cash transfers from Nairobi to northern Kenya because of increasing risks, cattle traders often invoke social relations to manage risks. Consequently, there has been an increase in risk reduction and management strategies among cattle traders in northern Kenya as the intensity of risk increases. In their efforts to minimise and where possible to eliminate risks, merchants act individually and collectively, thus utilising social capital at the local and regional levels. Individual and collective, as well as local and regional risk-reducing mechanisms not only curb recurrence of trading risks, but they also help improve cattle exchange at all levels of the trading chain and thereby enhance livelihood systems of businesspeople and other market actors. In essence there are four major types of risk management strategy in the trading circle: (a) trading partnerships, (b) communication, (c) cash transfer systems, and (d) establishment of trader associations. While the first strategy entails establishment of partnerships, the second strategy involves upgrading information dissemination, the third approach includes alternative local money transfer mechanisms and the last strategy is principally concerned with the establishment of trader associations in the trading corridor. Networking is best characterised as an individual-based arrangement, while involvement in trader associations is a collective effort. Trust and social relations are the foundation of these efforts and permeate all strategies.

Cattle Trading Partnerships

Partnerships in livestock trading have not been studied adequately in Africa in general and in the Horn of Africa in particular. A chain of livestock market actors, for example, has sometimes been referred to as 'trading partners'. The 'producer-broker-trader-butcher-consumer' chain of livestock marketing is a typical scenario in Kajiado district of Kenya and Oudalan and Seno provinces of Burkina Faso (Zaal 1998: 146; see Little 1992b and Nunow 2000 for a similar description of the livestock trading chain in Somalia and Garissa district of northeastern Kenya, respectively). In the cases of Kajiado (Kenya) and Oudalan and Seno (Burkina Faso), partnerships were determined by asking market actors to identify whom they buy livestock from and to whom they sell (Zaal 1998).

While risk evasion and working capital limitations can be some of the motives in cattle trader partnership formations in northern Kenya, there are also a host of other reasons that encourage the establishment of trading partnerships. In some partnerships in northern Kenya large-scale traders personally have huge working capital and enormous trading tasks. Such traders need trading partners to help with procuring livestock and securing markets. On the other hand, new entrants with minimal capital seek guidance and trading tips from experienced traders, and often serve as trading apprentices to the latter.

With the exception of apprenticeships, these partnerships entail relatively equitable forms of collaboration: equal work sharing, equal profit sharing, and a division of labour based on specialisation and expertise. The most frequent type of partnerships involves two traders, though it is not unusual for partnerships to involve more than two partners. The common residence pattern of partners indicates a dual model: one or more are based in Moyale, Kenya (cattle procurement area), while the other partner(s) are based in Nairobi (the terminal market).

Seventy-eight per cent of traders in northern Kenya have at least one trading partner, while 22 per cent have none. As mentioned earlier, animals from northern Kenya may only be transported to Nairobi for slaughter. This requirement has produced the emergence of two principal markets, that is, the source market (Moyale) and the consumption market (Nairobi slaughterhouses). Accordingly, partners must be based in both of these centers to facilitate trade. If the northern trading sphere had involved other Kenyan consumption markets, the resultant marketing scenarios would have produced a web of locations and trading relationships. The emergence of such settings is not completely being ruled out as other Kenyan towns and markets are growing. Mombasa and Nakuru are some examples. Moreover, the government is under concerted pressure from livestock traders, politicians, and organisations to permit northern traders to compete in other urban areas of the country. Policy amendments that allow northern Kenyan cattle traders to trade with the rest of the country would open diverse economic opportunities in northern Kenya in particular, and cattle trading in the Horn of Africa in general.

The primary duties of the partner stationed in Moyale are to monitor market conditions, including cattle prices, supply and condition of animals, and periodic trading constraints. After buying animals, he/she hires a truck and sends them to Nairobi, writes instructions on a piece of paper and hands it to the truck driver to deliver to his/her partner in Nairobi. These instructions are about the number of cattle purchased and loaded, buying prices, and cattle descriptions, including the colour, size, gender, and any peculiar features that would identify individual animals. Information about purchase prices is particularly important because trader profit margins are determined by how many animals are sold with respect to buying prices. The truck driver is responsible for animals until they are handed over to the dealer based in Nairobi. Traders also give money to the driver to pay for police 'tips', a form of rent that drivers pay to police. Other than the driver and the drover, no one else accompanies the cattle.

Soon after trucks depart from Moyale, the Moyale-based partner calls his/her counterpart in Nairobi on the latter's mobile phone[13] to alert him/her that animals will arrive in about 48 hours. In the context of northern Kenya, partnerships between traders are stronger than those between traders and other actors in the market. Individual traders establish trading partnerships for convenience and out of necessity. Most trading partners are members of the same family, such as father and son or brothers. Other partnerships are between members of the same clan or ethnic group, but not related by blood.

Ethnicity highly influences partnerships as can be seen in Figure 7.3, with same-ethnicity partnerships among merchants in northern Kenya being widespread. Merchants stress that trust is a fundamental requirement for the success of any

---

13   Several Nairobi-based and Moyale-based cattle traders from the north have mobile phones. They can make calls to and receive calls from Moyale on their mobiles.

Ethnic relations of trading partner (N=54)	
From the same ethnic group	96.3%
From a different ethnic group	3.7%
Total	100.0%

Figure 7.3   Cattle trading partnerships and ethnicity in northern Kenya
Source: Author's fieldnotes.

	Do you have a trading partner?			
	Burji (N=48)	Boran (N=18)	Gabra (N=10)	Garre (N=5)
Yes	90%	61%	60%	80%
No	10%	39%	40%	20%

Figure 7.4   Partnerships across ethnic groups in northern Kenya
Source: Author's fieldnotes.

business undertaking and trust is especially strong between members of the same group. Ethnicity was highly polarised during the period of this fieldwork, particularly with the run up to the third multi-party elections in Kenya. Ethnic tensions at the national level heightened local tensions in towns as remote as Moyale. In contrast to similar partnership structures within ethnic groups, there is a fundamental difference of trading partnerships between ethnic groups. This explains not only the relative importance of trading partnerships within different ethnic groups (see Fig. 7.4), but it may also determine vital future developments in cattle trading networks in northern Kenya.

Burji traders favour partnerships more than other groups, probably because they are heavily involved in the Nairobi market. For both, the Burji and Garre, establishment of partnerships is a key strategy because of questions surrounding their 'indigenousness' in Moyale. There is a long-standing feud between the Boran and the Garre on the one hand and between the Boran and Burji, on the other regarding the right 'to live and trade' in Moyale town. To overcome discrimination, harassment, and as a buttress against a risky and unpredictable trading environment of northern Kenya and Nairobi, Burji and Garre traders operate in ethnic clusters. While certain partnerships are forged between individual traders as has been examined in the preceding discussions, others are used to form clan alliances to facilitate trading.[14] Clan alliances in trading often emerge in violent and war-torn countries like Somalia.

Information Dissemination – Bridging the Communication Gap

Economic and political information is inherently crucial for residents of northern Kenya and for cattle traders as well. Changes in economic, political, and social

14   See Cassanelli (1982), Cohen (1969), Dalleo (1975), Little (1992b).

events profoundly affect commerce in the area. As such, effective trading requires that a trader is exceptionally familiar with current news. People from all lifestyles and economic occupations are constantly engaged in the search for information pertaining to political security issues, market information, and cross-border activities. In short, local residents continuously strive for any valuable information from within the area and outside. Information is precious and costly. It occupies a central position in trader's strategies and constitutes an important cost item.

The information needs of pastoral communities (i.e. livestock producers) are fundamentally different from those of livestock traders. Information needs of pastoral communities in East Africa and elsewhere have been discussed in some studies (e.g. in Nunow 2000 and Sandford 1983). As an imperative policy issue, African governments with a substantial pastoral population have attempted to keep this population informed about key policy issues that are of concern to them. States have often claimed that it has been a hurdle to 'keep in touch' with their pastoral communities because of the 'remoteness' of their locations and the lack of efficient and effective communication tools. As a result, vital national policy issues often do not reach pastoral areas. However, in Botswana, for example, radio broadcasts are used to disseminate information pertaining to pastoral populations; while in Kenya audio-visual tools in mobile units have been used to communicate policy issues to pastoral populations (Sandford 1983).

Specific forms of information exchange between traders, on the one hand, and between pastoralists and the government, on the other, are lacking. Nunow (2000), however, notes that the main source of information for Somali pastoral groups living in the Garissa district of northeastern Province of Kenya are government officials, local leaders, and travelers. It was further observed that radio is the most important mode of information dissemination in Garissa. This mode of information dissemination has two constraints. First, radio ownership among pastoral groups in Garissa is estimated at only 10 to 15 per cent of the population and, second, commodity prices that are announced through the national radio do not include livestock prices and market information (see Mahmoud 2006).

Merchants in northern Kenya are incredibly mobile and seldom have 'free time' to sit around and listen to radio broadcasts. The German Technical Assistance agency (GTZ) working in northern Kenya in the 1990s embarked on a program to transmit livestock price information in Nairobi to northern Kenya through the national radio (Chabari, personal communication). There were two major reasons why such a scheme did not work effectively in northern Kenya. Firstly, the market information transmission was made in the afternoon when traders were busy going about their business. Secondly, cattle prices in Nairobi are not static, but fluctuate on a daily basis, depending on the supply of cattle and season. Finally, traders view these transmissions as less credible. They prefer to rely on information related by friends and trading partners based in Nairobi, who also provide additional analyses of market conditions.

Among traders in the Horn of Africa the Somalis possess a complex information dissemination system. Garissa merchants make telephone calls to their counterparts in Nairobi, who in turn make radio calls to towns and trading centers in Somalia. This method is used to acquire and disseminate trading information in the region, in addition to being used for cash transfers as well (Little 2000).

Prior to the increase in telephone use as a medium for livestock market information dissemination between northern Kenya and Nairobi, traders relied on information transmitted by word of mouth. With subsequent expansion of telephone facilities in Moyale, particularly the introduction of mobile phone services in the town, traders are increasingly using these services to communicate with their counterparts in Nairobi. It is of vital necessity that traders have access to a medium of information in order to be aware of market conditions in Nairobi. While those traders who have partners in Nairobi acquire market information from their partners, solitary traders often have contacts with friends and clansmen.

Informal Cash Transfer Mechanisms in Northern Kenya

Cattle commerce in the north involves comparatively large amounts of money. Ordinary trucks can haul 18-22 head of cattle, 100-120 goats, or 10-12 camels. At average market prices, a truckload of cattle may be worth KES 180,000 to KES 300,000 (about US$ 2,308 to US$ 3,846). After cattle sales have been finalised in Nairobi, traders are confronted with the task of repatriating cash to Moyale for further animal purchases. Several means, with varying degrees of risk, are available to accomplish this. Each of the methods involves different fee payments and limitations. The seemingly easiest and obvious method of cash transfer is to carry the cash oneself, but this method has become increasingly unpopular because of the worsening security situation on the Isiolo-Moyale road. Traders have progressively become targets of bandit attacks. The Kenya Commercial Bank and the Postal Corporation of Kenya represent a conventional, but costly, system owned and operated by the state. The Somali cash transfer system, commonly known as *hawala*, is privately owned and operated, while the 'Burji system' is an institution developed exclusively for cash transfers. The latter system is discussed below as it is most commonly used by the traders studied and is a unique local response.

The 'Burji Cash Transfer System'   To avoid the hassle of existing cash transfer systems with their expensive fees, northern merchants have devised an indigenous system that is not only safe, but also fast and free of charge. In addition, trust and personal relations reinforce the system. It was devised and principally utilised by Burji cattle traders and merchants among whom the practice has grown considerably. This unique transfer system operates purely on the basis of mutual trust between merchants and cattle traders. Soon after receiving their cash from cattle sales in Nairobi, traders arrange to meet with Moyale wholesale, retail merchants, and used-clothes dealers. The cattle trader hands over their cash to a used-clothes dealer in Nairobi. Both traders (the used-clothes dealer and the cattle trader) have partners in Moyale and they call them for instructions on how to pay and receive the money that has just changed hands in Nairobi, respectively.

The used-clothes dealer instructs their partner in Moyale to pay the cattle trader's partner in Moyale giving out the identity details provided by the cattle trader, such as the recipient's full names and telephone number. At the same time, the cattle trader in Nairobi calls his partner to confirm whether they have received the stated amount of cash from the used-clothes dealer's partner in Moyale. The deal is over once the cash transfer in Moyale has been confirmed. The used-clothes dealer uses the cash to purchase used-clothes from Nairobi and transports the cargo to Moyale, while the cattle trader in Moyale uses the cash to procure another consignment of cattle. Cattle traders who do not have trading partners transfer their money to a merchant or a shop owner and collect it when they arrive in Moyale.

The advantages of this method of cash transfer are numerous. Most importantly, security risks are minimised since traders rarely travel with large amounts of cash between Nairobi and northern Kenya. Any amount of cash can be transferred as required by the needs of traders, while cash transfers are carried out on the same day by telephone. Cattle traders have highly praised the method for helping to reduce trader transaction costs and increase trader confidence for dealing with someone they know and trust, rather than dealing with strangers and a banking system that they barely understand. In short, informal cash transfer mechanisms are fast, free, fair, and flexible. According to traders, there have not been any serious fraudulent incidences in this system. Social relations and acquaintances help reduce occurrences of deceptive behaviours. As far as cattle trading is concerned, this system of cash transfer by cattle traders could be the only one of its kind in the Horn of Africa.

The key in this cash transfer system is trust and the relationships between the people engaged in the transfer. They are mostly members of the same ethnic group or close friends. In addition to these criteria, those involved must trust each other based on their previous business experiences.

Livestock Trader Associations

Livestock trader associations are rare in Kenya and particularly in the northern and northeastern parts of the country. In the recent past, however, there has been an escalation in the number of associations and for organisations in northern Kenya involved with pastoralism in general, and livestock and livestock products trade in particular. These associations can easily be identified in Moyale town from the names written on the doors of their 'offices' – with titles that include terms like herders, pastoralists, pastoral women, pastoral youths, cowboys, and brokers, just to mention the most prominent. Some of these associations sprung up almost overnight following workshops that NGOs in the area organised regularly. Some NGOs and other large scale organisations avoid Moyale because of security concerns and cross-border tensions. Moreover, the area has very poor roads and other vital infrastructure, which discourage NGO involvement.

The initiatives taken by stakeholders themselves to improve their own conditions present the best options. In a similar context, livestock traders in

Moyale and those based in Nairobi have initiated their own associations, formed and managed by them. These are the Moyale Livestock Traders and Northern/ Nairobi Livestock Dealer's Association.

Moyale Livestock Traders   The Moyale Livestock Traders was established in September 1998. It has 31 registered members, who meet on a monthly basis to discuss matters of importance to the association. Although membership is open to all interested livestock traders from northern Kenya, the association does not have any women members and the majority of the members are Burji. The association is financially and organisationally stable, being active in awarding monetary assistance to its needy members. For example, one member was awarded KES 50,000 (US$ 641) as general support, while another member was awarded KES 200,000 (US$ 2,564) from the association and a further KES 100,000 (US$ 1,282) from clansmen because of bankruptcy. Association members contributed KES 30,000 (US$ 385) as a good will gesture to a livestock trader who lost his cattle to thieves. These amounts are gifts and no repayment is required from the beneficiary, but reciprocity is expected in the future and the recipient should extend a helping hand to another needy member of the community.

The association has a long-term development agenda for its members and the community at large. The establishment of a school for the community is one aspect of this agenda. In 2002, the association purchased a prime plot in Moyale town for a school project and looks forward to building the institution in the future. As part of its long-term plan, the association would like to start a livestock-fattening project, similar to what Somali cattle traders are achieving in the Mombasa area (see Mahmoud 2006).

Northern/Nairobi Livestock Dealer's Association   The Northern/Nairobi Livestock Dealers' Association was officially registered in Nairobi on 1 June 2001. The group's membership is open to livestock traders from northern Kenya regardless of age, ethnicity, gender, or religion. However, membership is restricted to livestock traders who originate from northern Kenya and are currently based in Nairobi. The objectives of the association are broad-based and encompass many issues important to all members. For example, the association seeks to foster the socio-economic progress of its members and economically empower them through active participation. In addition, the association seeks ways and means to improve livestock marketing opportunities for its members in Nairobi; to provide advisory services to members in matters related to livestock diseases, livestock transportation, and marketing; and to seek professional training and advice on improved livestock production and management methods for the benefit of its members. Despite being in its infancy, the association has begun working on reducing the credit menace that is limiting marketing benefits for northern traders.

It will be a long and strenuous struggle to tackle a credit institution that has endured in the Nairobi market and that is backed by powerful businesspeople. Through these two associations, northern Kenya cattle traders – whether they are based in Moyale or Nairobi – appear to have found a potential mechanism in the absence of any other organisation, including the government, to assist them in periods of crises.

Conclusion

While livestock commerce in northern Kenya has had a phenomenal growth both in livestock movements to Nairobi and trader participation, the sector continues to face increasing constraints. Trading risks have increased in diversity, frequency, and intensity in all regions, from cattle producing areas in southern Ethiopia and northern Kenya to beef consumption areas in Kenya's capital, Nairobi. Traders increasingly use personalised relationships in the higher marketing chain (partnerships, information sharing, trust-based money transfer, and involvement in associations) to deal with increasing risks. On the other hand, in the lower marketing chain trading activities are mostly impersonal as shown by low levels of partnership, absence of trust-based money transfer systems and trader associations (Mahmoud 2008). While traders increasingly utilise trust and social relations to manage insecurities, institutional weaknesses continue to expose cattle trade to increased risks. There is a strong connection between cattle commerce constraints in northern Kenya and government policies that perpetuate them through apathy and negligence.

References

Almagor, U. 1978. *Pastoral Partners. Affinity and Bond Partnership Among the Dassanetch of South-West Ethiopia.* Manchester: Manchester University Press.
Barrett, C.B., Chabari, F., Little, P.D. et al. 1998. How might infrastructure improvements mitigate the risks faced by pastoralists in arid and semi arid lands? (Newsletter of the Small Ruminant/Global Livestock Collaborative Research Support Program (SR–CRSP)). *Ruminations*, 1(10), 12-13.
Barrett, C.B., Chabari, F., Little, P.D. et al. 2002. *Livestock Pricing in the Northern Kenyan Rangelands.* GL–CRSP–Pastoral Risk Management Project.
Behailu, M.S. 2001002E *Pastoralism and Cattle Marketing: A Case Study of Borana of Southern Ethiopia.* MA Thesis, Njoro: Egerton University.
Behnke, R.H. 2008. The economic contribution of pastoralism: Case studies from the Horn of Africa and southern Africa. *Nomadic Peoples*, 12(1), 45-79.
Bian, Y. 1999. Getting a job through a web of Guanxi in China, in *Networks in the Global Village: Life in Contemporary Communities*, edited by B. Wellman. Boulder: Westview Press, 255-277.
Bonfiglioli, A.M. 1993. *Agro-Pastoralism in Chad As a Strategy for Survival: An Essay on the Relationship Between Anthropology and Statistics.* Washington: World Bank.

Cassanelli, L.V. 1982. *The Shaping of Somali Society: Reconstructing the History of Pastoral People, 1600-1900*. Philadelphia: University of Pennsylvania Press.

Chabari, F. and Njiru, G. 1991. Livestock marketing, in *Range Management Handbook of Kenya, Volume II-1*, edited by H.J. Schwartz, S. Shaabani, D. Walther. Nairobi: Marsabit District Republic of Kenya, 111-129.

Cohen, A. 1969. *Custom & Politics in Urban Africa: A Study of Hausa Migrants in Yoruba Towns*. Berkeley: University of California Press.

Coppock, L. 1994. *The Borana Plateau of Southern Ethiopia: Synthesis of Pastoral Research, Development and Change, 1980-91*. Addis Ababa: International Livestock Centre for Africa.

Dahl, G. 1976. *Having Herds: Pastoral Herd Growth and Household Economy*. Stockholm.

Dahl, G. 1979. *Suffering Grass: Subsistence and Society of Waso Borana*. Stockholm: University of Stockholm, Department of Social Anthropology.

Dahl, G. 1981. Production in pastoral societies, in *The Future of Pastoral Peoples*, edited by D.A. Galaty, P. Salzman and A. Chouinard. Ottawa: IDRC, 200-209.

Dalleo, P. 1975. *Trade and Pastoralism: Economic Factors in the History of the Somali of Northeastern Kenya, 1892-1948*. PhD Thesis, Syracuse University.

Dietz, T. 1999. *Pastoral Commercialization: A Risky Business? Three Kenyan Case Studies: The Pokot, the Maasai and the Somali*. Amsterdam: University of Amsterdam, Amsterdam Research Institute for Global Issues and Development Studies.

Ensminger, J. 1992. *Making a Market: The Institutional Transformation of an African Society*. Cambridge: Cambridge University Press.

Fleisher, M.L. 2000. *Kuria Cattle Raiders*. Ann Arbor: The University of Michigan Press.

Fratkin, E. 1991. *Surviving Drought & Development: Ariaal Pastoralists of Northern Kenya*. Boulder: Westview Press.

Fratkin, E. 1992. Drought and Development in Marsabit District, Kenya. *Disasters*, 16(2), 119-130.

Fratkin, E. 1998. *Ariaal Pastoralists of Kenya: Surviving Drought and Development in Africa's Arid Lands*. Needham Heights: Allyn & Bacon.

Fratkin, E., Nathan, M.A. and Roth, E.A. 1996. *Sedentarization, Commoditization and Development among Rendille Pastoralists of Kenya. Development as Ideology and Practice: African Perspectives, American Anthropological Association Annual Meetings*. San Francisco.

Hjort, A. 1979. *Savanna Town: Rural Ties and Urban Opportunities in Northern Kenya*. PhD Thesis, University of Stockholm.

Hogg, R. 1986. The new pastoralism: Poverty and dependency in northern Kenya. *Africa*, 56(3), 319-333.

Holtzman, J.S. and Kulibaba, N.P. 1994. Livestock marketing in pastoral Africa: Policies to increase competitiveness, efficiency and flexibility, in *Living with*

Uncertainty: New Direction in Pastoral Development in Africa, edited by I. Scoones. London: Intermediate Technology, 79-94.

Homewood, K.M. 1993. *Livestock Economy and Ecology in El Kala, Algeria: Evaluating Ecological and Economic Costs and Benefits in Pastoralist Systems*. London: Overseas Development Institute.

Kassa, G. 2002. An overview of root causes of problems that currently affect Borana pastoralists of southern Ethiopia, in *Resource Alienation, Militarization and Development: Case Studies from East African Drylands*. (Proceedings of the Regional Workshops on East African Drylands), edited by M. Babiker. Addis Ababa: OSSREA (Organization for Social Science Research in Eastern and Southern Africa), 67-76.

Kerven, C. 1992. *Customary Commerce: A Historical Reassessment of Pastoral Livestock Marketing in Africa*. London: Overseas Development Institute.

Kitching, G.N. 1980. *Class and Economic Change in Kenya: The Making of an African Petite Bourgeoisie 1905-1970*. New Haven: Yale University Press.

Kynoch, G. and Ulicki, T. 2000. It is like the time of *Lifaqane*: The impact of stock theft and violence in southern Lesotho. *Journal of Contemporary African Studies*, 18(2).

Landa, J.T. 1994. *Trust, Ethnicity, and Identity: Beyond the New Institutional Economics of Ethnic Trading Networks, Contract Law, and Gift-Exchange*. Ann Arbor: University of Michigan Press.

Legesse, A. 1973. *Gada: Three Approaches to the Study of African Society*. New York: Free Press.

Little, P.D. 1992a. *The Elusive Granary: Herder, Farmer, and State in Northern Kenya*. Cambridge/New York: Cambridge University Press.

Little, P.D. 1992b. Traders, brokers and market 'crisis' in southern Somalia. *Africa*, 62(1), 94-124.

Little, P.D. 1996a. Conflictive trade, contested identity: The effects of export markets on pastoralists of southern Somalia. *African Studies Review*, 39(1), 25-53.

Little, P.D. 1996b. Rural herders and urban merchants: The cattle trade in southern Somalia, in *The Struggle for Land in Southern Somalia: The War Behind the War*, edited by C. Besteman and L. Cassanelli. Boulder: Westview Press, 91-113.

Little, P.D. 2000. *Cross-Border Livestock Trade and Food Security in the Somalia and Northeastern Kenya Borderlands*. New York: Institute of Development Anthropology, Binghamton.

Little, P.D. 2003. *Somalia: Economy Without State*. Bloomington: Indiana University Press.

Little, P.D., Teka, T. and Azeze, A. 2001. *Cross-Border Livestock Trade and Food Security in the Horn of Africa. The Case of Southern and Southeastern Borderlands*. (OSSREA Development Research, Report Series No. 1). Addis Abab: OSSREA (Organization for Social Science Research in Eastern and Southern Africa).

Lusigi, W. 1984. *Integrated Resource Assessment and Management Plan for Western Marsabit District, Kenya.* Nairobi: UNESCO–FRG–MAB Integrated Project in Arid Lands (IPAL).

Mahmoud, H.A. 2003. *The Dynamics of Cattle Trading in Northern Kenya and Southern Ethiopia: The Role of Trust and Social Relations in Market Networks.* PhD Thesis, University of Kentucky: Department of Anthropology.

Mahmoud, H.A. 2006. Innovations in pastoral livestock marketing: The emergence and the role of 'Somali Cattle Traders-cum-ranchers' in Kenya, in *Pastoral Livestock Marketing in Eastern Africa: Research and Policy Challenges,* edited by J.G. McPeak and P.D. Little. Rugby: Intermediate Technology Publications, 129-144.

Mahmoud, H.A. 2008. Risky trade, resilient traders: Trust and livestock marketing in Northern Kenya. *Africa,* 78(4), 561-581.

Mahmoud, H.A. 2009. *Breaking Barriers: The Construction of a New Burji Identity through Livestock Trade in Northern Kenya.* (Working Paper No. 113). Halle: Max Planck Institute for Social Anthropology.

McPeak, J.G. 1999. *Herd Growth on Shared Rangeland: Herd Management and Land Use Decisions in Northern Kenya.* PhD Thesis, University of Wisconsin.

McPeak, J.G. and Little, P.D. (eds). 2006. *Pastoral Livestock Marketing in Eastern Africa. Research and Policy Challenges.* Rugby: Intermediate Technology Publications.

Nunow, A.A. 2000. *Pastoralists and Markets.* (Research Report 61/2000). Leiden: African Studies Centre.

O'Leary, M.F. 1985. *The Economics of Pastoralism in Northern Kenya: The Rendille and Gabra.* (IPAL Technical Report No. F-3). Nairobi: UNESCO.

Plattner, S. 1985. *Markets and Marketing.* Lanham: University Press of America.

Sandford, S. 1983. *Management of Pastoral Development in the Third World.* Chichester: Wiley.

Smart, A. 1993. Gifts, bribes, and guanxi: A reconsideration of Bourdieu's social capital. *Cultural Anthropology,* 8(3), 388-408.

Smith, K. 1997. *From Livestock to Land: The Effects of Agricultural Sedentarization on Pastoral Rendille and Ariaal of Northern Kenya.* PhD Thesis, Pennsylvania State University.

Smith, K., Barrett, C. and Box, P. 2000. Participatory risk mapping for targeting research and assistance: With an example from East African pastoralists. *World Development,* 28(11), 1945-1999.

Teka, T., Azeze, A. and Gebremariam, A. 1999. *Cross-Border Livestock Trade and Food Security in the Southern and Southeastern Ethiopia Borderlands.* Addis Ababa: OSSREA/BASIS CRSP – Project on Cross-Border Trade and Food Security in the Horn of Africa.

Tonah, S. 2000. State policies, local prejudices and cattle rustling along the Ghana-Burkina Faso border. *Africa,* 70(4), 551-567.

Trager, L. 1981. Customers and creditors: Variations in economic personalism in a Nigerian marketing system. *Ethnology,* 20(2), 133-146.

Ufford, P.Q. van 1999. *Trade and Traders: The Making of the Cattle Market in Benin.* Amsterdam: Thela Thesis.

Zaal, F. 1998. *Pastoralism in a Global Age: Livestock Marketing and Pastoral Commercial Activities in Kenya and Burkina Faso.* Amsterdam: Thela Thesis.

Chapter 8

Market Spaces in a Globalising Periphery: Livestock Trade, Borders, and Liberalisation in Eastern Morocco

Ingo Breuer and David Kreuer

Introduction

This chapter examines the dynamics of shifting market spaces within a dryland region of eastern Morocco (Fig. 8.1).[1] It does so from a producer perspective, addressing the question how integration into international markets and commodity chains affects the livelihoods of a population that, until recently, derived almost their entire living from pastoral livestock production. Like all pastoral areas in the country, the eastern Moroccan high plateaus have been exposed to neoliberal policies of economic restructuring and evolving global market chains during the last two decades, when the country experienced several liberalisation programs, a gradual deregulation of economy and trade, and the implementation of free-trade agreements, the most important of which is with the United States, signed in 2004.

In the eastern Moroccan steppes, economic spaces of pastoral households are largely delineated by four dimensions, all of which are, in a liberalising and globalising context, undergoing profound changes (Fig. 8.2 shows the flows associated with these dimensions). First, the region's *Beni Guil* sheep traditionally have a well established position in the country's meat markets; the animals are mainly marketed to Moroccan cities, to which producers are linked through a multitude of trade chains. Second, the region's economy is characterised by informal cross-border trade with Algeria, involving the smuggling of sheep, fuel, and consumer goods. Third, livestock production has become dependent on large quantities of supplementary fodder, which is increasingly being imported from countries like the United States. And fourth, temporary labour migration into the European Union has become widespread among pastoral households; young men often work in the agricultural areas of southern Spain. Local livelihoods are thus embedded in transnational trade and not only affected by the newly negotiated

---

1 We will refer to this relatively homogeneous area – inhabited by Arab tribes – as the Eastern Moroccan high plateaus, steppes, or highlands, whereas 'eastern Morocco' denotes the whole administrative region (*al-jiha al-sharqiyya*, often referred to by its French translation, *l'Oriental*) that includes other landscapes as well.

Figure 8.1   Location of the study area in Morocco
Source: Gertel and Breuer (2007).

free-trade agreements, but also directly shaped by the concurrent and continuous making, un-making and re-making of highly selective border regimes between Morocco, Algeria, and the European Union (Berndt and Boeckler 2011).

This chapter analyses the transformation of these economic spaces and assesses consequences on local livelihoods. On a conceptual level, we depart from several assumptions and analytical starting points. Our analysis of the recent dynamics of livestock markets draws on 'real market' approaches in the sense of Alcántara (1993), Harriss-White (1999), and Mackintosh (1990), as well as – more implicitly – on commodity chain approaches (Gibbon and Ponte 2005, Raikes et al. 2000, Stringer and Le Heron 2008). While the former emphasise the (local) specificity of actually existing markets and the embeddedness of exchange processes in economic, political, and social structures and institutions, the latter focus on commodity-specific dynamics of exchange in studying the respective processes and transactions from primary production to consumption. We use the notion of a 'globalising periphery' to emphasise two aspects of how, in our view, changing

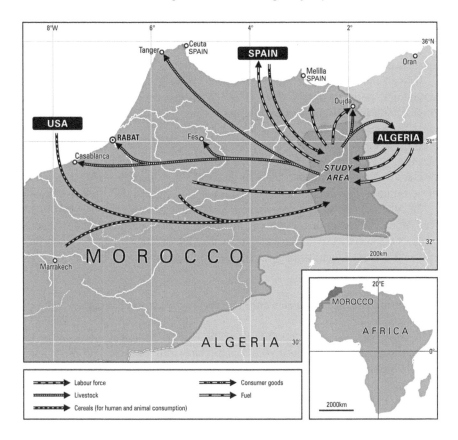

Figure 8.2   Economic articulations of the study area
Source: Fieldwork I. Breuer and D. Kreuer.

market spaces affect local livelihoods in eastern Morocco: On the one hand, we refer to Anthony Giddens' (1990) analysis of the consequences of a globalising modernity, claiming that local conditions are increasingly shaped by processes of time-space distanciation, disembedding, and reflexivity. On the other hand, using the term 'periphery', we adopt basic ideas from economic geography's core-periphery models (Krugman 1991), focusing however on unequal distributions of power within new (global) relations of exchange.[2] We argue that, in the colonial and post-colonial past, the eastern Moroccan high plateaus as a whole have experienced multiple developments of 'peripherisation', while current market integration triggers manifold exclusions of certain social groups, in the sense that decision making about their livelihoods' future is increasingly being removed from their scope of action. Both globalising and peripherising processes are, in our case study, visibly embodied in the emerging influence of global commodity chains

---

2   Cf. Castells (2008), Harvey (2005), Lee (1994), Smith et al. (1984).

on the local economy, in the selective integration of local labour force into transnational modes of accumulation, and in the power of international development discourses that increasingly shape the local conditions of pastoral production.

The chapter is structured in five parts. First, we give an overview of current agricultural policies, notably the 'Green Morocco' development plan, with its emphasis on liberalisation and world market integration of the Moroccan sheep meat, livestock and fodder sectors. Second, general properties of the national sheep meat market are summarised. Third, we discuss specific structures and dynamics of sheep production and, fourth, trade in the pastoral areas that constitute our case study. We then, finally, take the perspective of local producers, driven by the question of how future market transformations might impact their livelihoods.

In our analysis, we rely on data gathered during several years of work on and with Moroccan pastoralists,[3] material published by Moroccan colleagues,[4] and on a household survey David Kreuer conducted in 2009 among pastoral communities in the northernmost part of the high plateaus in the province of Jerada.

Liberalising Morocco's Agriculture: Impacts on Pastoral Systems

Morocco is situated in the north-west of the African continent, bordered by the Mediterranean Sea in the north, the Atlantic Ocean in the west, and the Saharan regions of Algeria and Mauritania in the east and south-east. With a population of about 32 million, a gross domestic product of 92 billion US dollars (CIA 2010), and a Human Development Index of 0.654 (UNDP 2009), Morocco can be classified as a developing country in the medium income range. As 44 per cent of the population live in rural areas, agriculture is a crucial element in the country's economy, accounting for 17 per cent of the GDP and employing some 45 per cent of Morocco's labour force (CIA 2010). Livestock production is widespread; most farms possess animals. Moroccans own about 17 million sheep and 5 million goats (FAO 2010), many of which are kept in pastoral or agro-pastoral production systems (Dutilly-Diane 2007).

The Moroccan sheep meat sector, which we describe in the next section, is currently almost completely sealed off from the world market. High tariffs still protect domestic producers from foreign competition, so there are no official imports or exports of sheep meat or live sheep of any noteworthy volume.

For several years, however, Morocco has witnessed a rush of bilateral and regional free trade agreements. Increasingly, these agreements encompass all economic realms, including agriculture, which had long been considered a 'sensitive' sector. While negotiations with the European Union have progressed only slowly, the agreement signed with the United States in 2004 concerns all sectors; all trade barriers are to be removed within 25 years starting from 2006. For

---

3   Breuer (2007a, 2007b), Gertel and Breuer (2007, 2011).
4   In particular Chiche (2007), Khalil (2007), Mahdi (2007), Rachik (2007).

agricultural products, a complete liberalisation is to take place within 10-15 years. The agreements include only two exceptions to this rule: quotas will be established for red meat and wheat imports into Morocco. The wheat quota is indexed, it will vary according to domestic production, whereas red meat imports will only be liberalised for high-quality products ('Hilton meat') that do not concern Moroccan producers. For standard meat, the protections will remain intact apart from a negligible quota. The EU is not likely to obtain more favourable conditions due to a clause in the US agreement that would automatically adjust its terms accordingly (Akesbi 2009 and personal communication).

Paralleling these free trade agreements, the 'Green Morocco' plan, drafted by the McKinsey & Co. consulting firm, was launched in 2008. The plan's authors do not elaborate on its theoretical underpinning. It is seen as a political instrument that aims to make Moroccan agriculture competitive on the world market. Although the plan has not been made available in any detail, it officially consists of two pillars: The quick development of modern agriculture through 'poles of growth' to become competitive, with high value added production, adapted to market rules, on the one hand; and the 'upgrading of vulnerable actors' and the fight against rural poverty by improving farm incomes through 'trickle-down' effects on the other (Agence pour le Développement Agricole 2010). Green Morocco is thus supposed to attract massive investments into the country's agriculture, to improve both productivity and rural incomes, to increase exports, and to create numerous additional jobs in the long run (Koné 2010). However, budgetary provisions reveal that the first pillar is given absolute priority by planners, while livelihoods of small scale producers are largely neglected (Akesbi 2009).

In the livestock sector, the plan aims to improve domestic supply in terms of meat quality and price, namely by increasing production; and within the second pillar, it strives to upgrade the sheep meat commodity chain to 'make it a social driving force' (whatever that may mean), and to double or triple the livestock producers' income (Bil-Khal 2009, Agence pour le Développement Agricole 2010). This is to be achieved through an 'integrated approach' following the model of the National Association of Sheep and Goat Breeders (ANOC) where small breeders' resources are pooled into larger productive units. Another axis of development is the opening up of export opportunities for niche products; moreover, a modernisation of slaughtering facilities and distribution networks is aimed for, notably the installation of cooling systems. In terms of export facilitation, the plans stipulate the gradual removal of tariff barriers (Bil-Khal 2009).

While it has been pointed out that the policies laid out in the Green Morocco Plan will likely have negative impacts for the rural poor in general (Akesbi 2011), the implications for pastoral production are mixed. Most producers already depend on supplementary feeding to some degree. When imported barley becomes available at low, but at times fluctuating prices, an even stronger conversion to feed-dependent production systems might be a first consequence, with increased profits for livestock breeders. On the downside, this dependence on the world market would expose them to price fluctuations and thus create new vulnerabilities, in addition to severe long term effects from undermining local grain production.

The Green Morocco Plan is only one in a series of modernisation projects targeted at Moroccan agriculture in the past few decades. Until now, these have not caused major disruption to pastoral systems. However, the new plan raises a number of concerns: (a) Through public-private partnerships, 80,000 ha of farmland throughout the country has already been leased to investors, with more acreage poised to follow (Koné 2010). This has so far only lightly affected pastoral areas, but a few large-scale olive plantation projects have already sprung up in the midst of the eastern Moroccan steppe. Should such policies be extended, an increasing fragmentation and territorial disruption of the available rangelands can be expected. (b) The Plan's emphasis on upgrading, modernisation, and quality improvement poses the question which kind of livestock producers will be able to keep up with such changes and who will benefit from them. Previous experience shows that, in the eastern Moroccan highlands at least, each wave of modernisation and innovation has exacerbated social polarisation among sheep breeders. The advent of motorised transportation, for instance, was accompanied by the emergence of a small group of large scale, mobile producers, while the majority of small scale immobile producers were gradually pushed out of pastoral livestock production. (c) The Green Morocco Plan suggests that the sheep meat sector develop export products with high added value, and concentrate on primary markets like the European Union – mainly butcheries that cater to the Muslim population – as well as on African and Arab countries (Agence pour le Développement Agricole 2010). In a related move, the ANOC is trying to establish its own label and get into certification by creating Protected Geographical Indications according to EU regulations (Fagouri 2009). Again, it is safe to assume that 'modernised', well-equipped producers might obtain privileged access to such opportunities, while the majority of small-scale producers would be excluded, particularly in view of the required standards for export markets.

The Moroccan Sheep Meat Sector: Rural-Urban Livestock Chains

The Moroccan sheep meat sector is characterised by significant spatial and temporal fluctuations in supply and demand. On the supply side, most sheep originate from extensive pastoral systems, where production depends on factors such as the animals' reproductive cycles, and regionally specific seasonal variations in precipitation. On the marketing side, national sheep meat remains, so far, shut off from competition by the global economy due to import tariffs of 304 per cent on live animals and processed meat (Agence pour le Développement Agricole 2010). The state further restricts rural-urban and interurban meat transport, limiting the availability of refrigerated transport equipment. Throughout the meat-marketing system, animals are thus transported alive and are generally slaughtered in the place of consumption.

For centuries, weekly markets have been the key interface in the traffic of goods in rural Morocco. Roughly 850 markets, held on one (or sometimes two) set days of the week, are spread over the country. Almost all of them have a livestock section. Individual markets, however, differ greatly from each other in many respects, such as size, the traded animal types, the suppliers and customers (Paulus et al. 1994, Troin 1975). Morocco's weekly livestock markets are characterised by a clear spatial hierarchy, where markets of different sizes assume different functions in the commodity chain. Several attempts have been made to develop typologies of these markets (Khalil 2007, Paulus et al. 1994, cf. also Breuer 2007b). Most often, three types are distinguished, namely: small rural livestock markets, located in remote community centres; middle-scale collector markets, located in regional towns; and middle- and large-scale distributor markets located in bigger cities. In many regions, as most studies indicate, the smallest markets play a subordinate role to interregional animal trade, functioning rather as food-provisioning markets for the rural population. The middle-scale markets, on the contrary, have developed into the central interface between pastoralists and interregional traders, especially as they are often located in regional towns experiencing substantial population growth.

Moroccan pastoralists' marketing opportunities are closely linked to demand and changes in the country's consumption structure. There are several dimensions to this. Generally, annual meat consumption per capita in Morocco is very low at 23.8 kg in 2005 (FAO 2010), and it has been at similar levels for decades. This is connected to the stagnating economic condition of large parts of the population. For the poor, both urban and rural – but especially in the countryside – meat remains a luxury item that is consumed rarely and in small quantities. However, Morocco has seen the emergence of a predominantly urban-based middle class with a higher meat demand (Chichaoui 2001). Hence, meat consumption is concentrated in the cities, and given the high rate of urbanisation, meeting urban demand with rural meat is increasingly important. Most importantly, consumption patterns and taste have changed. The share of sheep and other red meats in total meat consumption has dramatically dropped within the last four decades, from 82 per cent of meat weight in 1970 to 41 per cent in 2005 (FAO 2010). This decline is related to the state-subsidised development of a modern poultry sector, and the fact that poultry consumer prices are considerably lower than those of red meat. Despite the decline of its relative importance, the absolute volume of sheep meat production and consumption has increased during these last decades as a result of population growth.

There is also a strong cultural and religious dimension to Moroccan sheep meat consumption patterns. As geographer Jeanne Chiche puts it, male sheep are 'a prestige product, sold at high prices, ... reserved for special occasions' (2001: 266; our translation). In fact, it is difficult to overstate the relevance of religious festivals to the entire meat-marketing system. The event most strongly influencing sheep demand is the Islamic festival of sacrifice (*ʿīd al-aḍḥā*, locally *l- ʿīd l-kbīr*), where each household head who can afford to should slaughter a sacrificial animal. It is estimated that 50 per cent of all annual sheep slaughters occur at the *ʿīd* (cf. Brisebarre 2002). In November 2009, Moroccan newspapers reported an estimated demand of 5.1 million sheep and goats nationwide for the upcoming holiday, and total transactions were expected to exceed 7.2 billion Dirhams in the few days preceding it (Boukhalef 2009). About half of the demand is thus concentrated in one very short period of the year. Dates for *'īd al-aḍhā* follow the Islamic lunar calendar, in which a year has 354 days, and consequently the festival rotates backward through the Gregorian calendar. For example, while in 2005 it was celebrated in January, in 2010 it had moved to mid-November. Its temporal position in relation to animal production cycles in different Moroccan regions thus differs from year to year. For many pastoralists, *'īd al-aḍhā* is the single most important marketing season, although this does not seem to be the case in parts of eastern Morocco, as we see below.

It is difficult to evaluate the position of sheep meat in urban every-day meat consumption outside of the festivals. In Morocco, consumer preferences for certain kinds of meat are based on a great variety of cultural norms, which may differ according to an individual's region of origin and his or her socialisation (Chiche 2001). Likewise, different sheep breeds may receive variable ratings for taste and quality from distinct consumer groups. As Mohamed Khalil (2007) points out, animals with specific traits may be preferred in certain cities and at certain festivals. Throughout the country, there are thus numerous local niches for the marketing of sheep of defined age, size, breed, and taste. The *Beni Guil* breed of the high plateaus, for instance, enjoys a special reputation for its taste and is considered 'one of the best Moroccan meat breeds' (Brisebarre 2002: 113, our translation).

The Moroccan sheep meat market, in sum, offers considerable marketing possibilities for mobile livestock producers. Demographic change, domestic market protection, and consumption patterns tied to religious festivities have assured a meat demand that is, in the long run, relatively stable. In the short run, however, the Moroccan sheep meat sector is subject to considerable temporal fluctuations of supply and demand. Benefits from livestock marketing depend on the exploitation of price margins between different places or seasons, especially during Islamic festivals.

The Eastern Moroccan High Plateaus as a Globalising Periphery

Let us now introduce the eastern Moroccan highlands and the processes that have turned them into a 'globalising periphery'. Unequal economic ties and asymmetrical power relations are paralleled by this region's remoteness and physical distance from the political and economic core it depends on. And yet, modern globalisation has touched the area just like any other, and has begun irreversibly transforming it. We will now discuss these transformations as they relate to the livestock economy.

As Morocco is characterised by a wide range of semi-arid and arid zones, a variety of pastoral production systems can be found ranging from vertical transhumance to horizontal steppe pastoralism, and from semi-intensive agropastoral systems to more extensive forms relying almost entirely on natural pastures. These systems differ widely with regards to natural diversity, the degree of market integration, and state intervention. However, the main pastoral areas share several commonalities that qualify them as peripheral: they are prone to high inter-annual variability in precipitation; they are 'on the fringes' of the country, for example in mountain, steppe and desert regions far away from the major cities; and the prevalence of poverty and illiteracy is exceptionally high in these regions (see Gertel and Breuer 2007). The eastern Moroccan high plateaus, with over 5 million hectares of arid rangeland, are one of these pastoral areas. They are home to about 100,000 people and 2 million small ruminants (Mahdi 2007, USAID 2006).

Traditionally, the region was characterised by extensive, nomadic livestock breeding where people and their herds would move along routes of water points and forage availability (Guessous et al. 1989). Rangelands were in general collectively owned and used by tribal groups. However, a number of parallel developments – technical, political, socio-economic, and ecological – meant these old ways had mostly disappeared by the 1990s (see Rachik 2000). Basic tenets of former times, the nomadic mobility of livestock and people, as well as the collective access to land, have been widely modified or altogether abandoned.

The region's peripheral position has not prevented interventions by state and international development agencies that have regularly affected local pastoral livelihoods, inducing processes of social and spatial restructuring (Chiche 2007, Gertel and Breuer 2007). As a response to the perceived dysfunctional state of tribal institutions and to the administration's demand for solid governance structures, framed within a globalised development discourse, pastoral cooperatives were – based on ethnic lineages – established top-down throughout the high plateaus in the 1990s. Their main function was to ensure a regulated use of the associated pastures and the declaration of temporarily protected areas in order to fight degradation. This has, where enforced, encouraged a change in people's attitudes towards the rangeland: Access to the pastures is now members-only or has to be paid for (Mahdi 2007). Yet, 20 years after their creation, many cooperatives have failed to create the desired level of coordination.

Such institutional change has been paralleled by technological aspects of globalisation. From the mid-twentieth century onwards, trucks have been used to transport people, animals, water, and fodder rapidly over long distances, replacing the camel and making new pasture grounds accessible for those who can afford to buy or rent a vehicle. In combination with other technical innovations, prolonged droughts, and the growing tendency to 'fix one's tent for good' or move into a house in one of the towns, this has led to a polarisation between wealthy pastoral entrepreneurs on the one hand and impoverished ex-nomads on the other (Rachik 2000).

Another important aspect of globalisation concerns the opening up of international labour markets. Since the 1990s, there has been labour migration from the eastern Moroccan highlands on a large scale. According to data published by Mohamed Mahdi (2007), 2,088 workers from the 4,064 pastoral families surveyed had emigrated. Of those, 57 per cent were international migrants working in Spain as agricultural labourers, while the remaining 43 per cent work in the cities neighbouring the high plateaus. As Mahdi points out, the emigration is temporary, and emigrants do not cut their ties with their home communities; many of them go back and forth between Spain and Morocco several times a year. The flow of people and capital between the two countries is of course highly asymmetrical, so that the peripheral status of the steppes is reproduced on a global scale. It is thus in a double sense that our notion of a 'globalising periphery' should be read: A domestically marginalised area that takes part in globalising processes; and in turn, an unprecedented integration that does however not establish a more balanced power structure or a more advantageous economic position for the region.

Trading Sheep in the High Plateaus: A Market Perspective

In spite of all these changes, the region has remained mainly pastoral, producing sheep, and – to a lesser extent – goats. Thanks to its significance to the domestic sheep market, it is still well positioned when compared to other marginal regions, maintaining a key role in supplying Morocco's urban centres with meat. We now take a closer look at the market mechanisms in order to examine the supply flows. Figure 8.3 depicts the structure and volume of sheep commodity circuits and commercialisation in this zone; it is based on findings by Mohamed Khalil in the late 1990s (see Khalil 2007). Sheep traded within the region originate from both the Moroccan high plateaus (43 per cent) and, significantly, from the Algerian side of the border (57 per cent) This cross-border trade is related to different sheep breeds and will be discussed below. There is no evidence of any direct link between producers and consumers; one or more intermediaries are always involved in the marketing chain, although many of them are also livestock producers themselves. Khalil distinguishes three types of intermediaries (excluding local butchers who cater exclusively to markets within the study zone): local, regional, and external traders, who also buy and sell sheep amongst themselves. This explains why the number of sheep transactions (1,481,000) in the region exceeds the total number of animals traded (931,000).

Main weekly markets in the high plateaus are held in Tendrara (southern part), Ain Beni Mathar, and Tiouli (both in the northern part), and fall into the category of middle-scale collector markets, while the biggest hub linking the eastern Moroccan steppes to the rest of the country is the livestock exchange in Taourirt (USAID 2006). As Figure 8.3 shows, close to two thirds of the sheep are destined for consumption outside of the region of Eastern Morocco, while most of the remaining third goes to its urbanised areas – mainly the city of Oujda and the regions of Nador and Berkane. Due to the ban on meat transportation, they are trucked across the country alive. In this context, a second distinction needs to be made between two types of actors: middlemen who buy from the producers but do not slaughter themselves (*maquignons* in French), and urban meat wholesalers (*chevillards*) who deal with slaughterhouses directly and then sell the carcass to butchers (Khalil 2007).

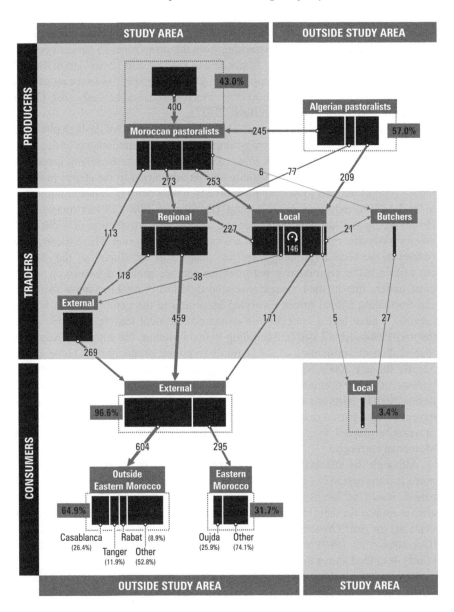

Figure 8.3   Sheep commodity chains in the eastern Moroccan steppes
Note: Numbers in 1,000 head.
Source: Khalil (2007).

Since consumption within the eastern Moroccan highlands is almost insignificant, urban demand shapes the provisioning chain. As Figure 8.3 reveals, the four most important urban destinations receive a large share of all sheep passing through the study zone. Most of this trade, once arrived in the city, is handled by *chevillards*. Of the overall *chevillard* trade originating from the study zone, 35 per cent goes to Casablanca, 16 per cent to Oujda, 15 per cent to Tangier, and 12 per cent to Rabat.[5] The relative importance of this supply chain from the perspective of the cities is as follows: Casablanca receives a third of its sheep meat from the study area, Oujda two thirds, and Rabat a quarter (Khalil 2007).

So what kinds of sheep are traded? The typical sheep breed of the high plateaus, the *Beni Guil*, is named after the most important tribal confederation of the area; locally, it is also referred to as *daghma* (dark-coloured). The *Beni Guil* breed has an excellent reputation 'for its rusticity and its adaptation to the steppe pastures [and] is one of the best Moroccan meat breeds, but its dairy qualities also allow it to lend itself to industrial crossbreeding' (Brisebarre 2002: 113, our translation). However, the past decades have seen the large-scale introduction of a different breed via Algeria, called *Ouled Jellal* after another tribal group. These sheep, *bīḍa* (white) in the local terminology, are less adapted to the climate of the steppe, but offer a better reproductive performance and are preferred by certain urban meat traders due to their slaughter-related qualities. Unstable supply structure and fluctuating animal prices – varying according to the enforcement of border controls – have lent a considerable dynamism to local trading activities since the mid-1990s (Khalil 2007). According to one observer, the entry of *bīḍa* sheep 'has shaken the principles of the region's livestock production' (Chiche 2007: 52). By the late 2000s, a crossbreed of *daghma* and *bīḍa* had become extremely widespread in the northern high plateaus, as we will discuss below. Its colloquial designation is *bargiyya* (gleaming, light-coloured). Moreover, local observers have noticed a partial reversal of the cross-border sheep flows in the past couple of years, prompted by exchange rate developments between the Algerian and Moroccan currencies.

Although the informal sheep trade with Algeria is fairly recent, it should be noted that the region has a long history of transnational trade. During the colonial period, *Beni Guil* lambs were primarily produced for export. They were sought after in France for their high quality meat, and were marketed there as 'the Small Oranian' (*le petit Oranais*). These exports, however, ceased in the early 1970s (Guessous et al. 1989) and it is only in the context of the current liberalisation efforts described above that they might be revived.

In conclusion, three points stand out about the area's sheep trade system: Firstly, most of the consumers are located in distant cities; secondly, this enables a complex network of intermediaries using fluctuations in demand and supply to enhance their profits; and thirdly, the high plateaus have become a major hub in trans-border livestock movements between Algeria and Morocco.

---

5   These percentages refer to the *chevillard*-handled portion of the trade only, therefore they are not comparable to the data displayed in Fig. 8.3.

Trading Sheep in the High Plateaus: A Producer Perspective

With these dynamics in mind, we now turn to the pastoral households living in this globalised periphery. Drawing on data from the 2009 survey, we take up the four dimensions of market relations established in the introduction: the intra-Moroccan sheep commerce with its fluctuations in demand and price; the cross-border trade involving Algerian *bīḍa* sheep; the producers' dependency on supplementary fodder; and their embeddedness in non-pastoral activities, especially European labour markets. The survey includes four rural municipalities in the northern part of the high plateaus that are associated with three tribes: Oulad Sidi Abdelhakem in the east, Beni Mathar in and around the town of Aïn Beni Mathar, and Oulad Sidi Ali Bouchnafa further to the west. Based on lists of all *douars* (ethnically and/or locally defined subgroups of each tribe), 484 heads of household were interviewed, covering 17 per cent of the households in the four municipalities. In our sample, about three quarters of the families engage in livestock breeding, with marked differences between *douars*. Other livelihood strategies that can be found include seasonal labour migration to Europe, day labour in the region, work in the construction sector, but also trade and a few more specialised professions. Overall, our sample appears representative of the study region although official records to establish this are not available.

In order to examine the pastoral society of the northern high plateaus, we propose four categories of livestock breeders in relation to the number of sheep they own. While this is, admittedly, a simplistic way of categorising pastoral livelihoods, in terms of the questions that concern us here, it works well and allows for comparison with Khalil's work, who uses similar categories of producers (Khalil 2007). If we take all 370 households from our sample that own livestock and divide them into four groups of about equal size, the results reveal some key characteristics of each quartile (see Fig. 8.4).

Price Formation in Sheep Trade

Generally, trade volumes fluctuate seasonally. In the summer, when pastures are exhausted and pastoralists sell animals in larger numbers, prices are expected to go down. During the winter, when vegetation is abundant and producers retain the maximum number of sheep, prices increase (Khalil 2007, see also Allali, Dalil and Mahdi 2002 for the neighbouring region of Missour). *ʿīd al-aḍḥā* has a special role regardless of season and is accompanied by a 'strong price increase for live animals' (Khalil 2007: 108).

Turning to the survey data, we start with these seasonal trade patterns. The total of animals sold by all respondents amounted to 10,004 sheep during the 2008 *ʿīd*; 10,757 in the winter of 2008-9; and 15,446 during the summer of 2009. In other words, the average number of sheep sold per pastoral household was: 27 before the *ʿīd*; 29 in the winter; and 42 in the summer. These figures underline the greater number of sheep sold during the summer months. But the often assumed

	Very small <31 sheep N=103	Small 31-67 sheep N=82	Medium 68-150 sheep N=98	Large >150 sheep N=87
Herd size (sheep only, averages)	19	49	103	488
Sales volume (annually in % of current herd size)	184%	137%	96%	57%
daghma share	52%	45%	49%	43%
bīḍa share	2%	1%	1%	1%
bargiyya share	37%	40%	45%	53%
Supplementary feed:				
always	51%	54%	66%	67%
most of the time	34%	41%	31%	32%
Practice of fattening	56%	57%	76%	90%
Sole reliance on pastoral income	32%	45%	37%	36%
Co-op member(s) in household	22%	33%	38%	43%
Wage laborer(s) in household	49%	40%	30%	37%
Migrant(s) in household	16%	17%	24%	28%

Figure 8.4   Pastoralist livelihoods in relation to production and marketing
Source: Fieldwork D. Kreuer.

importance of the festival of sacrifice is not confirmed by this data. For the average pastoral household in the northern plateaus, the *ʿīd* is apparently not the single big event where most of a year's transaction volume is realised. Instead, sales are distributed across the whole year. This makes sense given the frequently heard statement by herd owners with no significant second source of income: 'I sell one or two animals whenever I need some money.' Herd size does not seem to be an influencing factor: Small and large flock owners exhibit basically the same marketing patterns across the seasons.

Are the fluctuations in trade volume reflected in prices? In our sample, the mean prices quoted for *daghma* sheep are as follows: *ʿīd* 1,398 Dirhams (Dh), winter Dh 956, summer Dh 998. For the *bargiyya* crossbreed, the figures are: *ʿīd* Dh 1,443, winter Dh 1,031, summer Dh 1,062. The peaking demand at the *ʿīd* is clearly reflected in prices, and *bargiyya* are a little more lucrative than *daghma* sheep on average. Our low case numbers for *bīḍa* (*Ouled Jellal*) sheep prices do not allow a comparison. Interestingly, despite the larger sales volume in the summer, prices are slightly higher than in the winter, which runs counter to a simple supply-demand equation and received wisdom. Some other factors that determine animal prices must be at work. The annual influx of Moroccan emigrants who return home during the summer months and are comparatively wealthy certainly plays a role, as does the fact that marriages and the like are usually celebrated in the summer.

Other elements that, in our survey, have an influence on sheep prices, include a breeder's membership in a cooperative; and this, in turn, is related to the herd size as figure 8.4 demonstrates. While 43 per cent of the large herd owning households have one or more co-op members, only 22 per cent of the very small livestock keepers do. It must be noted that many of the pastoral cooperatives that were decreed around 1990 are currently defunct in the northern plateaus: Merely 32 per cent of livestock-breeding households declare having a member in one of them. This low rate is in stark contrast to official statements, according to which all pastoralists are formally enrolled. At any rate: In terms of sheep sales, those who state being in a cooperative tend to get slightly higher prices in summer and winter, but not during the *'īd*. Mobile nomads get lower prices than sedentary people (former and non-nomads). Those who hold a public office get far better prices than those who do not, except for the *'īd* – although case numbers are low so care needs to be taken in extrapolating.

The general impression is that people who command higher levels of social capital, expressed in access to networks and formal institutions, employ it to accumulate financial gains. They are less dependent on individual intermediary traders and thus have a stronger position in bargaining. In the days before the *'īd*, these systematic differences disappear and everybody seems to enjoy equal opportunities.

Speculation and Cross-Border Trade

When addressing the issue of cross-border trade, we should first note that reliable information is almost impossible to obtain as these activities are illegal; the Moroccan-Algerian border has officially been closed since 1994 due to diplomatic tensions between the countries. Yet, two indicators give hints as to who may be involved in this business: The percentage of *Ouled Jellal* sheep someone possesses, since they were originally brought from Algeria; and the proportion of animals sold each year in relation to the current herd size.

Starting with the latter, we observe that 39 per cent of those who own sheep have, throughout the past year, sold at least as many animals as they owned at the time of the interview. This phenomenon is most distinctly found among the very small owners (see Fig. 8.4) whose turnover is almost twice their entire stock, and decreases with growing herd size. Khalil, in his study, found essentially the same pattern and concluded that the smallest herd owners were the most active in sheep speculation and smuggling (Khalil 2007). There are however other possible explanations. Natural growth of the flock can lead to a surplus; or the majority of herds could simply have shrunk during the past year (since 2008-9 was a year with abundant rainfall, this is unlikely). It is quite common among the region's pastoralists to purchase a number of sheep, fatten them for a few weeks, and then re-sell them for a profit. Others buy and sell livestock within a single market day, trying to take advantage of intra-day price variations, and thereby become part of the intermediary network described above.

If 'taken with a grain of salt', trade volume numbers can serve as an indicator of cross-border trade activities, and the same is true for the share of *bīḍa* sheep in a household's animal stock since they are brought directly from Algeria. The survey suggests, however, that almost none of the interviewed pastoralists claim to own animals of the Algerian breed; *bargiyya* sheep, on the other hand, are very popular. Across the interviewed persons, the average herd size is 158 sheep, including 65 of the local *daghma* and 76 of the *bargiyya* crossbreed. A few other breeds can be found as well. Herd compositions are very heterogeneous, but the essential finding is that in the northern high plateaus, the Algerian breed is not present in large numbers, whereas the mixed breed has become more popular than the local one.

The larger the herd, the higher the proportion of *bargiyya* sheep tends to be (Fig. 8.4). This distribution is somewhat at odds with the non-linear distribution of sheep types reported by Khalil (2007), and if taken as an indicator of involvement in smuggling activities, plainly contradicts what we just concluded about sales volumes, namely that the smallest herders are most active in cross-border trade. Our interpretation is that, in the past 15 years and at least in the northern plateaus, the Algerian breed has become much less relevant than suggested by previous studies, while its cross-bred *bargiyya* offspring is now so common that its breeding can not be seen as an indicator of specific economic strategies. The *Ouled Jellal* and subsequent *bargiyya* introduction can, if this is true, be understood as a typical innovation process where the early adopters were to be found among small and medium herd owners (medium and large in Khalil's typology). Once the viability of the new crossbreed was proven, large owners entered into the business and massively increased their *bargiyya* stocks in order to maximise profits.

Fodder Dependency

In the 2009 survey, participants were asked to assess whether they 'always', 'most of the time', 'normally do not' or 'never' supplement. As Figure 8.4 demonstrates, two thirds of the medium and large herds completely depend on feed supplementation and the remaining third does so most of the time. The minuscule number of pastoralists who rely only on the natural pastures without using any extra fodder tend to be small herd owners. In this respect, the situation appears more uniform than in the 1990s (Khalil 2007). This could be due to extended drought and pasture degradation, which are cited by most pastoralists of the region as the main reasons for recurring to supplementary fodder. Another reason is of course fattening, which is practiced by nine out of ten large owners, but by only about half of the very small and small ones (see Fig. 8.4). Given these figures and the fact that most of the forage consists of barley and bran (USAID 2006), it is apparent that any change to the Moroccan fodder market would directly affect the vast majority of the pastoralists in the area as well as pasture ecosystems.

The loss of mobility in this formerly nomadic society is often seen as closely connected to drought and degradation. Many sheep breeders argue: 'Where am I supposed to go with my animals? The pastures are the same everywhere...' In our sample, 31 per cent of livestock-producing households still consider themselves nomadic, where the entire family is basically mobile. Thirty-four per cent say they used to be nomads, but have now turned to sedentary animal production; and 35 per cent say that they have been sedentary for all their lives, so if it ever existed in their family, nomadism was abandoned in earlier generations.

Non-Pastoral Activities and Labour Markets

At the same time, income diversification has become widespread, and pastoral production is intertwined with wage labour. Just 31 per cent of households derive all of their income from livestock, either as independent breeders or as paid herders – with the highest percentage among families who own 31-67 sheep (Fig. 8.4). Among the overall sample of 484 households, 18 per cent have no animal-generated income at all, and no less than half of today's families combine livestock breeding with other jobs, most frequently as agricultural labourers in Spain or as day labourers or construction workers in the region. Wage labour is most widespread among very small livestock owners and lowest, in this sample, among medium ones. The medium category seems to rely on freelance non-pastoral activities more frequently. Pastoralism thus continues to be the mainstay of the regional economy, but it is losing importance: Even among those who are not engaged in any livestock breeding, 80 per cent state that it was their parents' main activity a generation ago.

Currently, one fifth of the interviewed households have (or have had) at least one family member working in a Moroccan city or abroad, predominantly in Spain, mostly as seasonal migrant, and often in agriculture. When asked about migrants in their larger family, 55 per cent of household heads mention at least one. This variable seems, once again, to be directly related to herd size, which can be read in two ways: Either those who possess large numbers of animals can more easily afford to send one or two of their sons to Spain, or the migrants' remittances are invested in animals and lead to considerably larger numbers of livestock. Both statements are partly true according to locals: There is a certain investment threshold to be overcome before obtaining a Spanish contract, but once you are in the game, you can easily earn eight times the monthly wage that is commonly paid in the high plateaus.

The risks of integration into a distant labour market have been felt since 2008, when the global economic crisis hit Spain and the demand for seasonal agricultural labour went down. Many migrants have since stayed at home and concentrated on livestock breeding or other activities like house construction, slowly but surely fostering change in the steppes.

Conclusion

In the face of Morocco's integration into a global economy, pastoral livelihoods are increasingly determined by transnational commodity chains – both formal and informal – of livestock, labour and fodder, among others. Our empirical findings highlight the mechanisms that determine their living conditions within a region that we have characterised as being both 'globalised' and 'peripherised'. Making a living from sheep sales has become intricate: Price formation is variable and depends on the quality of the animals, the place and time of transaction, but also on an individual's social and institutional connections. Moreover, speculation with livestock has become a decisive livelihood strategy for many households with few animals. Access to international labour markets offers a chance of earning substantial non-pastoral income, but is highly selective, reproducing relations of power and dependence. All but the poorest pastoralists depend on supplementary fodder, a commodity where globalisation is particularly imminent. The future of pastoralists in the eastern Moroccan highlands thus hinges on the evolution of these market spaces.

Our findings indicate multiple processes of socio-economic polarisation happening in this context, many of which will probably be exacerbated by the current liberalisation measures. First of all, in the domestic sheep meat market, well-connected larger herd owners who take part in modernisation initiatives in the frame of the Green Morocco Plan or similar instruments are in a position to take the most advantage of changing demand structures. The economic aggregation promoted by the Plan is a further step towards concentration of sheep production in the hands of a few 'efficient', 'modern' and 'competitive' entrepreneurs. In trans-border trade, secondly, the same group is about to benefit most from new export opportunities: Livestock owners with sufficient monetary and social capital to allow them access to upgrading and certification programmes. Households with small herds are likely to be excluded from these opportunities. They may doubly suffer, for at the same time that export markets are being promoted, the Green Morocco Plan also includes the rubric of 'sanitary security of livestock and meat quality' under which it pledges to fight informal structures and smuggling – both decisive resources for the poorest pastoralists' livelihoods. Liberalisation measures and free trade on the fodder market, thirdly, will affect all of the region's livestock producers; although the exact implications for local livelihoods are yet unknown, we suppose that, with accelerated influx of foreign fodder, producers' exposure to price fluctuations on the world markets will sharply rise. Finally, markets for manual labour will critically determine the future of the eastern Moroccan high plateaus and their population's livelihoods, given the fact that increasing numbers of small livestock producers 'drop out of the system', becoming dependent on activities such as manual wage work in agriculture, informal petty trade, or even smuggling. For local employment markets, prospects are difficult due to ongoing demographic transition; and to which extent the Mediterranean border will remain selectively open to seasonal labourers depends on the European Union's needs and regulations. All in all, chances are that the new economic spaces will entail the creation of multiple borders – both visible and invisible – that govern access to these new market opportunities. They might, in the worst case, exclude everyone but a select elite.

Acknowledgments

This chapter presents results from the Collaborative Research Centre SFB 586 'Difference and Integration' project (www.nomadsed.de) funded by the German Research Foundation.

References

Agence pour le Développement Agricole. 2010. *Plan Agricole Régional. Région de l'Oriental.* [Online]. Available at: http://www.ada.gov.ma/uplds/pars/par08. pdf [accessed: 31 March 2011].

Akesbi, N. 2009. Un plan schématique et trompeur. *La Revue Economia*, 7, 39-43.

Akesbi, N. 2011. Marokkanische Landwirtschaft und Freihandel, in *Alltagsmobilitäten. Marokkanische Lebenswelten im Aufbruch*, edited by J. Gertel and I. Breuer. Bielefeld: Transcript (forthcoming).

Alcántara, C.H. de (ed.). 1993. *Real Markets: Social and Political Issues of Food Policy Reform.* London: Frank Cass.

Allali, K., Dalil, S. and Mahdi, M. 2002. Le marché des ovins dans la région de Missour: Structure, comportement et performance, in *Mutations sociales et réorganisation des espaces steppiques*, edited by M. Mahdi. Casablanca: Imprimerie Najah El Jadida, 91-109.

Berndt, C. and Boeckler, M. 2011. Mobile Grenzen, entgrenzte Orte und verortete Waren: Das Beispiel des Agrarhandels zwischen Marokko und der EU, in *Alltagsmobilitäten. Aufbruch marokkanischer Lebenswelten*, edited by J. Gertel and I. Breuer. Bielefeld: transcript (forthcoming).

Bil-Khal, 'A. 2009. Maḥāwir tanmiyat silsilat al-luḥūm al-ḥamrā' fī iṭār mukhaṭṭaṭ al-maghrib al-akhḍar. *Al-Kassāb*, 17, 8-9.

Boukhalef, A. 2009. Aïd Al Adha. Une offre de 6 millions d'ovins et de caprins. *Éco Plus (Le Matin insert)*, 6 (13 November 2009), 5.

Breuer, I. 2007a. *Existenzsicherung und Mobilität im ariden Marokko.* Wiesbaden: Dr. Ludwig Reichert Verlag.

Breuer, I. 2007b. Marketing from the margins: The Ilimchan pastoralists of the pre-Sahara, in *Pastoral Morocco: Globalizing Scapes of Mobility and Insecurity*, edited by J. Gertel and I. Breuer. Wiesbaden: Dr. Ludwig Reichert Verlag, 117-132.

Brisebarre, A. 2002. L'Ayd al-kabir : Un élément structurant de la production et de la commercialisation des ressources pastorales, in *Mutations sociales et réorganisation des espaces steppiques*, edited by M. Mahdi. Casablanca: Imprimerie Najah El Jadida, 111-124.

Castells, M. 2008. The rise of the Fourth World, in *The Global Transformations Reader*, edited by D. Held and A. McGrew. Cambridge: Polity Press, 430-439.

Chichaoui, H. 2001. *Etude d'un marché régional de la viande rouge: Cas de la région de Casablanca.* Unpublished thesis. Rabat: Institut Agro-Vétérinaire Hassan II.

Chiche, J. 2001. Les effets des programmes d'encouragement à l'élevage sur la production des ovins et des caprins au Maroc. *Options Méditerranéennes. Série A: Séminaires Méditerranéens (CIHEAM)*, 46, 55-64.

Chiche, J. 2007. History of mobility and livestock production in Morocco, in *Pastoral Morocco. Globalizing Scapes of Mobility and Insecurity*, edited by J. Gertel and I. Breuer. Wiesbaden: Dr. Ludwig Reichert Verlag, 31-59.

CIA. 2010. *The World Factbook. Morocco.* [Online]. Available at: https://www. cia.gov/library/publications/the-world-factbook/geos/mo.html [accessed: 31 March 2011].

Dutilly-Diane, C. 2007. Pastoral economics and marketing in North Africa: A literature review. *Nomadic Peoples*, 11(1), 69-90.

Fagouri, S. 2009. Stratégie de l'ANOC. *L'Eleveur*, 17, 10-11.

FAO. 2010. *FAOSTAT.* [Online]. Available at: http://faostat.fao.org/ [accessed: 31 March 2011].

Gertel, J. and Breuer, I. 2007. Introduction, in *Pastoral Morocco. Globalizing Scapes of Mobility and Insecurity*, edited by J. Gertel and I. Breuer. Wiesbaden: Dr. Ludwig Reichert Verlag, 3-9.

Gertel, J. and Breuer, I. (eds). 2011. *Alltagsmobilitäten. Aufbruch marokkanischer Lebenswelten.* Bielefeld: Transcript (forthcoming).

Gibbon, P. and Ponte, S. 2005. *Trading Down. Africa, Value Chains, and the Global Economy.* Philadelphia: Temple University Press.

Giddens, A. 1990. *The Consequences of Modernity.* Cambridge: Polity Press.

Guessous, F., Boujenane, I., Bourfia, M. and Narjisse, H. 1989. Sheep in Morocco, in *Small Ruminants in the Near East. Volume III: North Africa*, edited by FAO. Rome: Food and Agriculture Organization of the United Nations, 14-83.

Harriss-White, B. 1999. Introduction: Visible hands, in *Agricultural Markets from Theory to Practice*, edited by B. Harriss-White. London: Macmillan, 1-36.

Harvey, D. 2005. *A Brief History of Neoliberalism.* Oxford: Oxford University Press.

Khalil, M. 2007. Trading livestock: Eastern Moroccan sheep meat commodity chains, in *Pastoral Morocco. Globalizing Scapes of Mobility and Insecurity*, edited by J. Gertel and I. Breuer. Wiesbaden: Dr. Ludwig Reichert Verlag, 107-115.

Koné, I. 2010. *Agriculture/ Plan Maroc Vert: une promo pour attirer les entrepreneurs.* [Online]. Available at: http://www.yabiladi.com/articles/details/2429/agriculture-plan-maroc-vert-promo.html [accessed: 31 March 2011].

Krugman, P. 1991. Increasing returns and economic geography. *The Journal of Political Economy*, 99(3), 483-499.

Lee, R. 1994. Core-periphery model, in *The Dictionary of Human Geography*, 3rd Edition, edited by R.J. Johnston, D. Gregory and D.M. Smith. Oxford: Blackwell, 95-96.

Mackintosh, M. 1990. Abstract markets and real needs, in *The Food Question. Profits Versus People?*, edited by H. Bernstein, B. Crow and M. Mackintosh. London: Earthscan, 43-53.

Mahdi, M. 2007. Pastoralism and institutional change in the Oriental, in *Pastoral Morocco. Globalizing Scapes of Mobility and Insecurity*, edited by J. Gertel and I. Breuer. Wiesbaden: Dr. Ludwig Reichert Verlag, 93-105.

Paulus, I., Boueiz, M., Fischer, M. et al. 1994. *Le fonctionnement du marché ovin au Maroc – Approche méthodologique et résultats de l'étude pilote au Moyen Atlas*. Berlin: Humboldt-Universität.

Rachik, H. 2000. *Comment rester nomade*. Casablanca: Afrique Orient.

Rachik, H. 2007. Nomads: But how?, in *Pastoral Morocco. Globalizing Scapes of Mobility and Insecurity*, edited by J. Gertel and I. Breuer. Wiesbaden: Dr. Ludwig Reichert Verlag, 211-225.

Raikes, P., Friis Jensen, M. and Ponte, S. 2000. Global commodity chain analysis and the French Filière approach: Comparison and critique. *Economy and Society*, 29(3), 390-417.

Smith, J., Wallerstein, I. and Evers, H.-D. 1984. *Households and the World-Economy*. London: Sage.

Stringer, C. and Le Heron, R. 2008. *Agri-Food Commodity Chains and Globalising Networks*. Aldershot: Ashgate.

Troin, J. 1975. *Les souks marocains: Marchés ruraux et organisation de l'espace dans la moitié nord du Maroc*. Aix-en-Provence: Edisud.

UNDP. 2009. *Human Development Reports*. [Online]. Available at: http://hdr.undp.org/en/statistics/ [accessed: 31 March 2011].

USAID (U.S. Agency for International Development). 2006. *Promotion des viandes ovines à l'Oriental*. [Online]. Available at : http://pdf.usaid.gov/pdf_docs/PNADH517.pdf [accessed: 31 March 2011].

PART III
From State to Market Production: Post-Socialist Contexts

Chapter 9

Pastoralism in the Pamirs:
Regional Contexts, Political Boundaries, and
Market Integration in Central Asia

Hermann Kreutzmann

Introduction

Pastoralists utilising sparsely populated and marginal regions have often been affected by political decisions, made far away in the centres of colonial powers. The major transformations from the age of imperialism to the era of globalisation have had significant effects on the pastoral sphere and changed living and market conditions in the periphery. Central Asia is no exception and experienced, in addition, revolutionary interventions, which transformed the livelihoods of pastoralists and farmers alike. For an enhanced understanding of market integration in the Central Asian context it is important to analyse the effects of geopolitical interference, boundary-making, socio-economic reforms and revolutionary movements. The two major transformations of the twentieth century – the October and Chinese Revolutions on the one hand, and the replacement of collective strategies through market-oriented reforms and independence on the other hand – are followed by the end of the Cold War and the advent of modern globalisation. An investigation into historical and contemporary challenges by powerful external actors and a focus on local and regional responses provides insights into the effects of domination, resistance and reform in the spatial and economic periphery.

Ecological Structure and Spatial Utilisation

On the macro scale, Central Asia has been the sparsely settled periphery between Europe and Asia. Environmentally, the region is characterised by steppe, desert and mountains with arid conditions in the lowlands and precipitation and humidity increasing with altitude, resulting in snow-covered mountains, glaciation, high mountain pastures and scant forest. Given these assets, common utilisation patterns of ecological resources are related to a bi-polar approach. Extensive nomadism in the vast desert and steppe regions covers substantial sparsely vegetated areas. Animal husbandry, as a prime strategy, is enhanced by certain forms of mountain nomadism in the Hindukush, Pamirs and Tien Shan.[1] In contrast, agriculture is limited to oases in which intensive crop cultivation is linked to the demands of the bazaar towns and their surroundings along the traditional trade routes of the Silk Road network. Even more important than silk as an exchange product has been cotton, cultivated in major irrigated oases. Hydraulic resources for irrigation originate mainly from the glacier-fed rivers such as the Amu and Syr Darya issuing from the high mountain ranges within the desert-steppe environment. In remoter mountain regions we find different forms of combined mountain agriculture (Ehlers and Kreutzmann 2000: 15) in scattered mountain oases mainly supplied by gravity-fed irrigation schemes tapped from the tributary valleys of the main rivers. Niche production of valuable and marketable crops augments the general pattern of grain crop cultivation for basic sustenance.

Economically and politically there has been long competition between nomads and farmers over natural resources. While they compete in the production sector, political influence is mainly felt and contested in the urban centres of oasis towns. They were the prime target of all kinds of conquerors from Iran, Mongolia and China.[2] These conquests left their marks on the Central Asian socio-economic landscapes and prove the existence of an Eurasian exchange system over long periods of time.

In the nineteenth century Central Asia's role changed significantly. The *Great Game* started, creating a new landscape of power structures, stimulated by the two superpowers of that time: Russia and Britain (Fig. 9.1). Direct influence in the form of boundary-making and economic exploitation removed the former pattern of indirect control and tax collecting in a feudal system. For the understanding of the economic spaces of pastoral economies in Central Asia and the performance of independent states, the geopolitical dimension of the *Great Game* and subsequent territorial demarcations needs to be discussed in greater detail.

The *Great Game*

At the turn of the twentieth century the British Viceroy in India, Lord Curzon, identified the Central Asian countries and territories in his famous statement as 'pawns on a chessboard'. British India and Russia were the players who gambled their influence in Transcaspia, Transoxania, Persia, and Afghanistan (Fig. 9.1). But this battle was not solely about regional control, it was a contest about world domination of imperial powers. Great Britain had already achieved maritime supremacy, now the last land-locked area – Central Asia – came into focus. From

---

1   The specific utilisation patterns of high mountain pastures – such as observed in the 'pamirs' (see Kreutzmann 2003) – is characteristic for Central Asia and has repeatedly given scope for speculation about the economic potential of animal husbandry since Marco Polo's travels.

2   Bregel (2003), Christian (2000), Kreutzmann (1997, 2004).

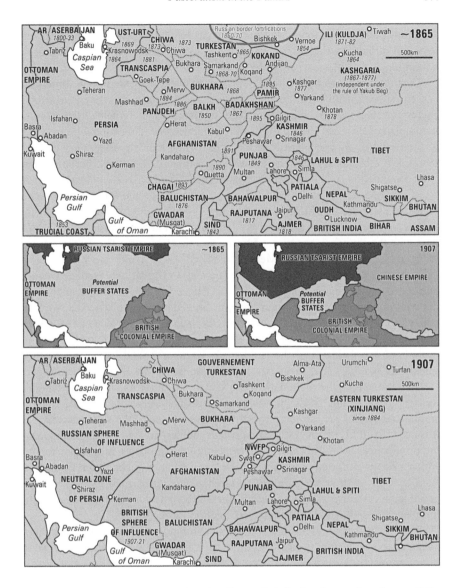

Figure 9.1   **Imperial expansion in Central Asia in the early phase of the**
                   ***Great Game* (approx. 1865) – Consolidation of British and**
                   Russian spheres of influence at the time of the Anglo-Russian
                   Convention (1907)

Source: Kreutzmann (2009a: 25-28), adapted.

a British viewpoint Central Asia was a 'buffer region' to protect the 'jewel of the
crown', the British possessions in India. From a Russian perspective expansion
towards the East and South had been pursued since Peter the Great stated in his

testimony that Russia's future was linked to Asia (Hauner 1989). Both superpowers expected sufficient wealth to be gained from Central Asia to pay for their exploring adventures and military expenditure.

During the nineteenth century both superpowers contested their supremacy in Central Asia. Both had literary celebrities justifying their cause and in both countries contemporary bourgeois debate highlighted the civilising mission to be accomplished. In Great Britain Rudyard Kipling was one of the foremost advocates of the *Great Game* and coined the term the 'white man's burden' (Kreutzmann 1997). With missionary zeal and state authorisation, civil society measures were to be promoted in Asia grounded in European standards. His Russian counterpart was Fjodor M. Dostojevsky who published an essay on the importance of Asia for Russia's future in which he justified the Asian conquest as a mission for the promotion of civilisation. Dostojevsky compared colonial expansion into Central Asia with the European conquest of North America (Hauner 1989, 1992).

In Russia the Gorchakov Memorandum of 1865 marks the beginning of the animated phase of the *Great Game* (Fig. 9.1). The British Premier Disraeli responded in his famous speech at Crystal Palace 1872 in which he announced imperial policies for further expansionism. Immediate results were the 'forward policy' in the Afghan borderlands and the subsequent crowning of Queen Victoria as Empress of India (1877). Russia and Great Britain fought this game in the remote mountains of the Hindukush, Karakoram and Pamirs where their spies-cum-explorers met in unexpected locations. At the same time there was competition among the diplomatic staff posted in Central Asian centres. Notably Kashgar became one of the hotspots of confrontation where a weak Chinese administration fell prey to the powerful representatives of the superpowers: The Russian Consul M. Petrovsky and his British counterpart George Macartney were the protagonists who informed their respective governments in detailed reports which give us historical evidence on the socio-economic conditions in Central Asia, as well as strategic and military intelligence during their rivalry. Nomads posed an uncertainty in their equations, therefore control of their movements and final sedentarisation were the attempted solution for the future.

The lower end of social hierarchies was composed of nomadic tribes who managed to evade strict control and followed a life-style which was perceived as one of utmost backwardness. Strategic control of vast tracts of deserts and steppes needed a coming to terms with mobile groups and their eventual domination. The second half of the nineteenth century experienced a heated debate in political and academic circles about the effects of the Anglo-Russian rivalry in Central Asia.

The *Great Game* in its narrow definition came to an end in 1907 without any military encounter or loss of lives. Russia and Great Britain negotiated the so-called Anglo-Russian Convention in which respective spheres of influence, buffer states and regions of non-interference were agreed upon (Fig. 9.1). Instrumental for the accord was the 'heartland theory' which drew geopolitical significance towards Central Asia.

The Geographer Mackinder formulated his 'heartland theory' in 1904 which became a hugely influential text on geopolitical debate. Mackinder drew prime attention towards Central Asia as he stated that the Tsarist regional dominance was linked to their equestrian tradition from nomadic Asian backgrounds. From the safe retreat of the Inner Asian steppe regions, conquests had begun towards Europe, Persia, India and China. He described European civilisation as the result of secular battles against Asian invasions (Mackinder 1904: 423). Imperial control of world trade by Great Britain, largely based on its naval predominance, was challenged by unfolding land-based transport infrastructure. The Russian railways were perceived as the successors of the equestrian mobile forces. Central Asia had become the arena of contest, all the more as a Russian-German and/or a Sino-Japanese alliance could contribute to a shift of world affairs to the 'heartland' of the Eurasian continent, which he perceived as a 'geographical pivot of history' (Mackinder 1904: 436). He predicted the transformation of Central Asia from a steppe region with little economic power into a region of prime geostrategic importance. Culture and geography would contribute to its key position. Mackinder identified four adjacent regions encompassing the heartland of 'pagan' Turan in the shape of a crescent and denominated by religious affiliations: Buddhism, Brahmanism, Islam and Christianity (Mackinder 1904: 431).[3]

Similar ideas of a Central Asian 'heartland' or a pivotal role stimulated Lattimore's perceptions in his book 'Pivot of Asia' (1950). Keeping the experiences of World War II in mind Lattimore drew a circle with a diameter of 1000 miles around Urumchi and identified Central Asia as a 'whirlpool' stirred-up by 'political currents flowing from China, Russia, India and the Middle East' (Lattimore 1950: 3). By following the same Central Asian-centred approach, Hauner shifted the centre in the 1980s to Kabul, drew a similar circle and identified a world of 'even greater contrasts' which 'touches upon the volatile and oil-rich region of the Middle East' (Hauner 1989: 7). The last statement has remained valid through the dissolution of the Soviet Union, the Taliban rule in Afghanistan, the aftermath of 9/11 and the Iraq crisis. The fact that Rashid (2000) subtitled his book on the Taliban as 'Islam, Oil and the New Great Game in Central Asia' is only one case in point for the reference to the *Great Game* connotation of contemporary geopolitical problems in the region.[4] The presence of American and Russian troops at airports and along borders in Central Asia proves the continuing geopolitical significance of the region and its linkage to contemporary crises zones.

What are the effects of certain lines of thought and resulting political actions on Central Asia and why do we still refer to the metaphor of a *Great Game* when discussing contemporary strategic interference and socio-economic transformations in geopolitical contexts? Boundary making, and its impact on nation-building and economic and political participation, severely influenced socio-economic developments in peripheral and nomadic areas of Central Asia.

---

3   With the passage of time Mackinder modified his theory due to the influence events during the First and Second World Wars had on him. He also influenced the thoughts of Karl Haushofer and other geopoliticians of his time.

4   Kreutzmann (1997, 2004, 2008, 2009a), Roy (2000).

Some cases in point need to be introduced to better understand the far-reaching consequences of imperial border delineations. First of all, the practical impact on trade relations and economic exchange need to be investigated.

The Aftermath of the *Great Game*

In Central Asia the *Great Game* resulted in the demarcation of international boundaries separating the spheres of influence of the super powers of the time. Great Britain and Russia executed direct control and established domination in the core areas of their empires while they created buffer states such as Persia and Afghanistan (Fig. 9.1) at the periphery. In their negotiations they excluded Kashgaria or eastern Turkestan which nominally was under Chinese administration. Trade between South and Central Asia was affected by this constellation and a rivalry had developed since British commercial interests entered this sector in 1874 (Davis and Huttenback 1987, Kreutzmann 1998). Both super powers competed for dominance on the valuable markets in urban oases of the Silk Road such as Kashgar and Yarkand. According to the theory of imperialism, the merchants of the industrialising countries tried to purchase raw materials such as cotton, hashish, and pashmina wool from nomadic production while, in exchange, textiles and manufactured products were offered in the bazaars (Kreutzmann 1998). Russia had some advantage, as access to markets and producers was easier. From the railhead at Andijan in the Ferghana Valley, which was linked to the Middle Asian Railway in 1899 the distance to Kashgar (554 km) could be covered in twelve marches via Osh, Irkeshtam, and Ulugchat by crossing only one major pass, Terek Dawan (3,870 m). On the other hand trade caravans from British India had to follow either of three trans-montane passages – the Leh, Gilgit, and Chitral routes – which were much longer and more difficult.

The competition for the Central Asian markets continued after the October Revolution (1917) which caused the closure of the Russian/Soviet Consulate in Kashgar from 1920-1925. This event affected Soviet commerce with Kashgaria detrimentally while the British share soared. Overall trade significantly declined due to the disturbances in Chinese Turkestan after 1935 and later due to World War II and the Chinese Revolution. Central Asian trade had become an important factor in cross-boundary relations affecting the economies in the regions traversed for a period of 40 years. The total annual volume of Indo-Xinjiang commercial exchange surpassed two million rupees for most of the era between 1895 and 1934.

At the end of the nineteenth century George Macartney, the British Consul-General in Kashgar, summarised the situation:

> The demand for Russian goods is without doubt ever increasing. Cotton prints of Moscow manufacture, as cheap as they are varied and pretty, are very largely imported. The bazaars of every town are overstocked with them, as well as with a multitude of other articles, amongst the most important of which may be

mentioned lamps, candles, soap, petroleum, honey, sugar, sweetmeats, porcelain cups, tumblers, enamelled iron plates, matches, knives and silks. These articles, with few exceptions, could, but for the competition, be supplied from India. But we have gradually had to relinquish our position in favour of Russia, until at last our trade has to confine itself chiefly to articles of which we are the sole producers and in which there is no competition.[5]

British interests in securing a substantial share in this commercial exchange governed their imperial designs and had an impact on the mountain societies involved. At the turn of the century Ladakh and Baltistan were dominated by the Maharaja of Kashmir, Gilgit had become an agency (re-established in 1889) under the joint administration of a British Political Agent and a Kashmiri Wazir-i-Wazarat. Principalities such as Hunza and Nager were affiliated after their defeat in the 1891 encounters, which were fought under the pretext of opening the Gilgit route for commercial purposes. At the same time the Mehtar of Chitral transferred his sovereignty in external affairs to a British Agent and was remunerated with an annual subsidy and a supply of arms.

This part of the region under study was controlled and de facto commercially incorporated in the British Indian exchange system. Trade with Afghanistan followed its own rules and became part of the special arrangements with the ruling Amir in Kabul. A major hiatus occurred in the aftermath of the October Revolution when a process of separation and isolation began. The economic relations of the Soviet-dominated Central Asian regions were re-directed and amplified towards Russia while, at the same time, international borders were sealed and became effective barriers for trade. This process took time and lasted until the mid-1930s. With growing alienation between the Soviet Empire and the Chinese-dominated part of eastern Turkestan all exchange relations between Tajikistan-Kyrgyzstan and Kashgaria came to a halt by 1930.[6] The undercutting of bazaar prices through the provision of cheaper commodities of the same quality in kolchoz shops – the collective farms provided goods of daily use at sub-market rates in state-run shops – led to the termination of trade in this sector. Nomadic production was confined to the limits of new territorial entities. Trade across borders nearly ceased to exist. Similar developments took place on the Soviet border with Afghanistan during the 1930s:

> During the past few years, the effect of Soviet policy has been to restrict, in an increasing degree, traffic, excepting state-controlled trade, from Soviet Central Asia across the Afghan frontier on the river Oxus ... more European Russian officers have been appointed to ensure that the frontier is effectively closed (IOL/P&S/12/2275, dated 13.10.1939).

---

5   Report of George Macartney of 1 October 1898, quoted from Captain K.C. Packman, Consul-General at Kashgar, 1937, Trade Report. India Office Library & Records. *Departmental Papers: Political & Secret Internal Files & Collections 1931-1947*. IOL/P&S/12/2354, 1.

6   Kreutzmann (1996: 179), cf. Kreutzmann (2009b).

The result was that border delineation and the establishment of different socio-political regimes affected a collapse of trade and exchange in this Central Asian region, which lasted for nearly 60 years until the end of the Cold War. With few exceptions traditional trade links and exchange routes were interrupted for two generations and are only becoming reanimated at a slow pace.

Boundaries in the Making

A few examples from the turn of the century illustrate how mountain and pastoral regions have been involved in the demarcation of spheres of influence. The contenders of the *Great Game* in High Asia agreed to lay down boundaries in the comparatively sparsely populated regions of the Hindukush and Pamir. Some-times these borders were described as natural frontiers, scientific boundaries and dialect borders. The Durand Line of 1893 separating Afghanistan from British India/Pakistan, epitomises such an effort and has continued to function as the symbol of colonial border delineation referred to as the 'dividing line' (Felmy 1993). In order to safeguard the physical separation of two imperial opponents, international borders were outlined and Afghanistan was created as a buffer state (Fig. 9.2). In this process local livelihoods and regional interests were neglected. The Pashtun settlement region was divided into two parts by an arbitrary line through the Hindukush ranges. This cut through traditional mi-gratory paths of seasonal nomads between the Central Afghanistan highlands and the Indus lowlands. Numerous clashes between tribal groups and imperial troops characterised political relations in the borderlands serving as a buffer belt on the fringe of the empire (Fraser-Tytler 1953). The Durand Line remains to this day a contested border and is a source of contempt to both Pakistan and Afghanistan – at the same time it is a battleground for Taliban and US forces, where drone attacks against suspected enemy strongholds are the present strategy of containment. The movement of nomads (*powindah*) and their herds now is severely affected and depends on the state of bilateral political relations and though restricted has not ceased. Even the numbers of nomads recorded in Afghanistan are the cause of major disputes (Tapper 2008). A recent 'National Multi Sectoral Assessment of Kuchi' fixed the number of nomads at more than 2.5 million for Afghanistan in 2004 (de Wejer 2005). Observers regard this as a purely political figure, which seems to be fivefold exaggerated (Glatzer 2006). Nevertheless, the report rightly stresses the further reduction of nomadic space and the limitations to mobility within the country and across borders (Kreutzmann and Schütte 2011).

Irredentism About Pashtunistan

Continuing border disputes and conflicts, like the claim for land by the irredentistic movement for 'Pashtunistan' (Fig. 9.3), are still thriving (one of the main squares in Kabul is named after this Pak-Afghan dispute ongoing since the 1960s). The

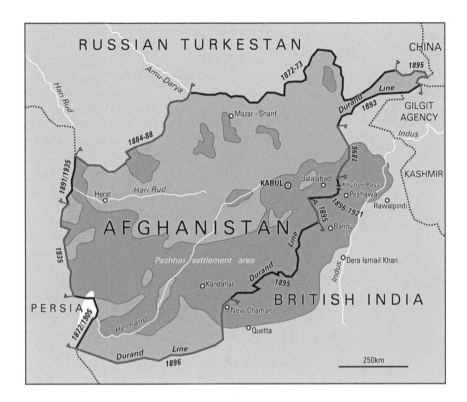

Figure 9.2   Demarcation of Afghanistan's boundaries and the partition of the Pashtun settlement region

Source: Kreutzmann (1997: 176).

Afghan demand for a territory named Pashtunistan and consisting of the Pakistan North-West Frontier Province and Baluchistan (including the tribal areas) is the result of the imperial design that led to the creation of the Durand Line and the referendum at the end of British Rule in India. Pashtun representatives use these incidents to politicise people towards the Pashtunistan cause. Imperial legacies and losses function as a measure of identity and supply the ideological platforms for charismatic leaders who mobilise their followership in order to re-write history. The Durand Line, as an acknowledged international boundary, has long been a cause for discontent and political crises between the neighbours Afghanistan and Pakistan, and will remain so in the future.

Wakhan as the Symbol of Division

The Wakhan Corridor of north-eastern Afghanistan symbolises colonial border delineation. The southern limit is formed by the Durand Line, while the northern part came into existence as a result of the Pamir Boundary Commission of 1895, in

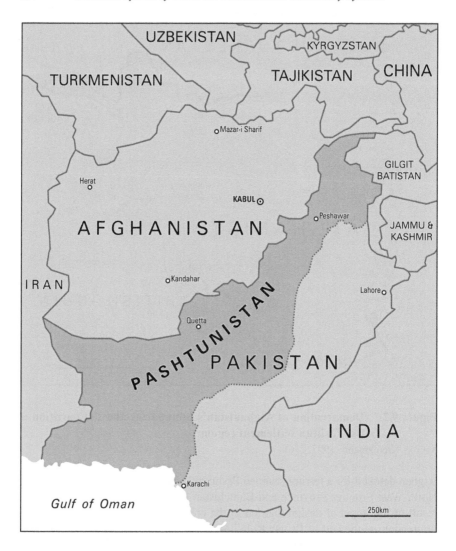

Figure 9.3   The spatial outline of Pashtunistan irredentism
Source: Abridged after Humlum (1959: 362).

which Russian and British officers negotiated the alignment, and Afghan officials
assisted in the demarcation (Fig. 9.4). This narrow 300 km-long and only 15-
75 km wide strip was created to separate Russian and British spheres of influence,
and fulfilled the function of averting direct military action between the two
superpowers. Part of the boundary follows the course of the Pjandsh (Amu Darya
River), according to the fashion of the time. The 'stromstrich' boundary followed
a role model tested in other regions of the world where rivers were accepted and
identified as separating entities (e.g. the Rhine).

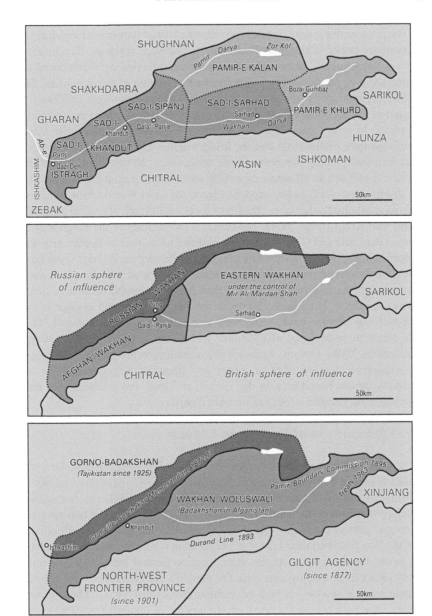

Figure 9.4   Concepts and realisation of the division of Wakhan: (I) Wakhan
prior 1883 (Wakhan four administrative districts: Istragh,
Kandut, Sipanj, Sarhad. Pamir-e Kalan and Pamir-e Khurd no
permanent settlement), (II) Plan for division and boundaries, (III)
Boundary delineation 1895 and existing international borders
Source: Kreutzmann (1996: 102), adapted.

The price of this colonial endeavour was the spatial partition of regional semi-autonomous principalities like Badakhshan, Darwaz, Wakhan, Shughnan, and Roshan. Subsequently both parts of each former principality experienced quite diverse socio-economic developments as part of greater political entities. Today we find regional units of the principalities in Afghanistan and Tajikistan. The creation of these boundaries resulted in immediate refugee movements by ethnic minorities. In recent years relatives separated by a century-old border have re-established their relationship and the bridges across the Pjandsh river in Langar, Ishkashim and Khorog symbolise those endeavours. Nevertheless, the effect of partition is still felt in all areas, especially when international borders are closed and strictly controlled, as has happened since the Cold War. Afghan Wakhan has suffered substantially due to its 'dead-end' location and lack of trading possibilities (Felmy and Kreutzmann 2004). The seasonal movements of the Kirghiz nomads living in the Little and Great Pamirs are confined to the narrow border strip within Afghanistan. Previous long-distance caravans to the commercial centres of Central Asia are only part of oral traditions, even the annual trade missions to the Kabul Bazaar have been interrupted for more than two decades. Nevertheless, itinerant traders (*saudegar*) have begun to reach the Pamirs again and engage in barter trade with the Khirgiz, exchanging basic goods and opium for livestock. Opium trade seems to be a source of additional income for transporters and smugglers (Kreutzmann 2008). Opium, weapons, and ammunition are highly volatile goods that require local knowledge of routes and transport arrangements for to be distributed across boundaries. In this context residents and pastoralists in remote mountain areas become a focus of interested parties.

China's Boundary with Afghanistan and Tajikistan

The missing link between both borders is the short Sino-Afghan boundary, which in itself is part of a disputed frontier. According to the Chinese, their border with Afghanistan and Tajikistan extends much further west, while the contemporary boundary is agreed on by China's neighbours (Fig. 9.5). All these borders formed an integral part of the major global divide after World War II. The frontlines of the Cold War followed their historical predecessors. Western and eastern alliances met in the Pamirian knot. At the same location neutral states like Afghanistan (up to 1978) lived side by side with the People's Republic of China, following an independent anti-Soviet path of communism (since 1958). Thus, a remote pastoral mountain region became a meeting-point of opposing political systems. The recent alleviation of this confrontation did not terminate any military action in the region. The Pamir boundary presently separates the newly independent state of Tajikistan (since 1991) from Afghanistan while their boundary was controlled by Russian troops for more than a decade after independence. The previous global confrontation has been replaced by regional conflicts. Nevertheless these examples are not singular cases. Nearly all borders of the Hindukush-Himalayan arc are under dispute by one or the other side. A symbol of contest and failing

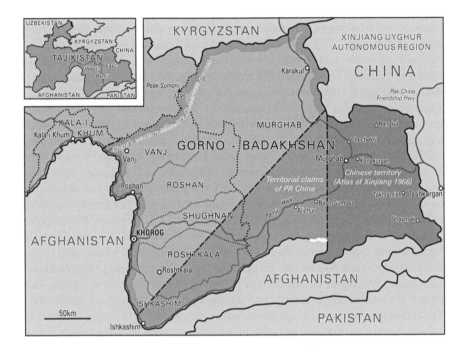

Figure 9.5   Chinese territorial claims towards Tajikistan
Source: Kreutzmann (2004).

reconciliation is the so-called 'system': The system identifies a demilitarised zone, which includes a 30 km-wide border strip with metal fences at both sides. Thus, communication and grazing across boundaries and within the 'system' is impossible. Kirghiz communities were restricted to their respective countries. In addition they lost valuable grazing grounds in the neutralised zone fenced by barbed-wire. Boundary disputes continue to hamper modern communication. The Kulma Pass Road, connecting Murghab in Tajikistan with Tashkurgan in China, epitomises this. The road is physically there, currently exchange and commerce are limited to the two neighbouring countries, with border markets the exception rather than the rule.

Border Disputes Within the Soviet Union and After Its Dissolution

Soviet policies attempted to create new republics, representing the ethnic groups of Central Asia in adequate spatial and administrative settings. Consequently by 1929, republics were created to represent Kazakhs, Kirghiz, Tajiks, Uzbeks, and Turkmens. The new republics did not have any boundaries in common with their predecessors, the Khanate of Khiva, the Khanate of Bukhara, and the Turkestan Governorate-General. If the term 'artificial boundaries' could be appropriate in any context, it would be here where Stalinist obsession with creating congruencies of ethnicity and territory grossly failed. Ethnic diversity did not allow for designing spatial expressions.[7] The newly defined republics consisted of a spatial nucleus, but very often they had in addition satellite territories of enclaves and exclaves within the territory of neighbouring republics (Fig. 9.6). While this phenomenon did not pose grave differences during the period of the Soviet Union – basically all territories were under the central command of the Kremlin and only international boundaries with neighbouring countries such as China and Afghanistan were of any importance and hermetically sealed – if it was another cause of dissent after independence in the early 1990s. Republican boundaries within the Soviet Union became international borders of sovereign states such as Uzbekistan, Tajikistan, and Kyrgyzstan. In a survey two years after independence and the dissolution of the Soviet Union, the Moscow Institute of Political Geography recorded 180 border and territorial disputes (Halbach 1992: 5). To illustrate the scope of conditions and demands a few cases are listed. The irredentistic movement in Turkmenistan expects Uzbekistan to 'return' the territory of the Khanates of Khiva and Khorezm. Tajik nationalists demand the 'return' of Samarkand and Bukhara. Uzbekistan lays claim on the eastern part of the Ferghana Valley, i.e. the Osh Oblast, the present-day economic and commercial centre of southern Kyrgyzstan. The Uzbek government does not permit colleagues from neighbouring republics to consult the archival material in Tashkent, which documents the boundary decisions from the 1920s. Rental arrangements and the production of natural resources in exclaves from Soviet times are under dispute, such as the Uzbek exploitation of oil and gas fields in southern Kyrgyzstan and the deviation of irrigation water from the Andijan reservoir towards the Ferghana Valley (Fig. 9.6). The Ferghana Valley alone contains seven enclaves through which major traffic routes pass. Freedom of travel is more restricted than before as new visa regulations have been introduced. Some of these measures have been justified in the aftermath of attacks from Afghanistan-trained rebels, who plundered Tajik and Kirghiz villages on their way to the Ferghana Valley in 1999 and 2000. The future of rented lands and exclaves that were created for the protection of ethnic minorities is at stake and neighbouring governments are discussing options for forced evacuation and migration to initiate population exchange. One example of this is the Tajik enclave of Sary Mogol in the Alai Valley, where additional fodder produced for the Kirghiz pastoralists of Murghab, had to be returned to Kyrgyzstan. The residents were offered a choice of citizenships, finally in 2004 they opted for Kyrgyzstan and continue to settle in Sary Mogol. Consequently, the pastoral system in the eastern Pamirs has changed. Households depend on local resource utilisation as much as on the additional income from migrants' remittances, entrepreneurship, services and trade.

---

7   Like in many other places on earth this approach failed. In the case of Central Asia the denominated and name-giving groups represented roughly two thirds of the respective republics' population even when assessed to Soviet standards of ethnic attribution (Landau and Kellner-Henkele 2001).

Figure 9.6   Isolated exclaves in the borderlands of Tajikistan, Kyrgyzstan and Uzbekistan

Source: Kreutzmann (2004).

Future Prospects and Conflict Resolution

The hope of friendly relations and mutual understanding occurring has suffered several setbacks in recent years. All negotiating partners are interested in obtaining the most advantages possible to their side. On a regional scale there is some hope in the Shanghai Cooperation Organization (SCO) which was founded in 1996 as the Shanghai-5 (Russia, PR of China, Kazakhstan, Kyrgyzstan, Tajikistan) and became a fully-fledged organisation under the name of SCO in 2001 when Uzbekistan joined.[8] Its mandate is to improve mutual relations and to improve Central Asia's economic competitiveness in a globalised world. Therefore the SCO has supported the opening of new trade corridors between the PR of China and Kyrgyzstan (Irkeshtam Road) and Tajikistan (Khulma Road) respectively. The two major regional players – Russia and PR of China – have cooperated with the European Union to link the Central Asian republics with Europe through a road network (TRACEA route) via the Caucasus. Participation in regional and international trade may be a prime stimulus to overcoming the legacies of geopolitical interference. Trade will also reflect the economic interests of the big economic players of today in the future of Central Asia.

Nevertheless, the region suffered not only directly from Cold War confrontations, but also from regional problems, which have developed into conflicts between neighbours, an unfortunate colonial legacy. After more than 50 years of independence India and Pakistan are still engaged in military confrontation

8   The SCO became an internationally acknowledged organization in 2004 and operates a secretariat from Beijing.

that affects economic exchange tremendously and keeps the mountain regions of the Karakoram and western Himalaya in a state of dispute and uncertainty.

Conclusion

The starting point of our deliberations was the external interest by colonial powers in the Central Asian periphery, with long-lasting implications for the livelihoods of people. The major impact has been the delineation of international boundaries and internal borders. Most of the mountain and pastoral regions became even more peripheral after border demarcation and lost their economic value as a transit region for traders and herders. The deadlock situation between competing powers has partly changed since the end of the Cold War, but not into a great style of regional cooperation.

The second external intervention had even greater impact, especially on Kyrgyzstan and Tajikistan. About 70 years ago a major transformation of socio-economic conditions took place as the Soviet modernisation project changed lifestyles and civil rights. To quote contemporary sources on the contents of the project:

> The CPC [Communist Party Committee] of the Tajik S.S.R. is drawing up a plan for agriculture in the Pamirs, the idea being thereby to transform the migrant tribes into stationary inhabitants and to encourage them to grow their own food instead of importing it. A biological station on the Pamirs, at a height of 4,000 metres above sea-level, is just being started (Pravda 7.5.1934, quoted after IOL/ P&S/12/2273).

The 'Pravda' records nothing less than the truth: Modernisation meant the sedentarisation of nomads which was executed with great force and rigour. The effects of settlement and the introduction of 'modern' animal husbandry can be observed in all areas north of the Amu Darya, while on the southern bank of the river 'traditional' forms of livestock-keeping prevail.

Similar developments can be observed in people's organisation, education and agriculture. To quote again a source from 1934:

> Khorog is the capital town of the Soviet Pamir, and there has been held there the 5th congress of the Soviets of the mountainous Badakhshan region. On foot on horses, on yaks, on donkeys, along mountain tracks hanging over precipices, the delegates come from the distant Murghab, Borgang [Bartang], Bakhan [Wakhan], and other places in the S. and E. edges of the U.S.S.R. that border with Afghanistan, India and western China. The 110 delegates elected were 78 Tajiks, 16 Kirghiz, and 16 Russians. In the conference hall were many women in their white garments of homespun silk. Khorog is now lit with electricity that was started and first seen by the Pamir people in the spring of this year.

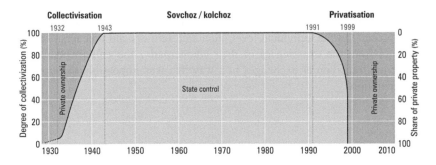

Figure 9.7   Transformation of land ownership and property rights in Tajikistan

Source: Kreutzmann (1996: 173), modified.

The president of the congress, Faisilbekov, spoke of the wonderful things that have taken place in the Soviet Pamir. Aeroplanes are flying over inaccessible mountain ranges, a splendid automobile road has been made from Khorog to Osh, 700 km long, that now links the Pamir with the rest of the U.S.S.R. Formerly there was only 1 school in the whole of the Pamirs – now there are 140, and a training school for teachers: instead of dark smoky earth huts or skin tents, European houses are now being built: collective farms are established in the Pamirs, and they are growing and getting good crops of wheat, millet and beans; and now they know how to manure their fields and be sure of good crops (Izvestia 29. 11.1934, quoted after IOL/P&S/12/2273).

It is the irony of history that now a transformation process has started which attempts to revert these reforms and to privatise collectivised property again (Fig. 9.7), in which households return to the farming practices of their grandfathers, and in which the traditional knowledge of neighbouring countries is adapted as a measure to overcome food crisis situations and to minimise risks. In that respect the external interference in Central Asia is a failed attempt to implement modernisation theory, while in many other aspects it succeeded. The transition, beginning with the independence of sovereign nation states in Central Asia, has failed so far to continue the path of modernisation in a post-Soviet society. Pastoralists utilise natural pastures as a coping mechanism for survival in times of economic and political crises. Just across the border in the PR of China, modernisation strategy is continued in an attempt to resettle Kirghiz pastoralists in centralised winter townships with social infrastructure and external support.

The lesson to be learnt from geopolitical interventions in peripheral mountain areas is that decisions made in the core of empires always affect the livelihoods of people who have not been involved in the decision-making process. Socio-political interference led to the creation of an arena of confrontation in the Pamirs, Hindukush and the Himalayas during the Cold War, which was one of the least permeable frontier regions in the world. Present developments might result in an evening out of living conditions, income patterns and indicators of human development. On the individual level, mountain farmers and breeders especially can learn from the experiences of their counterparts, and entrepreneurs might profit from trans-border exchanges in a way that was impossible for more than two generations.

References

Bregel, Y. 2003. *An Historical Atlas of Central Asia.* (Handbook of Oriental Studies, Section 8: Central Asia, Vol. 9). Leiden/Boston: Brill.

Christian, D. 2000. Silk Roads or Steppe Roads? The Silk Roads in world history. *Journal of World History*, 11(1), 1-26.

Davis, L.E. and Huttenback, R.M. 1987. *Mammon and the Pursuit of Empire: The Political Economy of British Imperialism, 1860-1912.* Cambridge: Cambridge University Press.

Ehlers, E. and Kreutzmann, H. 2000. *High Mountain Pastoralism in Northern Pakistan.* (Erdkundliches Wissen 132). Stuttgart: Franz Steiner-Verlag.

Felmy, S. 1993. The dividing line. *Newsline*, 5(5-6), 72-78.

Felmy, S. and Kreutzmann, H. 2004. Wakhan Woluswali in Badakhshan. Observations and reflections from Afghanistan's periphery. *Erdkunde*, 58(2), 97-117.

Fraser-Tytler, W.K. 1953. *Afghanistan. A Study of Political Developments in Central and Southern Asia.* London, New York and Toronto: Oxford University Press.

Glatzer, B. 2006. Wie viele Nomaden gibt es in Afghanistan? *Afghanistan-Info*, 58, 12-14.

Halbach, U. 1992. *Ethno-territoriale Konflikte in der GUS.* (Berichte des Bundesinstituts für ostwissenschaftliche und internationale Studien 31-1992). Köln: Selbstverlag.

Hauner, M. 1989. Central Asian geopolitics in the last hundred years: A critical survey from Gorchakov to Gorbachev. *Central Asian Survey*, 8, 1-19.

Hauner, M. 1992. *What is Asia to Us? Russia's Heartland Yesterday and Today.* London: Routledge.

Humlum, J. 1959. *La géographie de l'Afghanistan.* Copenhague: Gyldendal.

India Office Library & Records. *Departmental Papers: Political & Secret Internal Files & Collections 1931-1947.* IOL/P&S/12/2273, 2275, 2354.

Kreutzmann, H. 1996. *Ethnizität im Entwicklungsprozeß. Die Wakhi in Hochasien.* Berlin: Dietrich Reimer-Verlag.

Kreutzmann, H. 1997. Vom 'Great Game' zum 'Clash of Civilizations'? Wahrnehmung und Wirkung von Imperialpolitik und Grenzziehungen in Zentralasien. *Petermanns Geographische Mitteilungen*, 141(3), 163-186.

Kreutzmann, H. 1998. The Chitral Triangle: Rise and decline of trans-montane Central Asian trade, 1895-1935. *Asien-Afrika-Lateinamerika*, 26(3), 289-327.

Kreutzmann, H. 2003. Ethnic minorities and marginality in the Pamirian Knot. Survival of Wakhi and Kirghiz in a harsh environment and global contexts. *The Geographical Journal*, 169(3), 215-235.

Kreutzmann, H. 2004. Ellsworth Huntington and his perspective on Central Asia. Great Game experiences and their influence on development thought. *GeoJournal*, 59, 27-31.

Kreutzmann, H. 2008. Kashmir and the northern areas of Pakistan: Boundary-making along contested frontiers. *Erdkunde*, 62(3), 201-219.

Kreutzmann, H. 2009a. Geopolitical perspectives on cross-border exchange relations, in *Experiences with and Prospects for Regional Exchange and Cooperation in Mountain Areas*, edited by H. Kreutzmann et al. Bonn: InWEnt, 20-57.

Kreutzmann, H. 2009b. Transformations of high mountain pastoral strategies in the Pamirian Knot. *Nomadic Peoples*, 13(2), 102-123.

Kreutzmann, H. and Schütte, S. 2011. Contested commons: Multiple insecurities of pastoralists in Northeastern Afghanistan. *Erdkunde*, 65(2), 99-119.

Landau, J.M. and Kellner-Heinkele, B. 2001. *Politics of Language in the Ex-Soviet Muslim States. Azerbayjan, Uzbekistan, Kazakhstan, Kyrgyzstan, Turkmenistan and Tajikistan*. Ann Arbor: University of Michigan Press.

Lattimore, O. 1950. *Pivot of Asia. Sinkiang and the Inner Asian Frontiers of China and Russia*. Boston: Little, Brown & Co.

Mackinder, H.J. 1904. The geographical pivot of history. *The Geographical Journal*, 23(4), 421-444.

Rashid, A. 2000. *Taliban. Islam, Oil and the New Great Game in Central Asia*. London/New York: Tauris.

Roy, O. 2000. *The New Central Asia: The Creation of Nations*. London/New York: Tauris.

Tapper, R. 2008. Who are the Kuchi? Nomad self-identities in Afghanistan. *Journal of the Royal Anthropological Institute*, 14, 97-116.

Wejer, F. de. 2005. *National Multi Sectorial Assessment of Kuchi. Main Findings*. [Online]. Available at: www.mrrd.gov.af/vau/nmak/htm [accessed: 30 September 2005].

Chapter 10

Mongolian Pastoral Economy and its Integration into the World Market Under Socialist and Post-Socialist Conditions

Jörg Janzen

Mobile Livestock Keeping

A considerable section of the population in Mongolia relies heavily on mobile livestock keeping for a living;[1] livestock production and marketing continues to play a crucial role, even in post-socialist times. Despite difficult conditions it is the most important source of income in the agricultural sector. It not only feeds the livestock producers and many of their urban-based relatives but also provides an adequate living for most of them through the sale of animals and animal products. This chapter analyses the driving factors of realigning commodity chains and their consequences, alongside world market integration of Mongolian livestock producers during the socialist and post-socialist period. It argues that new market structures foster social differentiation that shapes the Mongolian pastoral economy of today.

Changes in Mobile Livestock Keeping[2]

After seven decades of dependence on the Soviet Union (USSR) and virtual isolation in political and economic terms, the reforms introduced in Mongolia since 1990 to establish democracy and a market economy have brought about far-reaching changes in the society and all sectors of the economy. This is particularly

---

1   Cf. Finke (2004), Janzen (2002, 2005a, 2005b), Janzen and Bazargur (1999, 2003a, 2003b), Müller (1994, 1999a, 1999b), Müller and Janzen (1997).

2   The term 'mobile livestock keeping' is used as the 'traditional' nomadism that originated in the arid zones of the Old World Dry Belt – the 'Sozio-ökologische Kulturweise Nomadismus' (Scholz 1995) – has, in most cases, lost its original character, and indeed in many places has declined or no longer exists. Still, mobile livestock keeping continues to play an important role. Its significance in some countries has even grown. This is not only true for Mongolia during the 1990s but also for the countries of Sub-Saharan Africa including Somalia, where agro-pastoralism (mobile livestock keeping in combination with farming) seems to have increased (Janzen 2001).

true in the sphere of mobile livestock production. In the 1990s the rural sector was neglected by the government of the day and left to fend for itself in the free-for-all of a developing market economy. Cancellation of state subsidies corroded the quality of infrastructure and services. Many of the population's needs could not be adequately met. The result was a noticeable deterioration of living and production conditions for most of the rural population. Livestock production, formerly largely export market oriented, was reduced to little more than a simple subsistence activity. Only wealthier herder households continued to produce for the market; notably cashmere hair from goats, sheep wool, hides, and skins.

When the collectively owned herds of the livestock production cooperatives (*negdel*) were privatised, the flocks comprising sheep, goats, cattle and yaks, horses and camels were distributed amongst its members. Because *negdel* employees from all walks of life received animals from the cooperative herds, many families, unable to find non-pastoral employment, abandoned their settled existence and took up a new life as herdsmen. This caused the number of mobile livestock keeping households, now with mixed herds, to nearly triple across Mongolia compared to during the socialist era. However, many of these 'new nomads' (Müller 1997) lacked experience in livestock keeping. Subsequently this group is much more vulnerable to adverse conditions than the experienced herdsmen from the former *negdel*. As a result of deteriorating grazing conditions, and in the wake of natural disasters, the number of these 'new nomads' has been in decline since the late 1990s. By the end of 2009 however, there were still 170,142 herdsmen households in the country, which equalled 27 per cent of total Mongolian households (NSOM 2010: 208).

Mobile livestock keeping in Mongolia is conducted under difficult climatic conditions. Natural phenomena such as drought (*gan*) in the summer season and heavy snowfalls accompanied by very low temperatures (*zud*) in the winter are common. Large areas of west and central Mongolia (Fig. 10.1) were exposed to these extreme conditions over three consecutive years from 1999 and once again in winter 2009/2010, causing heavy stock losses (Fig. 10.2, Fig. 10.3). But the vicissitudes of climate alone cannot be blamed for the death of millions of animals. They were also caused by the mobile stock keepers' grazing practices, which have altered under the new conditions of the transformation period and are often not ecologically sound. In addition, agriculture – both farming and livestock producing – and rural infrastructure were severely neglected in the 1990s, despite rural development projects. This caused increasing rural-urban disparities, and fostered heavy migration from the land into the cities, foremost to the capital Ulaanbaatar (Janzen et al. 2002, 2005).

The out-migration of mainly poorer livestock keepers from rural areas has increased the share of wealthier and rich herders. The owners of large herds are generally experienced livestock breeders who did not lose as many animals during the natural hazards as the smaller and often unskilled herders. Having large herds the wealthy animal keepers are forced to make frequent and long-range movements with their livestock, which is ecologically more viable. Often households with few

Figure 10.1   Mongolia: Cities and administrative boundaries on *aimag* level
Source: ALAGAG 2004. Geographic Atlas of Mongolia: 10/11, Ulaanbaatar.

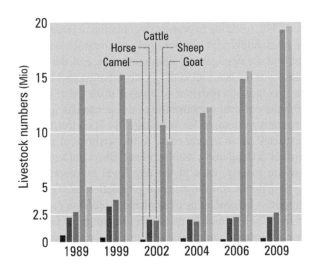

Figure 10.2   Mongolia: Livestock numbers between 1989 and 2009
Source: NSOM (2000-2010).

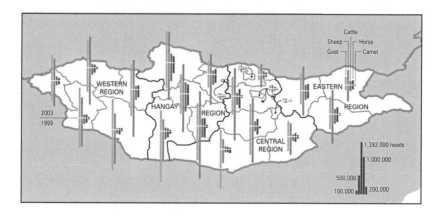

**Figure 10.3   Mongolia: Number of livestock on *aimag* level in 1999 and 2003
– Before and after *Gan* and *Zud***

Source: NSOM (2004).

or no livestock work for the herders with large flocks (Batbuyan 2003). By the end of 2005 only 1,570 households nationwide (0.7 per cent of the total number of Mongolian herder households) kept more than 1,000 head of livestock (NSOM 2006: 177).

Spatial Organisation and Seasonal Pasture Use

The frequency and distance of pastoral movements of many herder households have decreased since 1990, resulting in pasture degradation in large parts of the country. Many pastoral households with smaller flocks of livestock want to avoid the high transport costs, and are negligent or simply not aware of the fact that a high rate of pastoral mobility is the basis for ecologically adapted pasture management. The administrative structure of a province (*aimag*) is made up of a number of districts (*sum*). Although the political organisation of a *sum* has changed since the days of the planned economy system, as far as pasture use is concerned the herdsmen are still largely restricted, as in *negdel* time, to the territory of their respective *sum*. Each *sum* is further divided into a number of sub-districts (*bag*). The arrangements for pasture use are mainly made at the *bag* level in cooperation with the *bag* governor, *bag* parliament and the pastoral population as well as representatives of informal groups or communities of herdsmen (*malchin buleg*: loose unions of related and/or friendly livestock keepers). These groups of herdsmen consist of a number of *khot ail* (camp communities) who usually live grouped together in a particular area, often in a valley or around a water source. Each *khot ail* consists of several *ail* (households) that have joined forces for reasons of efficiency. In contrast to the socialist era, movement across territorial boundaries has increased and is regulated by oral and written agreements between the respective governors. 'Illegal' border crossings often result in disputes over pasture and water.

Ideally, the four seasonal pasture areas (winter, spring, summer and autumn) are in different locations, divided into smaller units, and used by the different groups of livestock keepers. The winter grazing area is of major importance and centres around permanent winter sheds or shelters made of wood and stone that usually date from the *negdel* era. Whereas the grazing land belongs to the state and the herdsmen only have use rights, these buildings are the registered and transferable property of a livestock keeping household. Occasionally herders also own shelters in the spring pasture or use their winter shelters in spring as well. However, only about 70-80 per cent of households have a winter shed. The other 20-30 per cent find accommodation with relatives and friends or move into a *sum* centre for the winter, where most of the animal herders own land in the form of *khashaa* (rectangular compounds surrounded by wooden fences) with a *ger* (yurt) and/or a wooden house.

During the other three seasons the herder groups ideally use different specified grazing areas whose limitations remain more or less constant over longer periods of time. Migration is supposed to follow a schedule agreed on between the herdsmen groups and the administration. If it is not adhered to conflict can arise during movement through the pasture areas of other livestock keeping groups. Depending on the spatial distribution of ecological zones, long-distance horizontal movements of 100 km and more can occur between the summer and winter pasture. This applies in particular to those groups of herdsmen whose areas of activity extend between the steppe zone of the foothills of the high mountain range (*khangai*) and the desert steppe of the lowlands (*gobi*). By contrast, in the mountain valleys of the forest steppe, vertical migration from one season's grazing area to the next is only a few kilometres. When feed and/or water is short within one seasonal grazing area wide-ranging moves, so-called *otor* migrations, are undertaken to peripheral (reserve) pastures, often located in border areas.

The migration patterns of the mobile livestock keepers are determined not only by grazing conditions but also on the availability of water for animals and humans. During the winter, and especially at high altitudes, water supply is guaranteed by snowfall, but in low snowfall areas and from spring to autumn the situation is more difficult. Bearing in mind that at present the majority of the *negdel* era wells have been either destroyed or are out of order, it is understandable that when there is no snow and little rain the herders keep their animals in the vicinity of working wells. In any event, stock and humans tend to concentrate near natural springs, watercourses, lakes, and permanent settlements.

Rural-Urban Migration

The difficult living and production conditions in rural Mongolia, the attractions of urban life and the growing gap between rich and poor have resulted in a strong internal migration directed towards Ulaanbaatar, where migrants mainly settle in the yurt-quarters at the fringe of the capital (Janzen and Bazargur 2003a and 2003b). Out-migration from the rural areas began in 1993 as a result of livestock privatisation and dissolution of the *negdel* and accelerated by end of the 1990s. Many of the rural-urban migrants are herders who owned only small flocks and who lost their animals through natural hazards (*gan* and *zud*) between 1999 and 2003. The major areas of out-migration in Mongolia are the far away Western and the Khangai (Hangay) Region of the country (Fig. 10.3), with Uvs, Zavkhan, Gobi Altai, Khovd, Khuvsgul and Arkhangai *aimag* (Fig. 10.1) being the main provinces of origin. While migration flow is low from the Eastern Region, a large number of people also came to urban areas from the Central Region, especially from Tuv *aimag*. Continued pasture degradation, often accelerated by locusts and rodents, resulted in a considerable increase in the number of rural poor. This in return caused another rise in the number of rural-urban migrants many belonging to the group of the 'new nomads'.

The majority of migrants also left their home because of the poor technical and social infrastructure, insufficient medical, as well as educational facilities and services. Lacking alternative income opportunities, many livestock keepers have only survived in the countryside by selling their livestock. After all animals were sold these former livestock keepers were forced to move to the large towns and look for waged work (Janzen et al. 2002 and 2005, PTRC/NUM 2001). To some extent new income opportunities arise from new settlements along major traffic routes. Formerly mobile herdsmen now live a half-settled life, adding to their earnings from livestock keeping by running small restaurants and kiosks.

The exodus from the countryside implies a strong skill and brain drain leading to a lack of qualified workers and highly educated specialists. Consequently, the lack of well-trained personnel negatively influences the quality of services and livestock production in the rural settlements. Population density in western Mongolia has decreased considerably. This is especially true along the Russian border where theft of livestock, organised by Tuva-Mongolian gangs, has increased general insecurity. A high percentage of the border population has already left their homelands. However, out-migration to towns has not led to a marked decrease in livestock numbers, because most of the migrants' animals stay in the countryside, either sold to other herders or left with relatives who take care of the livestock. At the same time the pre-existing problems of overstocking and overgrazing around the large cities of central and northern Mongolia have worsened because of livestock migration and ecologically inappropriate, mostly unorganised pasture usage.

Overstocking and Changing Markets

The rapid rise in animal numbers, combined with a shifting flock structure towards goats for cashmere production (Fig. 10.2) caused considerable over-grazing problems in large parts of the country. Some of the most common reasons for this development are, firstly the herdsmen's ambition to own as large a herd as possible as security against the risk of loss through natural disaster and disease. Secondly the lack of at least one seasonal pasture area, above all spring and/or autumn grazing, in the majority of the *sum*, resulting in longer grazing periods on available seasonal pastures and ecological damage. This is compounded by the non-existent or insufficient water supply due to the decay of the well network established during socialism, and increasing drought drying up natural springs and pockets of permafrost. The situation is further aggravated by the fact that administrative boundaries dating from the socialist era (*aimag* and *sum* boundaries are also grazing boundaries) hinder long-distance movement of herds. Hence, the concentration and duration of herd visits at the best locations (good pasture, water supply, favourable conditions for marketing and supplies, on major roads, rivers and lakes, and close to permanent settlements) has led to severe over-grazing in many places.

Experience suggests that the national animal population should not exceed its 1989 figure of 24.7 million (under the socialist regime, Fig. 10.2) to ensure a viable future for livestock production through sustainable use of the unique pasturelands of Mongolia. However, after years of natural hazards (1999-2003) rainfall once again has been sufficient to allow herders to increase their herds. By the end of 2009 the total national livestock population had reached more than 44 million (Fig. 10.2), two thirds of which were concentrated in the Khangai and Western Regions (NSOM 2010: 193) (Fig 10.3). This high number of animals also reflects lost livestock meat exporting opportunities, which during the Socialist period had helped maintain a reasonable national livestock population.

Marketing of Mongolian Livestock Products

The integration of Mongolia's pastoral economy into the world market has undergone great changes since the end of socialism. Mongolia was a satellite state of the Soviet Union (SU) and was obliged, as a member of the COMECON (Council for Mutual Economic Assistance), to provide livestock, meat, other raw animal materials, and processed animal products, to the socialist brother states, particularly the USSR. Since 1990 and the beginning of the transformation to a market economy, this has changed from a SU oriented flow to PR China directed marketing. Figure 10.4 reveals the marketing chain of livestock products within the centrally planned Mongolian economy. It started at the lowest level of the *brigade* continued through the *sum* to the *aimag* and from there to the national capital. Livestock from all parts of the country was driven on the hoof to the outskirts of Ulaanbaatar, where the animals were held to increase weight before being slaughtered. From there deep frozen meat and other animal products were transported by rail to the USSR.

During the Socialist period Mongolia was a very important provider of meat and other raw animal materials to the Eastern Bloc. A closer look at the development of Mongolian meat exports shows that the country was a large supplier of sheep and cattle meat. After a peak in 1970, the total amount of exported meat oscillated between approximately 220,000 and 400,000 tons per year during the following two decades. Chevrette (high quality goat leather), cattle leather and sheep skins were also exported in large quantities. Between 1965 and 1989 the number of

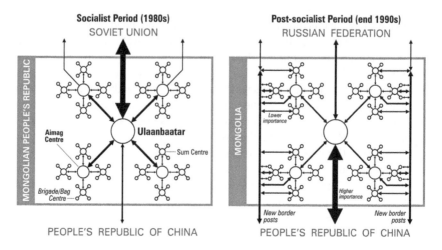

Figure 10.4   Marketing flows of Mongolian animal products
Design: Janzen.

exported pieces of cattle leather ranged between approximately 120,000 (1973) and 310,000 (1989) pieces. Sheep and camel wool were also important. An annual average of up to 110,000 tons of sheep wool and about 95,000 tons of camel wool were exported between 1965 and 1975. Between 1975 and 1980 sheep and camel wool, and cashmere exports sharply declined to nearly zero, when the raw materials were needed in the young Mongolian processing industry.

After the break-up of the USSR the direction of marketing, the kinds and quantities of animal products sold has changed considerably (Fig 10.5). Meat exports to the USSR almost disappeared. The PR China did not begin importing Mongolian meat, but became the most important importers of Mongolian raw livestock materials, such as cashmere, sheep wool, hides, skins and other animal products.

Reasons for this fundamental change were Mongolia's opening up to the world market, the collapse of the Eastern Bloc, the termination of its existence as a major meat supplier for the USSR, and the break-down of the large Mongolian combined collectives processing raw animal materials. Becoming a member of the WTO, Mongolia had to follow new international laws and regulations. High standards of quality and hygiene were often difficult to fulfil. Although Mongolia continued to export small quantities of deep frozen meat, for example, horse meat to Japan and cattle and sheep meat to the Russian Federation, the country did not succeed in finding new markets for large-scale meat exports. Attempts to develop mutton exports to the oil-rich Arabian Gulf States were not successful because the taste of Mongolian meat did not meet the taste of Arab customers. Expensive shipment of meat by aeroplane from a landlocked country like Mongolia compounds these problems.

A further major limitation on meat exports is the regular outbreak of diseases, such as foot and mouth disease in parts of Mongolia. Thus the country, treated as

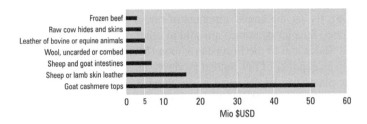

Figure 10.5   Exports of products of animal origin 2005
Source: MIT (2006).

a whole, cannot meet international standards of quality and hygiene. As the PR China produces enough meat by itself, Mongolia has no opportunity to export meat to its southern neighbour. The sale of products of animal origin contributed only 9 per cent to the total value of Mongolian exports in 2005 due to very high prices for copper and gold on the world market and a strong increase in mineral sales (mining products contributed to 70 per cent of export earnings, MIT 2006). Nevertheless exports of animal products still play a very important role for the Mongolian economy and the rural-pastoral economy in particular. During the transformation period, cashmere production became the major source of income for pastoral households. The cashmere price has remained high in recent years associated with a further increase in goat keeping (Fig. 10.2, Fig. 10.5). In spring 2010 up to US$ 40 was paid on average per kilogram. This is especially true for the arid Gobi desert and desert steppe where in several *aimag* the goat population has more than doubled. The number of goats also increased to a lesser extent in the dry steppe zone, and has become significant in the forest steppe zone, where previously they played only a minor role in livestock keeping.

Alongside the realigning marketing chains since 1990 the export of sheep skins played a major role (MIT 2006). Two peaks in 1993 and 2000 can be explained by an increased slaughtering rate for livestock during privatisation and the high death rate of animals during the first *gan* and *zud* year in 1999/2000. To a much lesser extent skins and hides of goats, cattle and horses were exported. Exports of intestines showed high export quantities at the beginning and the end of the 1990s and decreased at the start of the new millennium. Amongst other export commodities meat and sheep wool ranked highest. Of special interest is the increase of meat exports to the Russian Federation between 1995 and 2002 and decline since 2003 due to animal diseases such as foot and mouth. Although the world market price for sheep wool has gone down during the past decade, 5,000 to 10,000 tonnes of wool were still exported annually.

Figure 10.5 represents the share of total export value in 2005. About 56 per cent of exports in 2005 (or US$ 51.3 Mi.) were cashmere, followed by sheep skin leather (17.7 per cent or US$ 16.3 Mi.). Leather of large animals, sheep and goat intestines, and uncarded or combed wool ranged between 7.5 and 5.5 per cent of total export value. Raw cow hides and skins and frozen meat had a share of less than 5 per cent

(MIT 2006). By far the largest amounts of raw animal materials are exported to the PR China. Smaller quantities are sold to other industrialised countries such as the United States and to the European Union (MIT 2006). Although Mongolia is still mainly an exporter of raw animal products, an increase of raw animal material processing plants can be observed in the three large urban centres of Mongolia, especially in Ulaanbaatar. Hopefully this trend will continue in the years to come in order to increase the value of industrial production and decrease Mongolia's role as simply an exporter of cheap raw materials mainly to PR China.

The Spatial and Social Structure of Trade

After 1990 export trade to the PR China occurred through eight border posts. The most important was Zaminuud in central southern Mongolia (Ereen on the Chinese side). The other seven were newly established at the beginning of the 1990s. This was the shortest way from the rural areas of Mongolia to the PR China (cf. Fig. 10.4, Fig. 10.6). However, since the end of the 1990s this situation has changed. Now Mongolians increasingly market their animal products internally to the large Mongolian cities of Erdenet, Darkhan, and Ulaanbaatar, where raw materials are either processed locally in new factories, or exported abroad by Mongolian and to an increasing extent by Chinese wholesalers. The main reason for changing the spatial trading pattern is that the former exchange of goods took place on Chinese territory near the new border posts where the Mongolian pastoralists and business people often did not receive a fair price for their products.

Today, the best access for pastoralists to the market is in the surrounding areas of the large cities of Ulaanbaatar, Erdenet, and Darkhan, in the vicinity of *aimag*-capitals and along major traffic routes. Herder households located in the periphery and lacking their own motor transport facilities, which is still the case for the majority of less wealthy herders with small flocks, thus often face considerable difficulties in marketing their produce for reasonable prices and in acquiring needed supplies for a low price. This is mainly true for the remote areas of western Mongolia with distances of more than a 1,000 kilometres from the major market of Ulaanbaatar. The real winners within the mobile livestock keeping economy of Mongolia are the small group of wealthy and rich herder households with more than 500 animals (3.6 per cent of the total number of Mongolian pastoral households at the end of 2005). Approximately 80 per cent of herder households, which possessed less than 200 livestock at the end of 2005, can be classified as ranking between less wealthy and poor. Their production is (except Cashmere) only to a limited extent market-oriented and mainly aims at satisfying their subsistence needs (NSOM 2006).

Unlike during the socialist period, when animal products were centrally marketed, now under market conditions livestock and raw animal materials are purchased from herders by migrant traders and trading companies, who sell the raw materials to wholesalers and processing factories in the big cities. Only a

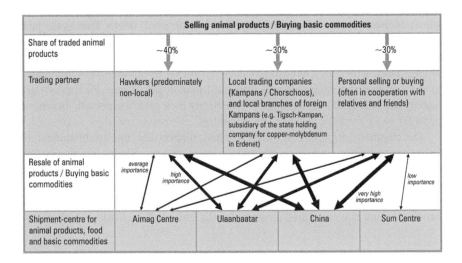

Figure 10.6   Market chains in the post-socialist period at the end of the 1990s
Design: Janzen.

small share of animal products is transported by pastoralists themselves directly to trading companies in rural and urban centres. Migrant traders are men who belong to different professional and wealth groups. Most of them are urban based, are often drivers by profession who possess a car of their own, and carry out this business on a seasonal basis.

Milk and dairy products are mainly sold to the domestic markets during the summer and autumn seasons and to a lesser extent in a deep frozen form during the winter. The main marketing period for cashmere is in April/May, sheep wool in July/August, livestock in September/October and meat, as well as hides and skins during the main slaughter period at the start of the cold season, in October/ November. Currently, only small quantities of meat are exported, instead most of livestock and meat are sold domestically. Annually, between five and six million livestock are slaughtered in Mongolia (in 2006 about 2.9 million sheep, 2.3 million goats, 0.4 million cattle, and 0.3 million horses; NSOM 2007: 184).

Due to the lack of state organised marketing facilities and often weakly developed group oriented self-help organisations, the majority of herders do not market their animal products on a larger scale, for example, together with neighbours, but instead sell individually to migrant merchants. Mutual mistrust often exists even among relatives and friends of the same neighbourhood, and explains why herders tend to not market their animal products collectively. This is counterproductive as herders could make a much greater profit selling their raw animal materials in bulk together. The same is true for the purchase of provisions, which are normally bought after the sale of animals and raw materials. Another negative aspect of post-socialist times is that most of the raw animal materials are sold un-cleaned and unsorted, and in relatively small quantities, because of the subsistence orientation of less wealthy herders with small flocks. Due to the lack of cash in the countryside, barter is still the usual method of payment. As most of the smaller pastoral households with less than 200 livestock have debts with high interest rates, which they have to pay back as quickly as possible, they are unable to store animal raw materials and wait until market prices rise. Pastoralists are heavily dependent on migrant traders who dictate their prices, especially in remote areas far away from the *sum* and *aimag* centres,

Development organisations offer limited support for the establishment of marketing facilities and/or multi-purpose cooperatives. There are also a few examples throughout Mongolia where former *negdel*-directors are running livestock cooperatives on a private basis in a similar way as in socialist times. Complaints from members of these cooperatives show however that the economic benefits are not equally shared amongst the directors and the herder members.

Conclusion

This chapter investigates the transformations of Mongolia's pastoral economy since 1990. It reveals three key developments. Firstly, during the early capitalist years many of the 'new nomads', those who received animals after the privatisation of the former socialist production units, contributed to and enlarged the number of the new poor due to heavy livestock losses caused by natural hazards and degrading pastures. Subsequently they were forced to migrate to the city and to settle under often difficult circumstances in constantly increasing yurt-settlements at the urban fringes, predominately in Ulaanbaatar. Secondly, orientated towards Moscow for decades, meat commodity chains almost completely broke down. However, marketing chains for pastoral produce were (re)constituted and realigned towards China. Alongside this spatial reorientation of demand, a product reorientation took place; instead of meat, only raw materials such as hides and wool are now in demand. Thirdly, while the trade value from marketing sheep, cattle, horse, and camel products lost of importance, the sale of cashmere hair from goats became the major source of income for Mongolian herders.

The future of mobile livestock keeping in Mongolia however largely depends on government policy that will have to meet the needs of the herder households and open new export markets for meat. Market orientation is necessary to reach a higher take-off rate for livestock and keep the stocking rate at an ecologically sound level. The introduction of new legal provisions for ecologically suitable pasture management (land law amended June 2002) specifically precludes the privatisation of pasture land and is crucially important for the mobile livestock keeping sector. The granting of long term user rights for winter/spring grazing areas to existing or new groups of livestock keepers is now possible and desirable, providing that the pasture is used in an ecologically sustainable manner and the carrying capacity is not exceeded. The ongoing discussion about territorial reform aiming at the creation of larger administrative units, which would allow more long-range, ecologically more adapted pastoral migrations and a more economical use of the available public budgets, will hopefully have a positive outcome.

Finally it has to be emphasised that sustainable rural-pastoral development can only be achieved through a holistic approach. Major goals are to keep the number of animals at a reasonable level, the introduction of further legal regulations for securing group-oriented ecologically adapted pasture and water management, the establishment of small and medium enterprises processing raw animal materials, as well as the improvement of technical and social infrastructure in rural areas. The improvement of animal breeding and health, as well as the provision of better marketing opportunities for herders, are important prerequisites if Mongolia wants to regain its importance as an export nation for meat and livestock. Taking these recommendations into consideration mobile livestock keeping in Mongolia can have a viable future.

Acknowledgements

The research results presented in this chapter are based on fieldwork carried out in western Mongolia since the mid-1990s. The author thanks DFG, GTZ, CIM, ADB and SDC for their generous financial support for carrying out various research projects focused on the mobile livestock economy and rural development in Mongolia.

References

Administration of Land Affairs, Geodesy and Cartography (ALAGAC) 2004. *Geographic Atlas of Mongolia*. Ulaanbaatar: ALAGAC.

Batbuyan, B. 2003. Verarmung von Nomaden in der Mongolei nach 1990. Sozialgeographische Probleme eines Transformationslandes. *Geographische Rundschau*, 55(10), 26-32.

Finke, P. 2004. *Nomaden im Transformationsprozess. Kasachen in der postsozialistischen Mongolei*. (Kölner Ethnologische Studien, 29). PhD Thesis, University of Cologne. Münster: Lit.

Janzen, J. 2000. Present state of development and perspectives after ten years of transformation, in *State and Dynamics of Geosciences and Human Geography of Mongolia. International Symposium 'Mongolia 2000'*. (Berliner Geowissenschaftliche Abhandlungen, Reihe A, 205), edited by M. Walther, J. Janzen, F. Riedel, and H. Keupp. Berlin, 149-152.

Janzen, J. 2001. *What Are Somalia's Development Perspectives? Science Between Resignation and Hope?* (Proceedings of the 6th SSIA-Congress, Berlin, 6-9 December 1996). Berlin: Das arabische Buch.

Janzen, J. 2002. *Mobile Livestock Keeping and Pasture Management in Mongolia.* (Gasarzuin Asuudal II/156). Ulaanbaatar: National University of Mongolia, Faculty of Geography and Geology.

Janzen, J. 2005a. Mobile livestock-keeping in Mongolia: Present problems, spatial organization, interactions between mobile and sedentary population groups and perspectives for pastoral development, in *Pastoralists and Their Neighbors in Asia and Africa.* (Senri Ethnological Studies 69), edited by K. Ikeya and E. Fratkin. Osaka, 69-97.

Janzen, J. 2005b. Changing political regime and mobile livestock keeping in Mongolia. *Geography Research Forum*, 25, 62-82.

Janzen, J. and Bazargur, D. 1999. Der Transformationsprozess im ländlichen Raum der Mongolei und dessen Auswirkungen auf das räumliche Verwirklichungsmuster der mobilen Tierhalter – Eine empirische Studie, in *Räumliche Mobilität und Existenzsicherung.* (Abhandlungen – Anthropogeographie, 60), edited by J. Janzen. Berlin: Dietrich Reimer Verlag, 47-81.

Janzen, J. and Bazargur, D. 2003a. Wandel und Kontinuität in der mobilen Tierhaltung der Mongolei. *Petermanns Geographische Mitteilungen*, 147(5), 50-57.

Janzen, J. and Bazargur. D. 2003b. The transformation process in mobile livestock keeping and changing patterns of mobility in Mongolia. With special attention to Western Mongolia and Ulaanbaatar, in *The New Geography of Human Mobility – Inequality Trends?* (IGU, Vol. 4), edited by Y. Ishikawa and A. Montanari. Roma: Home of Geography, Societá Geografica Italiana, 185-222.

Janzen, J., Gereltsetseg, D., Bolormaa, J. et al. 2002. *A New Ger-Settlement in Ulaanbaatar, Functional Differentiation, Demographic and Socio-Economic Structure and Origin of Residents.* (Research Papers 1). Ulaanbaatar: National University of Mongolia, Center for Development Research.

Janzen, J., Taraschewski, T. and Ganchimeg, M. 2005. *Ulaanbaatar at the Beginning of the 21st Century: Massive In-Migration, Rapid Growth of Ger-Settlements, Social Spatial Segregation and Pressing Urban Problems. The Example of the 4th Khoroo of Sanginokhairkhan Duureg.* (GTZ Research Papers 2). Ulaanbaatar: National University of Mongolia, District-, Center for Development Research.

Janzen J.; Enkhtuvshin, B. (eds.) 2008. Present State and Perspectives of Nomadism in a Globalizing World. Proceedings of the International Conference: 'Dialog Between Cultures and Civilizations: Present State and Perspectives of Nomadism in a Globalizing World', held at National University of Mongolia, August 9-14, 2004, Ulaanbaatar.

MIT (Ministry of Industry and Trade). 2006. *Trade Statistics of Mongolia.* Ulaanbaatar.

Müller, F.-V. 1994. Ländliche Entwicklung in der Mongolei – Wandel der mobilen Tierhaltung durch Privatisierung. *Die Erde*, 125(3), 213-222.

Müller, F.-V. 1997. New nomads and old customs – General effects of privatization in rural Mongolia. *Nomadic Peoples*, 36/37, 175-194.

Müller, F.-V. 1999a. *Der unverbesserliche Nomadismus – Sesshaftigkeit und mobile Tierhaltung in der Mongolei des 20. Jahrhunderts.* Habilitationsschrift, Berlin: Freien Universität Berlin, Geowissenschaften (unpublished).

Müller, F.-V. 1999b. Die Wiederkehr des mongolischen Nomadismus – Räumliche Mobilität und Existenzsicherung in einem Transformationsland, in *Räumliche Mobilität und Existenzsicherung.* (Abhandlungen – Anthropogeographie, 60), edited by J. Janzen. Berlin: Dietrich Reimer Verlag, 11-46.

Müller, F.-V. and Janzen, J. 1997. Die ländliche Mongolei heute. Mobile Tierhaltung von der Kollektiv- zur Privatwirtschaft. *Geographische Rundschau*, 49(5), 272-278.

NSOM/National Statistical Office of Mongolia. 2000-2010. *Mongolian Statistical Yearbook, 1999-2009.* Ulaanbaatar.

PTRC/NUM (Population Teaching and Research Center/National University of Mongolia). 2001. *A Micro Study of Internal Migration in Mongolia.* Ulaanbaatar.

Scholz, F. 1995. *Nomadismus. Theorie und Wandel einer sozio-ökologischen Kulturweise. Stuttgart.* (Erdkundliches Wissen, 118). Stuttgart: Franz Steiner Verlag.

Chapter 11
Nomads and their Market Relations in Eastern Tibet's Yushu Region: The Impact of Caterpillar Fungus

Andreas Gruschke

This chapter investigates pastoral livestock production in the Yushu Region on the Tibetan highlands, which is under increasing pressure from decreasing pasture areas and deteriorating livestock markets. While the larger part of Yushu still depends on livestock production, it is argued that pastoral production only persists due to an unfolding niche economy, built on a single resource: the caterpillar fungus. The marketing of the fungus makes up to 50 per cent and sometimes even up to 80 per cent of cash income of households. This resource allows many Tibetan pastoralists to subsist with their livestock rather than give up animal husbandry, especially after local pastoralists were recently able to assert greater participation in the marketing chain of this special commodity.

The Yushu Region

The Tibetan plateau extends across a vast area of approximately 2.5 million km², but only a small portion of this land is arable. The majority is however fit for mobile pastoralism. The Yushu[1] Tibetan Autonomous Prefecture (TAP), one of six autonomous prefectures of Qinghai province, is almost entirely within the purely pastoralist sphere of the Tibetan highland. Some 300,000 inhabitants populate this region, of whom two-thirds are pastoralists, representing more than 80 per cent of the rural population. The prefecture belongs to the poorer and least developed provinces in China, lagging behind in both productivity and income. Within Qinghai Yushu TAP is considered to be one of the least developed and poorest

---

1   The name Yushu is used in different contexts: (a) as designation of a sub-provincial district, the Yushu Tibetan Autonomous Prefecture; (b) as name of one of the six counties of that prefecture; and (c) as designation for the seat of both that county and the prefecture. Although the latter's name is Gyêgu, the old Tibetan town of Jyekundo (*skye rgu mdo*), it is common use to call this modern town by the same name as the prefecture. In this chapter, Yushu will always refer to the prefecture, Yushu TAP; while the county will be called Yushu county and the town Gyêgu.

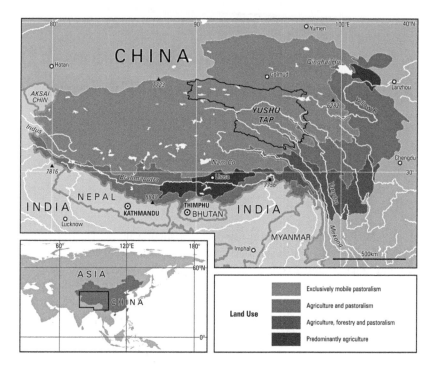

Figure 11.1   Land use in the Tibetan Plateau
Source: Author.

areas. It covers an area of 197,000 km², more than a quarter of Qinghai province.[2] Rangeland used by pastoralists makes up 56 per cent of the entire Qinghai province and 78 per cent of Yushu's surface.[3] It, therefore, is only slightly smaller than Great Britain, while its population of 297,000 inhabitants compares to a medium-sized European city. The Yushu prefecture includes the territory of the sources of the Yellow River in the north, the upper reaches of the Yangzi and Mekong in the west and south, and the border areas to the neighbouring Tibet Autonomous Region in the south and to Sichuan province in the east. It consists of six counties – Qumarlêb, Zhidoi, Zadoi, Chindu, Yushu and Nangqên, with the prefectural capital Gyêgu located in Yushu county. The counties differ considerably in size, with Zhidoi covering more than 80,000 km² and Nangqên only about 12,700 km². Three of the counties are completely pastoralist, while the other three feature agriculture on a very limited scale: In 2004, cultivated land in the prefecture amounted to less than three per cent of the cultivated land in Qinghai province (Fig. 11.1).[4]

---

2   Surface data for the Yushu TAP vary greatly: Yushu Survey (1985: 2) notes 197,000 km², QPSB (2005: 46) 197,953 km², YZT (2006: 59) 210,300 km².

3   Data for 2004, acc. to QPSB (2005: 42-46).

4   QPSB (2005: 213).

Household type	Household	Population	
	number	number	%
Pastoralist	47,522	202,787	68.3
Peasant	23,288	48,921	16.5
Non-rural	-	45,296	15.2
Total	70,810	297,004	100.0
Non-registered	-	50-60,000	17-18

Figure 11.2   Population of Yushu TAP in 2005

Note: Data for non-rural households and for non-registered households are not available
Sources: YZT (2006: 63f) and local informants.

The terrain gradually rises from the southeast, where it never falls below 3,510 m above sea level, to the northwest where the average altitude is 4,200 m to 4,500 m. The climate of Yushu is one of extremes: the average annual temperature varies between -4.3°C (Qingshuihe) and 5.2°C (Nangqên). In winter temperatures can drop as low as -43°C (1975, Qingshuihe) while in summer, temperatures can rise to +28°C (Yushu and Nangqên) during the day. The precipitation decreases from southeast to northwest, with Nangqên and Zadoi counties receiving 550-635 mm annually, Zhidoi and Qumarlêb only 380-500 mm.[5] Not only are the weather conditions unpredictable, but climate variability is especially high.

Winter pastures and winter houses in Yushu can be found at altitudes of up to 4,800 m. Above this level, but in more favourable locations also at lower altitudes, the nomads' summer area of activity extends. There they set up their tents and use the adjacent meadows up to 5,000 m altitude as pastures. Arable field cultivation is only possible in the zone below 4,000 m. Today, the highest fields in Yushu are no higher than 3,900 m above sea level. A limited population of farmers and semi-nomads have fields and permanent houses there. On the whole, animal husbandry is the prevailing sector of production, while less than a fifth of Yushu's population is involved in field cultivation (Fig. 11.2),[6] the extent of which varies regionally.

Livelihood Situation of Yushu Nomads

During the last 50 years Yushu has seen considerable fluctuations in both population and livestock numbers (Fig. 11.3).[7] The rural human population has

---

5   YZT (2000: 50, 2006: 58).

6   The non-registered population (the so-called floating population, *liudong renkou*, for example both long and short term residents without local household registration, *hukou*) are, of course, not well represented in the statistics. An estimate of 50,000-60,000 people, communicated by well-informed local government employees, matches what was in evidence during the field research. Approximately 30,000-40,000 of the non-registered individuals (Tibetans, Han and Hui) are said to reside in Gyêgu.

7   Total livestock numbers include horses, which traditionally played a great role in Yushu, and goats. Although, due to well-known problems with official statistics, especially

Year	Total Population	Rural Population		Non-rural Population		Livestock ('000)	
	number	number	%	number	%	Yaks	Sheep
1950	123,110	105,333	85.6	17,777	14.4	832	974
1959	168,005	155,027	92.3	12,978	7.7	379	645
1969	123,071	114,425	93.0	8,646	7.0	1,171	2,781
1979	178,935	160,102	89.5	18,833	10.5	1,656	3,848
1989	224,071	196,286	87.6	27,785	12.4	1,578	2,301
1999	252,696	217,596	86.1	35,100	13.9	895	1,742
2004	283,144	238,903	84.4	44,241	15.6	868	1,949
2005	297,004	257,859	84.8	39,145	15.2	908	1,950

Figure 11.3   Development of population and livestock in Yushu
Sources: YZT (2000: 66-109; 2006: 64-88), QPSB (2005: 68, 223).

more than doubled, while problems of environmental degradation are increasing. Natural disasters are frequent, grassland 'carrying capacity' is decreasing and the impact of climatic and ecological changes are expected to be severe.[8] According to Nori (2004: 13) almost 10 per cent of Yushu's total herd (1.32 million heads) were lost in the winter of 1981/82, while the winter 1984/85 cost another 990,000 heads, representing 17 per cent of the (remaining) total herd.

The development of the population-livestock ratio in Yushu TAP reveals the diminishing livelihood basis of the pastoralists. The livelihood situation can be assessed by calculating sheep units (SU). This is a reference unit to compare different types of livestock. The basis for the SU is one adult female sheep. Calculated on the assumption that one SU requires 4 kg of hay per day, other animals are usually converted as follows: 1 yak = 5 SU; 1 horse = 6 SU (Miller 2001, Yan et al. 2005: 37) or 7 SU (Goldstein 1996). 25 Sheep Units per person is considered to be the break-off point for poverty in Tibetan nomad areas. Families with less than 25 SU per person are not able to meet their basic needs (Miller 2001). In 1950 the average rural household in the Yushu TAP owned 48.8 SU (horses and goats not counted)[9] per individual (see Fig. 11.4).

---

during the highly ideologised periods of Communist China 'government statistics on livestock populations are fraught with problems' (Costello 2008), these numbers need to be read with caution, they allow – in combination with historical and natural data – interpretations about times of political disturbance, secularisation, economic policies and natural disasters. The numbers for 2004 were adopted and generated from the QPSB's data. Numbers for 1999 differ slightly (between 2.75 and 0.16 per cent) from Wiener, Han and Long (2003/06: 265).

8   For climatic and ecological changes see Wu Ning (1997), G. and S. Miehe (2000: 298-306), and for the effects and increase of snow disasters see Nori (2004: 12-13). For a discussion of the controversial carrying capacity of the grassland see Wiener, Han and Long (2003/06).

9   Horses and goats are not separately displayed in Fig. 11.3 and Fig. 11.4. If their figures (273,900 goats; 36,800 horses) are included, there were 53.4 SU per person for Yushu's rural population, and 45.7 SU per person for the total Yushu population in 1950.

Year	Rural Population	Yaks per Person		Sheep per Person		SU per Person	
	number	TP	RP	TP	RP	TP	RP
1950	105,333	6.8	7.9	7.9	9.3	41.7	48.8
1989	196,286	7.0	8.0	10.3	11.7	45.5	51.9
1999	217,596	3.5	4.1	6.9	8.0	24.6	28.6
2005	257,859	3.0	3.5	6.6	7.6	21.6	25.2

Figure 11.4   Development of livestock distribution in Yushu

Note: TP = based on total population; RP = based on rural population
Source: Calculated from data in YZT (2000, 2006).

Following the disbandment of the people's communes and the beginning of a 're-nomadisation', large livestock numbers continued to exist. The year 1989 represents a situation in which an average rural individual in Yushu owned almost 52 SU and thus a livestock number which was twice as high as the poverty level. This was even though two serious snow disasters occurred in Yushu during the 1980s (1981/82, 1984/85). Ten years later, the numerical value of SU/person had decreased to 28.6 and has plummeted to such an extent that today, on average, rural inhabitants of Yushu live just above the poverty line. This is further compounded by the fact that the grassland of Yushu is stressed much more than before: in 1950 there were 5.14 million SU. In 2005 there were 6.49 million SU for the district. Hence, 25 per cent more hay was needed from the grassland, although at the same time the basis of the (remaining) pastoralists' sustenance in animal husbandry was reduced by almost 50 per cent. Miehe and Miehe (2000: 306) claim: 'The natural decrease of the carrying capacity[10] in consequence of climatic desiccation does not regulate the herd size any more where sedentarism and supplementation of feed are offered. On the contrary, naturally shrinking pastures face rapidly increasing herds of people who increase in number and welfare.' This is not to say that every household's herd is growing, but rather the district's total livestock number. Most herders in Yushu, with medium-sized or larger herds complain about the shortage of grassland. More and more pastoral people are left without pasture, and will gradually fail to subsist from the rapidly decreasing number of their animals.

Changing Features of Tibetan Nomadism

Yushu's physical infrastructure within the rural areas is in the course of being upgraded. By 1999 there were 2,562 km of roads in Yushu prefecture (YZT 2000: 204) and investment in road infrastructure has increased tremendously since.

---

10   Since 1950, the area of usable pastureland in Yushu has decreased by approximately 25 per cent. Furthermore, the biomass per unit area of pastureland has also considerably decreased – for example the actual average carrying capacity per unit area is, even without further livestock growth, lower than in 1950 (YZT 2000; cp. also various sources in Gruschke 2009: 188-189, 197-198).

Metalled roads connect the county seats of the prefecture, and all community townships are now accessible by a network of improved gravel roads. Compared to other parts of the Tibetan Plateau, pastoralist regions in eastern Tibet may thus, due to shorter distances to the big markets of Chinese cities, be considered as having favourable conditions for marketing animal husbandry products. However, new links between the local economy and external markets have not unfolded. Subsistence economy still, or again, largely prevails. Thus, public policies aim at developing, or rather transforming, the habitual nomadic pastoralism into modern livestock farming in order to improve livelihoods. Both Chinese and Tibetan government officials, who in general are not at all familiar with pastoralist livelihoods and its pressures, believe Tibetan livestock production can be revitalised by increasing market push-factors. Foreign observers, however, are more sceptical. Their assessments rather criticise market integration policies as damaging, if not destroying, traditional structures that are rather well adapted to a difficult environment, and consequently consider such projects as major challenges for the pastoralists' livelihoods.

Yet, even without the policies of integration aimed at the once far-off nomads living in the core of the Tibetan Highlands, these pastoralists have already been touched by forces of the world market. Yushu's economy has always been complemented by trade and was thus connected to supra-regional markets. Nomad merchandise was wool, fur, hides, yak and sheep skins and horses; in modern times meat (live yaks and sheep), wool and leather are more important.

The livestock market, however, plays a diminishing role for Tibetan pastoralists nowadays. The same is true for commodities produced from animal husbandry. This was different during the 1980s when, after the start of economic liberalisation policies, nomads with their exchange commodities of meat, dairy products, wool etc., could offer relatively high-priced products – as compared to Tibetan farmers who had to compete with the agricultural products of the whole of China, a major agricultural producer. Furthermore, the 1980s still offered a marketing infrastructure run by the state, be it profitable or not. The privatisation push started in the 1990s, and fostered the closure of many state businesses for marketing and processing. While, time and again, Buddhist values are also given as a rationale for decreasing deliveries of products, they seem, more often than not, to be pretexts in cases where people either *can afford not* to sell, or *cannot afford to sell* animals.[11]

There are however multiple reasons why traditional pastoralist products are marketed to a lesser extent than before – at least intra-regionally. Among the major explanations for meat and dairy products we have to consider that they are increasingly consumed at home in the pastoralists' households. While formerly

---

11   The same is true for the common explanation of some Tibetan nomads' practice of killing a yak by suffocation: a leather bag is bound around the mouth of the animal so that it 'dies by itself'. While it is generally said that people do this in order to avoid accumulating bad karma by the straight action of killing, many informants in Yushu maintained that it is done as this kind of death makes the yak meat more tasty.

very little meat was eaten in summer, it now has become very common to eat meat throughout the entire year. Secondly, local markets in Yushu (Gyêgu and county seats) have developed a higher demand for meat and dairy products. Finally, they are produced in lower quantities as productivity has decreased by 20 to 30 per cent during the last two decades.[12] Simultaneously wool prices have seriously dropped, due to China entering the world market and the downward trend of international wool prices.[13] Tibetan sheep wool is, due to its coarse structure, also not as competitive as the high quality Merino wool of Australia and New Zealand. Although China has the world's largest sheep population, Australia produces more wool than any other country. China is also the biggest purchaser of Australian wool, with exports to China valued at US$ 1.3 billion in 2004/2005.[14]

Furthermore, we have to acknowledge that the marketing of livestock products in peripheral areas like Yushu has a number of intrinsic difficulties. In more remote yak-raising areas, the marketing of yak products is very limited – there are only a few, remote market outlets, and due to weak communication, transport costs are high. The reliance on individual traders with poor management and financial capabilities further aggravates the situation and does not provide a good basis for large-scale marketing.[15] Without market intelligence, it is also difficult to evaluate trends in the market or to observe price changes. The major obstacle for Tibetan pastoralists, however, is the combination of population growth and decreasing pasture areas. The latter has shrunk by approximately 20 to 25 per cent since 1950, as has the grass yield per area unit.[16] Consequently, individual pastoralist families have – on average – a smaller stock of animals available and, thus, command over a weaker base of subsistence, meaning that fewer animals or animal products can be sold. The subsequent higher local demand for animal products, thus, obstructs any aim of integrating the pastoralists' livestock economy into a supra-regional market.

Given the difficulties of Tibetan nomads in taking advantage of marketing their livestock products, the larger part of Yushu is nevertheless a region where pastoral livestock production persists. This is – as will be argued – largely on account of a niche economy built on a unique resource: caterpillar fungus. It is this resource that allows many Tibetan pastoralists to subsist with their livestock rather than give up animal husbandry.

---

12   See He and Li (2004: 95). Since the start of economic liberalisation in 1980, livestock numbers in Yushu have steadily decreased: from 1980 to 1990 by 30 per cent; then more or less levelled off for half a decade; and after 1995 decreased again by 27 per cent until 2005 (see data in YZT 2006: 91-92). This adds to the decrease in productivity.

13   See AIECE-Prognose (2000/2001).

14   See AWI (2005).

15   Op. cit. He and Li (2004: 95).

16   Yushu statistics show a decrease of pastureland by 25 per cent (YZT 2000). In the neighbouring districts of Nagqu in the TAR and Garzê and Aba in Sichuan, this decrease amounts to 20-24 per cent (He and Li 2004: 95).

Caterpillar Fungus – A New Niche Economy Widening the Economic Space of Yushu Pastoralists?

Unlike nomad areas outside 'High Asia', many Tibetan pastoralists have access to a traditional resource and medicinal plant called caterpillar fungus (*cordyceps sinensis*).[17] The gathering and marketing of this fungus, while not related to animal husbandry, has become more important for most Yushu Tibetan nomad livelihoods than any other income opportunity. It enables many of them to participate in modern consumption, and thus often makes households appear better off than their vulnerable pastoralist livelihood situation would actually allow. Bringing significant amounts of capital and assets into the Yushu region, as well as into its remote nomad households, it has become one of, if not *the* major commodity of pastoralist households, notably in the north-eastern part of the Tibetan plateau.

This caterpillar fungus, locally known as *yartsa gumbu* (*dbyar rtsa dgun 'bu*) 'is a fungus parasitizing the larvae of a moth of the genus Thitarodes (Hepialus), which lives in alpine grasslands of the Tibetan plateau ..., but the market is driven by Chinese consumers, who know it as chongcao (dongchong xiacao), a highly valued tonic in Traditional Chinese Medicine' (Winkler 2004: 69).

When economic liberalisation began in the 1980s, the price of caterpillar fungus started to rise, at first only gradually, then, due to the growing demand in Inner China, Southeast Asia and Japan, it was pushed up significantly in the last few years (Fig. 11.5). The digging for, collecting and trading of caterpillar fungus thus allows even poor pastoral households with only a small stock of animals to safeguard their livelihood, and even allows for extra consumption.

With the establishment of a cash economy on the highlands, and, concurrently, the decreasing surplus of livestock products, the collection of tradable plants and fungi, such as medicine, tonic or food, has gained more and more weight – especially for rural populations who until recently hardly participated in the new cash economy. It enables them to pay for consumer goods, medical treatment, transport or high school fees,[18] or even to invest in business ventures. Collecting medicinal plants has an age-old history in Tibetan culture, be it for personal use or for trade, especially with the Chinese inland provinces. Although this tonic also seems to be of significance within traditional Tibetan medicine, the nomads themselves do not use it. As in past centuries, when it was an important bartering good to obtain tea and other goods from China, it is again traded out of the region to the adherents of traditional Chinese medicine – in China proper, but also to international markets in South East Asia (Thailand, Singapore, Malaysia), Japan, Korea and the United States.

---

17   For details see Gruschke and Winkler (2010).

18   Winkler (2004: 69) still lists taxes and school fees in general. In 2006/07 land and/ or animal taxes had been abolished, as were primary school fees. As for secondary and high schools, the regional policy may vary, most likely depending on the subsidies received from the Chinese central government.

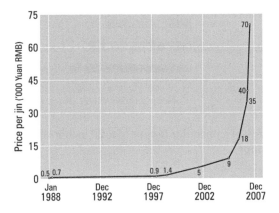

Figure 11.5   Trend of caterpillar fungus prices in Yushu (1988-2007)
Source: Gruschke (2009).

An extended household survey[19] conducted in summer and winter 2006/2007 reveals the economic importance of caterpillar fungus for pastoral households. Almost 85 per cent of households used this resource. As only about 60 per cent of (rural) households still own animals (171 HH) this is particularly crucial. The low animal ownership is partly explained by the fact that the survey includes families who had accepted state programmes for resettlement in the county town. According to the contract conditions, re-settlers have to waive the use of their pastureland for five years and therefore have to sell their herds. However, this contract does not automatically deprive them of their land. If they have a leasehold contract, a relatively new and locally uncommon right, they are allowed access to their pastures to collect caterpillar fungus, a right of major importance for the resettled pastoralists. The situation of the pastoral economy is further aggravated by the fact that many livestock owners do not own enough animals to subsist on them. If among the surveyed households livestock was their only means of subsistence, 62 per cent of them (106 HH) would not be able to maintain a livelihood. Apart from animal husbandry and collecting caterpillar fungus, the district does not offer many job and income opportunities. Traditionally, there are also other medical herbs, which can be collected, and eight villages in Nangqên county can mine salt in mineral springs. Arable farmland is however so rare in Yushu, that semi-nomads, who own fields, can barely produce enough for their

---

19   The household interviews (N = 296) have been conducted in three different areas of Yushu (TAP): a) In Zadoi, a county that is totally shaped by its pastoral economy, with a dynamically developing district town. b) In Nangqên, this is a region with some additional agriculture and semi-nomadism. Here arable land is limited to near the county seat at Xangda and the nearby valleys of the Mekong river and its tributaries. c) In Yushu county, including its urban centre Gyêgu, an extraordinarily dynamic centre with traditional and modern extra-regional trade connections (the town was largely destroyed by an earthquake in 2010).

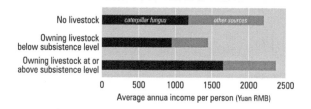

Figure 11.6   Share of caterpillar fungus of total income in different household groups

Source: Author's survey (2007).

own consumption. Only genuine peasant households manage this. Traditionally, trade played a considerable role for nomads, yet the door to business generally remains closed to those who do not have enough income to invest in goods and transportation. In addition, pastoralists often live some distance away from towns where construction sites and retail trade offer low-wage job opportunities. Figure 11.6 reveals how little income is derived from these other sources and how important cash income from caterpillar fungus is.

The widespread dependency on caterpillar fungus is, however, not a guarantee for a substantial income. Analysing earning from caterpillar fungus sales is thus crucial. To contextualise the findings the regional context of the research layout is important (see Fig. 11.7): The majority of sample households do not live in the most renowned places of caterpillar fungus recovery. These are mostly located in the prefecture's southwest, notably in Zadoi county and a few spots are in Nangqên and Zhidoi. Nor are the sample households from the few Yushu districts that do not collect fungus. The sample hence represents households with mediocre access to caterpillar fungus, given their territorial setting. Yet, as Figure 11.8 reveals, even in this sample, the average income derived from collecting and selling the fungus is significantly higher than any other locally yielded profit. There is only one very particular exception in the trade sector; namely, the breeding and marketing of a certain kind of dog, the Tibetan mastiff boosts earnings.[20] However, if the three cases of dog breeding households who stated their profits are considered separately from the trade and shop category, then the income type remains below average earnings from caterpillar fungus.

According to a semi-official website,[21] Qinghai province yields between 20 and 50 tons of caterpillar fungus per year. Given the prices at the time of this survey this

---

20   This dog has become a fashion among rich people in the west and in mainland China, where the newly rich purchase many of these dogs to show off. In Yushu, the demand for this breed has initiated a kind of 'dog breeding fever' since about 2004/2005. Due to a widespread readiness to get into debt in order to start the very profitable breeding business, combined with the dogs' susceptibility to diseases (distemper), the business is extremely risky.

21   Available at: http://www.qhei.gov.cn/qhly/gytc/mgyc/t20060420_203261.shtml [Accessed at: 03 August 2007].

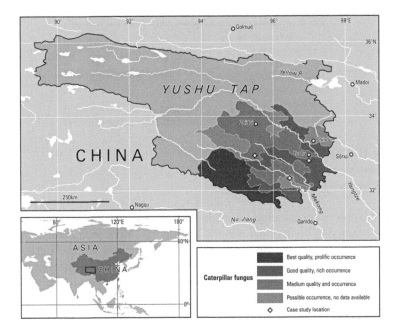

Figure 11.7   Map of caterpillar fungus recovery in Yushu
Source: Author's survey (2007).

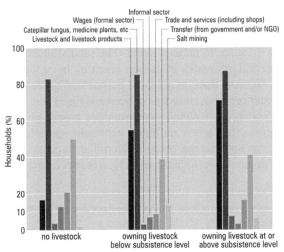

Figure 11.8   Use of different income opportunities in Yushu
Note: N = 296 is 100%
Source: Author's survey (2007).

What do you consider the most successful strategy to secure your livelihood?		
Strategies *(N=281)*	*number*	%
To get more livestock	208	74.0
To collect caterpillar fungus	196	69.8
To get more pastureland	123	43.8
To get a job (regular or irregular)	67	23.8
To do business	32	11.4
Other	29	10.3

Figure 11.9   Livelihood strategies
Note: Multiple answers possible.
Source: Author's survey (2007).

translates into an annual turnover of 1 to 2.5 billion Yuan (approximately US$ 350 million) earned by traders and pastoral households. Most of the fungus-yielding regions are situated in the Tibetan prefectures of Golog and Yushu.[22] Cash income earned from caterpillar fungus in these two major caterpillar yielding prefectures is thus roughly five times higher than the official annual budget of the entire Yushu TAP, subsidies from Beijing and Xining included. Due to the high vulnerability of the present-day pastoralists, and due to the political sensitivity of Tibetan areas, pastoralists in Qinghai – unlike other primary producers in China – are usually not taxed. All this explains the outstanding significance of caterpillar fungus in Tibetan nomad areas.

The importance of the income from collecting caterpillar fungus is also reflected in people's views on the best strategies to secure their livelihood. Two answers are almost of equal importance: A slight majority believes the best strategy remains getting more animals. This reflects the fact that most households do not possess enough livestock to make a living. The second most important strategy is 'to collect caterpillar fungus', an answer given by more than two thirds of all households and far more than 'to get more pastureland'. Alternative strategies of getting a job or starting a business are not favoured (see Fig. 11.9).[23]

Alongside decreasing livelihood opportunities based on animal husbandry, caterpillar business offers a new unfolding economic space for Tibetan nomads. This, of course, can only happen because caterpillar fungi are (a) typically found in parts of their pastureland; (b) because it requires special experience to find it, a

---

22   Interestingly enough, official provincial and prefectural statistical yearbooks do not even mention caterpillar fungus. It is likely that the figure listed in the website is calculated from the caterpillar fungus processed and packed in Xining city and the prefectural seats of the province. Hence, we have to assume that the actual output of caterpillar fungus in the province is much higher, since middlemen of the trade (Tibetans and Muslims coming to Yushu after the end of the digging season) are also in direct contact with wholesalers (so called *da laoban*, 'Big Bosses') in Chengdu, Guangzhou, Shenzhen and other places.

23   Several answers were accepted by the interviewers as the interviewees did not want to restrict themselves to a single choice.

skill which is more often found among nomads than among other people;[24] and (c) because the financial threshold to enter the business is low, instead knowledge of high altitude pastoral areas are required.

To make a long story short: In the ideal set of circumstances, pastoralists simply go to their pastures and dig for caterpillar. At the end of the season, collecting is followed by trading caterpillar fungi to middlemen or consumers. Caterpillar has become the most important mobile source of cash income for the majority of pastoral Tibetan households in Yushu (Fig. 11.8). Those who use caterpillar fungus as resource for cash income outnumber stock-raising families by 46 per cent.[25]

Market Relations

Income from caterpillar fungus enables the pastoralists of Yushu to adapt to the consequences of natural, demographic and economic changes in their living space. It allows many pastoralist households to widen their range of economic activities and to develop new forms of social mobility. In this, the use of modern means of communication for marketing this commodity has become a matter of course – be it the use of modern transport, mobile phone or the internet. Thus, the digging for and trading of caterpillar fungus has markedly changed pastoral strategies in Yushu's rural areas and established marketing chains that obviously reach beyond regional borders. This allows nomads to not only safeguard their habitual pastoral livelihood, but also to invest into small private enterprises (like mini buses/taxis, trucks for transport and trade, dog raising), and to participate in the intermediary trade and thus start to take control of (parts of) the market chain.

So what are their market perspectives? While for yak and mutton there is a limited market perspective, such as participating in the 'green food' business, medium-term prospects for pastoralists to sell caterpillar fungus are far more favourable. The ecological basis of this commodity is rooted totally in Tibetan nomad areas and, therefore, puts them in a position to profit from a monopoly situation – as long as they are able to gain control of, or participate substantially in the market chains – not only as basic suppliers, but also as traders who reinvest the money earned in Yushu proper.

This had not been the case in the meat market chains. Before the dissolution of the commune system, meat, wool and other pastoralist commodities were almost exclusively marketed through the state. Economic liberalisation a priori entailed

---

24   The number of caterpillar fungus pieces an individual can find each season varies greatly by individual and natural conditions of the pastureland. Children are often said to be especially good at finding caterpillar fungus. Interviewees in Sulu xiang and other parts of Zadoi county noted that 20-50 pieces a day per individual is normal in Sulu, while in Golog Costello (2008) was told of averages between 600 and 1,000 in a season.

25   Only 57.8 per cent of the pastoralist households surveyed still owned livestock while 84.5 per cent used caterpillar fungus as resource for cash income (Gruschke 2009).

some surplus which was subsequently sold through a system of regional and intraregional market chains. Typically these markets are controlled by outsiders. Pastoralists do not act as intermediaries in meat commodity chains – except for some local rural-rural and rural-urban market chains. The supra-regional rural-urban and the urban-urban chains are dominated by external stock traders. Those oriented towards the north are mostly controlled by Hui Muslims from Xining (the provincial capital) and sometimes from neighbouring Gansu province. As the Moroccan case study also reveals (Breuer and Kreuer in this volume), marketing profits deriving from temporary and spatial price differences, are not realised by the pastoralists, but by the external intermediaries. As for the 'southern market chain' aiming at the TAR, notably Lhasa, both buyers (who in the rural-rural case might be pastoralists as well) and traders (interregional rural-urban chain, Lhasa city dwellers, among others) are TAR Tibetans. As the Yushu nomads' 'wait-and-see' sale strategies are essentially based on herd reproduction cycles and on road conditions they cannot be considered to be trade market oriented to maximise profits. Instead a subsistence orientation, aiming to spread economic risks, guides economic action. The same is true for other livestock commodity chains, such as sheep's wool, yak hair, skins and dairy products.

Correspondingly, we have to ask: Are caterpillar fungus commodity chains comparable to meat and livestock market chains? Until recently the answer would have been a clear 'yes'! Year after year, in May, June and July the central plazas, major roads or community centres in Yushu's counties, gradually fill with Hui Muslims of northern Qinghai, of Gansu and Ningxia.[26] For a decade or two, they also came in large numbers to dig for caterpillar fungus, but those in the main streets and on the market-like open spaces in towns came predominantly to trade the 'gold of the grassland'. Caterpillar fungus trade has given rise to the special occupation of small, medium and big intermediary traders who connect with the *da laoban* – the 'big bosses'. The numerous middlemen's task is to buy single pieces and small sets of caterpillar fungi, and resell them according to the local supply and their own financial resources. Thus, the fungi may be resold several times in the same community and town before being transported to a city and resold in larger quantities. The 'big bosses', generally residing in big Chinese cities or Hong Kong, are never seen in the streets, and only buy big deals. While they are often said to be Han Chinese, almost all the middlemen active in Yushu and Golok were Hui Muslims – until about 2005.

The following year this changed dramatically. In July and August 2005, about 80 per cent of the intermediary caterpillar fungus traders on Gyêgu's big central square (*da guangchang*) were, by all appearances, Muslim men. By the summer of 2006, this situation had been reversed: ten to 20 per cent were Muslim traders, with the rest being Tibetans. Among the latter are salespeople from other east Tibetan towns like Qamdo, Ngawa, Kangding or Qabqa (Hainan) and even Lhasa.

---

26   While Muslims dominate the scene in Yushu and other Qinghai nomad areas, in neighbouring Sichuan the intermediary traders are said to be mainly Han Chinese.

The bulk, however, were local Tibetans, more pastoralists than peasants,[27] who had made their way into the caterpillar fungus business.

What enabled Tibetan nomads to take over from the formerly omnipresent Muslim agents? During the last decade, with ever-rising profits from resource turnover, more and more outsiders entered Yushu (and other respective areas) during the collecting season – April through June. After a number of local conflicts over land access, a major escalation between villagers from Jiduo and Sulu communities in 2005 led to a week-long clash at a bridge across the river Chi Chu in Zadoi county. The villagers barred outsiders – east Tibetans from the TAR and Sichuan and Muslims from Qinghai and Gansu – from entering their pastures to collect fungus. As those barred had paid a fee to the authorities to get access, they started a fight for what they saw as their right. During the clashes at least one person was shot dead.[28] Authorities could only stabilise the war-like situation after armed police were deployed and outsiders returned home. Tension persisted and led to the semi-official barring of outsiders from Yushu during the caterpillar season in 2006. Since then, access in the whole of Yushu has been restricted during the collecting season; only Yushu people are allowed to go 'fungus hunting'. In 2006, more than 50 checkpoints and road-blocks were – semi-legally, as it were – organised by the community people and their leaders, tolerated and, later on, quasi-legalised by the higher administration.

As collectors from outside were locked out during the season, so were the intermediary traders who ordinarily came in as soon as the trading season started. Could a better situation exist to let Tibetan pastoralists familiar with the system, jump at the chance to catch hold of the marketing chain? By the time outsiders were allowed in again in July 2006, much of the intermediary dealing of the fungus had already been done. Although intermediary trading takes place over the whole year (at least in the prefectural seat Gyêgu), traders who are present in the most important initial phase of the collecting season, gain the highest profits as prices are low due to large supply. Prices of the commodity do not increase evenly: They are at their height before the season starts, since by then the markets are relatively empty, and at their lowest when the nomads pour into the county towns to sell their fungi, get cash and begin their procurement of annual provisions, consumer goods and durables. This time is also most attractive for other merchants and artisans, dealing with and processing jewellery, clothing and any kind of electric appliances.

Pastoralists engaged in the caterpillar marketing did not restrict themselves to the simple intermediary trade. Among fungus dealers, certain hierarchies have evolved, hence sale methods and strategies have diversified and improved.

---

27   In 2006 there were also some monks and women among the new intermediary traders.

28   The number of casualties reported grew with the distance: In Gyêgu I was told that three or four were killed, while a Xining taxi driver talked about two dozen. Nomads of Sulu, who were engaged in the fighting, assured me that there was one casualty among the outsiders – shot by the people's armed police.

Apart from a multitude of small scale traders who buy single pieces and small quantities of caterpillar fungus, more and more Tibetans have developed direct business contacts with partners in the consumer markets – in the far-distant Chinese megacities. Since the *da laoban* are never present in Yushu, the exact sequences of their part of the commodity chain still needs further investigation. It is essentially reminiscent of a western stock market: The dealers in Yushu – generally pastoralists in business suits at the big plaza – just observe the hustle among the sellers and buyers. Armed with their mobiles they are in permanent contact with their business partners in Chengdu, Guangzhou, Shenzhen, Hong Kong or other cities. These keep them up-to-date with price developments. Like stockbrokers, the new pastoralist-businessmen buy large deals of the fungus if the price difference becomes promising. For fast transport, local cars are hired – ordinarily new Japanese or joint venture jeeps – and the freight is transported to Xining within one day. From there, the commodity's distant place of destination is reached by aeroplane.

A turnaround of pastoralists' share in the commodity chain has occurred. Within only two years, they managed to change their relatively passive, and at most, secondary role as caterpillar fungus dealers, to become major stakeholders within and outside of Yushu. Their formerly weak presence in the markets has metamorphosed to their absolute dominance over the previously omnipresent Muslim traders. And their business connections, meanwhile, reach far beyond the region, beyond Xining into the Chinese mega-cities on the wealthy coast. There, as informants related several times, one can also find Tibetan 'big bosses' who are active on all levels of the commodity chain. Towards the top level, they certainly 'thin out', yet formerly they were not even close to this. There are about a dozen Nangchenpa – pastoralists from Nangqên – among them. This is not unexpected, since business in China and among Tibetan pastoralists as well, depends very much on long-standing personal contacts. This gives them an edge in the distant markets, which are also partly international markets. Hence it is not the characteristics of a former and still-nomad society – animal husbandry and its commodities – that relates them to the world market. Instead, it is the peculiar niche market of a very special tonic endemic to the pastoralists' living space that finally enabled them to hold catch of and be present on all levels of a commodity chain originating in their midst.

Social Impact of Caterpillar Fungus

Caterpillar fungus business certainly entails a further differentiation of pastoralist society, since not everybody has easy access to the fungus yielding pastures, and not all the pastures produce the same amount and quality. There is, however, enough reason to believe that pastoral communities able to subsist from animal husbandry even in the new cash economy, have similarly shaped social differentiation. If it is difficult to judge the impact of caterpillar fungus on inequity, it is clear that the new business prevents the social gap between countryside and urbanised towns from dramatically widening.

However, the subsistence-oriented marketing system of animal husbandry has been replaced, or at least supplemented, by the intensive use of a new resource entailing considerable changes in the pastoralists' role in the commodity chains. Although the caterpillar fungus trade is not a product of economic liberalisation,[29] what is new is its dimension and implication on nomadic society. It is easy to imagine that this change of economic spaces has manifold effects.

Due to natural and demographic restrictions, a pastoralist livelihood at subsistence level is no longer possible for most of Yushu's nomads. The speeding-up of rural out-migration would be inevitable. This has already increased. Survey data reveals that unsolicited migration has been larger than resettlements triggered by state programmes.[30] Both outward migration and staying in the pastoral countryside are linked to caterpillar fungus. People who would otherwise have to give up animal husbandry are able to continue their way of living, since economic deficits deriving from livestock production are compensated by caterpillar cash crop. As for migrants to urban areas, the bulk of them are young and middle-aged who use their income from caterpillar fungus to develop new economic opportunities and lifestyles.[31]

Still, nomad areas in Tibet, and more so in eastern Tibet's caterpillar fungus yielding areas, have witnessed a growing number of conflicts. While in the 1980s, such conflicts were addressed as disputes over grazing areas, in recent years they have been regularly related to collecting caterpillar fungus. This is less the case in communities where output and quality of the resource are low, than rather in areas where it abounds.

Funds of formerly unknown size, to which pastoralists *may* now gain access – with most of them still remaining on a precarious financial level – allow new forms of mobility. Social mobility is nothing new to Tibetan society, partly owing to the Buddhist system and the implicitness of every household sharing in the monastic system. Yet, social mobility goes far beyond the religious sphere now. The high spatial mobility pastoralists are accustomed to, is combined with a new financial, and consequently economic mobility and adaptability. This certainly owes more to the caterpillar fungus business widening the economic space than to any other economic or administrative development measures – be it by the state or foreign NGOs.

---

29   Chinese socioeconomic investigations of the 1950s included records of the 1930s in which caterpillar fungus was listed among the 20 major commodities traded out of Yushu (Personal communication of Tashi Gonchab, Yushu, August 2006).

30   I know that there are other regions, such as Golok, where forced resettlement happened more often than not – which by nature of the Chinese central government's conception of the programme should be considered illegal.

31   Cp. Gruschke (2008 and 2009).

Acknowledgements

This chapter represents a result from the work of the Collaborative Research Centre SFB 586 'Difference and Integration' (www.nomadsed.de) that is funded by the German Research Foundation.

References

AIECE-Prognose. 2000/2001. *Weltrohstoffpreise.* (ifo Schnelldienst 31/1999). [Online]. Available at: http://www.cesifo-group.de/pls/guest/download/ F70561/SD31TEXT.HTM [accessed: 12 September 2007].

AWI (Australian Wool Innovation Ltd.). 2005. *AWI Woolfacts.* [Online]. Available at: http://www.wool.com.au/attachments/Education/AWI_WoolFacts.pdf [accessed: 12 September 2007].

Costello, S. 2008. The flow of wealth in Golog pastoralist society: Towards an assessment of local financial resources for economic development, in *Tibetan Modernities. Notes from the Field on Cultural and Social Change.* (PIATS 2003: Proceedings of the Tenth Seminar of the IATS, Oxford, 11), edited by R. Barnett and R. Schwartz. Leiden/Boston: Brill, 73-112.

Goldstein, M.C. 1996. *Nomads of Golog: A Report.* unpublished manuscript. [accessed: August 2006].

Gruschke, A. 2008. Nomads without pastures? Globalization, regionalization and livelihood security of nomads and former nomads in Kham, in *In the Shadow of the Leaping Dragon: Demography, Development, and the Environment in Tibetan Areas.* (Proceedings of the 11th Seminar of the IATS, IATS 2006, JIATS 4), edited by K. Bauer, G. Childs, A. Fischer and D. Winkler.

Gruschke, A. 2009. *Ressourcennutzung und nomadische Existenzsicherung im Umbruch. Die osttibetische Region Yushu (Qinghai, VR China).* PhD Thesis, Universität Leipzig.

Gruschke, A. and Winkler, D. 2010. *Yartsa Gunbu and Yak Herders. Cordyceps Sinensis and the Economic Transformation oft he Tibetan Plateau.* (Interdisziplinäre Betrachtungen zu Themen in Asien, 1). Freiburg.

He, A.X. and Li, L. 2004. Study on the relation between yak performance and ecological protection, in *Yak Production in Central Asian Highlands.* (Proceedings of the Fourth International Congress of Yak), edited by J. Zhong, X. Zi, J. Han and Z. Chen. Chengdu: Sichuan Publishing House of Science & Technology, 93-95.

Miehe, G. and Miehe, S. 2000. Environmental changes in the pastures of Xizang, in *Environmental Changes in High Asia.* (Marburger Geographische Schriften 135), edited by G. Miehe and Y. Zhang. Marburg, 282-311.

Miller, D. 2001. *Poverty among Tibetan Nomads in Western China: Profiles of Poverty and Strategies for Poverty Reduction.* Paper prepared for the Tibet Development Symposium: Brandeis University. May 4-6, 2001 (online).

Nori, M. 2004. *Hoofs on the Roof. Pastoral Livelihood on the Qinghai-Tibetan Plateau. The Case of Chengduo County, Yushu Prefecture.* ASIA. [accessed at: May 2005].

QPSB (Qinghai Province Statistics Bureau). 2005. *Qinghai Statistical Yearbook.* Beijing: Zhongguo Tongji Chubanshe.

Wiener, G., Han, J. and Long, R. (eds). 2003/06. *The Yak.* 2nd Edition. Bangkok: FAO, Regional Office for Asia and the Pacific.

Winkler, D. 2004. Yartsa Gunbu – Cordyceps sinensis. Economy, ecology & ethnomycology of a fungus endemic to the Tibetan Plateau, in *Wildlife and Plants in Traditional and Modern Tibet: Conceptions, Exploitation and Conservation.* (Memorie della Società Italiana di Scienze Naturali e del Museo Civico di Storia Naturale di Milano, 33(1)), edited by A. Boesi and F. Cardi, 69-85.

Wu, N. 1997. *Ecological Situation of High Frigid Rangeland and Its Sustainability. A Case Study on the Constraints and Approaches in Pastoral Western Sichuan/ China.* (Abhandlungen Anthropogeographie, 55). Berlin: Dietrich Reimer Verlag.

Yan, Z., Wu, N., Yeshi Dorji and Ru, J. 2005. A review of rangeland privatisation and its implications in the Tibetan Plateau, China. *Nomadic Peoples,* 9, 31-51.

Yushu Survey = Yushu Zangzu Zizhizhou Gaikuang Bianxiezu [Survey of the Tibetan Autonomous Prefecture of Yushu Compilation Group]. 1985. *Yushu Zangzu Zizhizhou Gaikuang [A Survey of the Tibetan Autonomous Prefecture of Yushu].* Xining: Qinghai Renmin Chubanshe.

YZT [Yushu Zangzu Zizhizhou Tongjiju (Yushu Tibetan Autonomous Prefecture Statistics Bureau)]. 2000. *Yushu Zangzu Zizhizhou Tongjiju Nianjian 1950-1999 [Yushu Statistical Yearbook 1950-1999].* Yushu: Yushu Zangzu Zizhizhou Tongjiju.

YZT. 2006. *Yushu Tongjiju Nianjian 2005 [Yushu Statistical Yearbook 2005].* Yushu: Yushu Zangzu Zizhizhou Tongjiju.

Chapter 12

Capitalism in the Tundra or Tundra in Capitalism? Specific Purpose Money from Herders, Antlers, and Traders in Yamal, West Siberia

Florian Stammler

Velvet antlers, or in Russian *panty*,[1] have a long tradition as a medicine in Asia. According to old Chinese sources, the extract of velvet antlers 'reduces hot temperedness, dizziness, strengthens male kidneys and testacles, cures involuntary ejaculation of male semen during sexual intercourse with a ghost during the sleep' (Iudin 1993: 3). This sounds like healing from the medieval idea *sacubus*, a female ghost lying on a sleeping man. Today many people in Asia believe in the mysterious qualities of velvet antler extract, for male sexual performance and strengthening a person's overall condition. In this chapter, I draw connections between these 'believers' as potential customers and the reindeer herders as the producers of this mysterious medicine, *panty* in Siberia, who mostly are 'non-believers'. Based on an ethnography conducted in 2001 and shorter pre- and post-studies in 1995 and 2010, I show that the growing importance of national and international trade in this commodity affects the way of life of arctic reindeer nomads, and how they react to integration into a network of worldwide exchange. I argue that it is because of the interest in this commodity, *panty*, that we can talk about the globalisation of reindeer herding. I investigate the history and impact of *panty*, arguing that herders' increasing connection to world markets has not resulted in significant marginalisation of reindeer herding, nor did it turn upside down the other factors in reindeer herding, such as migration, meat production, and subsistence. Subsistence livelihoods based on reindeer and fish remain extremely important. Rapidly changing developments in the globalised economy are seen as a welcome addition that can usefully supplement the core economy and socio-culturally significant practices of the tundra.

---

1   Although being aware of the English meaning of this term, I would like to encourage the reader to accept this as the Russian 'terminus technicus' throughout this contribution.

Fresh Velvet Antlers: Historical-Biological Background

Originally, *panty*/fresh velvet antlers, were not a Northern commodity. Before Perestroika, the southeast Asian markets were satisfied with raw material from New Zealand, and from poached animals of different origins. Within Russia, the most important source animals for this raw material are marals (*cervus elaphus maral*) in the Altai mountains. We know that harvesting *panty* from marals occurred from the early 1930s on, although the quantity produced was minimal (Iudin 1993: 5). Soviet state planners had an ambiguous relation to the animal product. On the one hand, in the Russian Far East, state- or collective farms started to produce reindeer *panty* in 1971, approximately 8 tons per year (Iudin 1993: 6). In Soviet laboratories in Magadan and Yakutsk, research had been carried out to prove the medicinal effects of *panty* extract, 'on the basis of experience of Tibetan and Chinese traditional medicine' (Iudin 1993: 5). This is true for both reindeer and for maral *panty*. The relevant ingredient in fresh antlers is *pantocrine*, and sometimes for reindeer *rantarine*. *Panty* from northern reindeer has identical ingredients as *panty* from other deer and marals, with the only difference being that they have half the saccharin of more southern antlers. *Panty* are rich in amino-acids and many other elements, analysed in detail by Russian scholars (see Iudin 1993: 63-69). However, research about medical effects still is very unsatisfactory. So far we know that giving *pantrocrine* to mice over a long period increases testosterone, and protein in the liver. We know that in humans, performance on a velo-ergometer increases when *pantocrine* is consumed over a long period (Iudin 1993: 66). Research has also shown, according to Russian sources, that *panty* can indeed reduce sleeping problems, headaches, and dizziness (Iudin 1993: 73). On the other hand, Soviet authorities were very sceptical about the *panty* trade, arguing that selling *panty* to Chinese is sharlatanery (Iudin 1993: 6). However, economic reforms in China created a greater demand for this kind of product, and Perestroika in the Soviet Union permitted the establishment of extensive trade networks between South East Asian countries and Russia. It was only after Perestroika that *panty* became significant as a source of income for reindeer herders. In contemporary Russia, *pantrocrine* is sold as having similar effects as *echinacea*, strengthening the overall condition of the human body.[2] As well as being used for medicinal purposes, *panty* are also consumed as dried chips in expensive restaurants in China and other Asian countries.

The North as a Producer of the Raw Material

Harvesting velvet antlers from northern reindeer has only recently been on a large scale. Although the first Chinese bought reindeer *panty* as early as 1910

---

2   See for example the instructions for use for 'cigapan', which can be purchased in pharmacies in Russia.

(Iudin 1993: 62), real production only started in the 1970s, mostly in state farms in the far east of Russia, and in rather small quantities. The real boom began in the beginning of the 1990s, when the limits on entrepreneurial activities in the Soviet Union ceased, and simultaneously the need for additional income increased because of the lack of state support. West Siberia was the only region in the former Soviet Union to experience growth of domestic reindeer after the end of the planned socialist economy, and it was here where building up a velvet antler trade network was most promising. The Yamal-Nenets Autonomous Region, where I did my fieldwork, today has the world's largest herds of domestic reindeer, with 651,100 animals (Ministry for Agriculture 2009). This is also the number one gas producing region in Russia. In comparison to the income from gas exports, reindeer herding has close to no economic importance. However, for the approximately 15,000 mostly indigenous people directly occupied with herding reindeer on the tundra, it is their basic means of subsistence and their main means of income, as well as an important symbol of ethnic identity. All reindeer herding is done in a highly mobile way, with herders performing extensive migrations varying from 100 to more than 1,000 km a year (see Fig. 12.1). Whereas in Pre-Soviet and Soviet times, reindeer herding was mainly for subsistence, meat production, and transport, today many reindeer herders say that producing *panty* has become at least as important as meat production for their income. This changing economic orientation of the reindeer herders influences their day to day interactions with their animals, as I show below.

Velvet Antler Production: A 2001 Ethnographic Account

Velvet antler (*panty*) production is closely interlinked with the seasonal nomadic cycle of reindeer herders. In spring most reindeer herders migrate with their herds towards the summer pastures in the North of the Yamal peninsula (see Fig. 12.1). This is the season when the antlers begin to grow. Reindeer (*rangifer tarandus*) is the only animal where both male and female grow antlers, and also bulls and castrated bulls. This makes reindeer, unlike other deer, an exploitable raw material for *pantocrine*. Depending on the weather and the quality of pasture, the antlers reach their full size, up to 50 cm, towards the middle/end of June. There is then a four-week optimal period for cutting the fresh antlers. However, it is usually done a little later than this ideal time suggests, from mid-July to mid-August. The process of cutting antlers is very time consuming, and has turned summer, which was previously a calm season for both reindeer and herders, into a very busy period. The deer are driven into a corral in the tundra, and the *panty* animals chosen and separated from the rest of the herd, either by lassoing them or by rounding them up in front of the nomadic camp. Although all reindeer produce antlers, most herders choose to cut them only from castrated bulls or those bulls that are not considered to have very good reproductive capacities. Reproductive bulls are considered to be weakened unnecessarily by *panty* cutting, and females

Figure 12.1   The Yamal-Nenets Autonomous Okrug, Northwest Siberia, Russia

Source: Author.

need their antlers to compete for the best pastures in autumn and winter, also, they should not be unduly stressed, as their main task is feeding their calves, who are usually two to three months old during the *panty* harvest. The cutting itself is done by a team of several herders per deer, since one or two men have to keep the animal calm while one is sawing the antlers. Ideally, after the cut, the wound which is normally bleeding is treated with a bandage and iodine to stop bleeding and keep the wound clean.

The most important challenge in the process is getting the raw material for the commodity *panty* within the time schedule. The success of the whole production depends on the interaction between the reindeer herders and the enterprise collecting the cut antlers. The later *panty* are cut, the more bone they contain, which means a decrease in quality and a reduction in the price. Therefore, the younger antlers are better. Secondly, *panty* must be collected as soon as possible after cutting (in one or two days) or they begin to rot. This is why the 'producers' (for example reindeer herders) and the 'collectors' (for example traders) have to agree on an exact date for cutting and collecting the *panty*.

I will give one example of a herding family where I stayed in the summer of 2001 to illustrate this process: Anniko and his family have a midsize reindeer herd of slightly more than 1,000 reindeer in North Yamal. In June, the antlers are already quite well developed, and they know that soon the helicopter will come. The collecting enterprise normally flies by helicopter to each reindeer herder camp initially to agree on an exact date for the collection of *panty*. Imagine the local knowledge of the land possessed by those who direct the helicopters: They have to know where several hundred nomadic camps are situated in the tundra on any given day. So they fly to Anniko to ask him how many kilogrammes of *panty* he wants to cut this year. He answers 300 kg. Then Anniko chooses from the goods in the helicopter offered by the trader: fresh bread of the day, tea, tobacco, noodles, spices, newspapers, books, rubber boots, and other items. During this first flight, the enterprise brings to the tundra a variety of the most common goods, which reindeer herders normally buy. Anniko and all herders order the goods in advance, before having cut the antlers. The entrepreneur writes the price of the purchase in his account book, and when the *panty* are collected, he counterbalances the given *panty* (payment) against the sold goods, and sees whether there are debts or profits. The first visit of the trading helicopter is also when herders may put in an order for other items. The entrepreneur records both the order and the prices.

This first flight is necessary because most private camps are not connected by any means of telecommunication. Only through personal communication does an entrepreneur know what his customers would like to buy this summer, how many kilogramme of *panty* they want to sell, and on exactly which date he will be able to collect the *panty*. Cash at this stage does not play any role. The whole pricelist of goods is calculated in kilogramme of *panty*. During the first visit of the trader, no *panty* are collected, since they are not yet cut. The second helicopter flight collects the *panty* one or two days after the reindeer herders have cut and packed the *panty* in linen sacs. Only then will Anniko know whether he sold more or less *panty* than goods he received in advance. With the second flight, global culture enters the tundra. Before the *panty* is loaded on the helicopter, the herders receive the goods that they ordered. This time Anniko got a spare piston for his snowmobile, batteries for his Chinese tape recorder, some tapes with recent Russian disco music for his sons, canvas for his *chum* (nomadic tent), and ammunition for his gun. Moreover, he got some more food since he had run short of bread and tea. Herders insist that the *panty* is weighed with scales before they are loaded on the helicopter to avoid cheating. Anniko sold 300 kg, which equates to 2/3 of the price of a *buran* (snowmobile) (he already has two that do not work very well). Normally, Anniko does not know the actual price of the goods he ordered and purchased. He trusts the entrepreneurs, because he knows them, and they know him. They are his only interface to the outside world, because only they know how the herders migrate and where they will be during *panty* cutting time. If one entrepreneur charges dishonest prices for Anniko, he will work with another one next year. This time it turned out that he received less in product from the helicopter than he sold *panty*, which means that the trading enterprise has a debt to Anniko. He can take goods equal to this sum in early winter when he migrates to the trading post. He thinks that in recent years *panty* has become as important as a means of income as meat. This is why he slaughtered less male reindeer, resulting in a slightly higher share of males in his herd, as he does not want to cut antlers from his females. The fact that many herders made similar husbandry decisions lately resulted in what Anniko's neighbour calls 'the overcrowded tundra' (*tundra perepolnilas*). Many young people think that living in the tundra as a reindeer herder, producing *panty* and meat, promises a better economic return than living in a village with a badly paying job.

Once the *panty* has been collected, Anniko and his family are happy, because the rest of the summer is a quiet period, until slaughtering begins in October. However, his son Nikolai mentions how the growing market competition for *panty* collection heavily affects their lives. He reported that in the space of one month in 2000, eight helicopters flew to his camp in the tundra to ask for *panty*. Anniko decided to give smaller amounts of *panty* to each of these enterprises. Nikolai welcomes this development. He says in the best cases, the price of goods decreases as a result of more competition. Anniko, his father, however, complains about how much work it is, so that he and his wife do not succeed in repairing sledges, tent covers, clothes, harnesses and the like. In general, the summer has become too hectic, which is why they decided to give *panty* only to two enterprises next year.

How *Panty* Enter the Global Economy

Once the *panty* are collected, the entrepreneur flies them to his base village and puts them in a natural freezer, which is an underground ice chamber in the permafrost. Here, the *panty* are kept for 20 days. After this period, Anniko's and other's *panty* are shipped to one of three centers where the headquarters of the vertically integrated reindeer enterprises are located. When the *panty* are cut, a herder such as Anniko gets the equivalent of around 150 Russian Rubles/kg or US$ 5 (in 2002). There are numerous ways the material reaches its East Asian destination market. The collecting enterprise has several options to achieve this:

- shipping panty fresh-frozen to Moscow (frozen wagon/container), receiving a maximum price of US$ 30/kg from Chinese or Korean businessmen, who take over and organise the drying and shipping to their home country;
- transporting it only as far as the village, and selling it to the first middleman for US$ 10-15/kg, who will ship it by train to Moscow, and receive US$ 30/ kg from Korean or Chinese traders;
- organising transport to the regional centre Salekhard, drying and processing the panty, and selling it for US$ 60/kg to Asian traders who come directly to Salekhard to organise their own shipping to China. This is also the way

panty was originally traded in 1910. The Chinese came to the source of panty arriving from the tundra and bought them for cash. The local trader can also organise transport of dry panty to Moscow and sell it to Asian traders for US$ 60/kg. Both these options promise more profit, as initially processed material is sold instead of raw material. Recently a second 'unofficial' drying facility opened in Salekhard to cater for this.

I want to draw attention to the fact that cash money only enters the antler business after the material has been shipped to Salekhard. As soon as the *panty* leaves the control of the collecting enterprise, the raw material becomes a commodity. In Moscow, the *panty* business is highly flexibly and unstable. The enterprise bringing *panty* to Moscow calls the phone number of a private house, normally where Koreans or Chinese live. He announces that he has a certain amount of *panty* to sell right on the spot. He gets a call back from a Korean or Chinese businessman, who then comes to meet him and pays for the *panty* in cash. This businessman normally does not disclose his identity, nor does he give his own phone number. He organises shipping to his home country, along unofficial trading routes in order to avoid problems with taxes and border guards (bribing increases the costs). In China or Korea he sells the *panty* again either as extract for medicine, or chips for restaurants. Unfortunately, from the point when the *panty* arrives in Moscow, we know almost nothing about further trade lines. These are dominated by mafiotic structures, which change very quickly.

The Nature of Entrepreneurship in the *Panty* Business

In this section I look in a more abstract way at the categories of entrepreneurs in this business. I suggest that this is a case, where a new kind of native entrepreneurship develops, which engages in all trading areas of the business. Caroline Humphrey (1999) introduced a typology of post-Soviet traders and entrepreneurs, which provides a good framework to understand how various people who make money from *panty* differ. A general characteristic of trade in post-Soviet provincial Russia is the importance of what Humphrey called 'trust networks' (1999: 45). These imply that informal social contacts are more important than official business relations, although this is often hidden from the outside world. Since *panty* trade works with very broad ranges of profit, and is mostly on the fringes of legality, these networks have a crucial impact on the *panty* economy. The common term in Yamal and elsewhere in Russia for traders is *kommersant*. The only over arching definition applicable everywhere in Russia is that 'traders' are 'those who aim to profit from middlemen activities involving goods and services' (Humphrey 1999: 24). Humphrey (1999: 38) thinks of *kommersanty* as 'trader retailers', who can afford transport, buy local products in the villages (or in the tundra) and sell vodka and other goods. In Yamal these traders make up only a small portion of the phenomenon which is referred to as *kommersant*. The main type of traders is defined as 'shuttlers' by Humphrey, non-locals travelling to foreign countries to bring in all kinds of products for sale (Humphrey 1999: 19). The source of their profit is the price difference between the place where they obtain their products and where they sell them (Humphrey 1999). Additionally, Humphrey introduces four other categories of traders in Russia: the small scale resellers (*perekupshchiki*, inside the borders); entrepreneurs (*predprinimateli*); brokers (former *snabzhentsi*), who have an intimate knowledge of the production and producers, having a rather modest profit range; and businesspeople, who can also invest in production, but are typically engaged in buying and selling waste, metals, cars and the like (Humphrey 1999: 34-38). Humphrey's main point concerning all these categories is that traders controlling the market always come from outside the region, and have no interest in the region as such. Their only incentive is the price difference available between locations.

In Yamal, a kind of *kommersant* developed with the *panty* economy that does not fit with this pattern, but combines all of these categories into one, namely because these traders are local, they are indigenous, most of them have direct kinship ties to the tundra reindeer herders, and they are engaged in the 'civil society of reindeer herders'. When I did fieldwork, there were 3-4 enterprises that controlled the Yamal *panty* economy completely. All their bosses are well educated Nentsy, having previously worked in the *sovkhoz*[3] headquarters, as leading zootechnicians, the person responsible for husbandry decisions and for the health of the reindeer herds. This job has proven to be an ideal starting point for a career as a local *kommersant*. They were not satisfied by the poor performance of *sovkhozy* after Perestroika, felt themselves overqualified for a job with few prospects. Having worked with reindeer herders all their working life, in younger years spending entire seasons on the tundra with the herds, these people have all the knowledge required concerning quality of reindeer products and the organisation of the nomads' mobility. On the other hand, having worked in the *sovkhoz* headquarters, they have also built up networks with markets in cities which are interested in reindeer products, offering goods in exchange. This explains why locals/indigenous people are able to control the reindeer herding business.

For these people, the definition of a trader, implying that they are not engaged in production (Humphrey 1999) does not work. The line between 'trader' and 'producer' is not at all strict, and frequently the bosses of these enterprises are rich reindeer owners themselves. Equally blurry is the border between the barter

---

3   Sovkhoz is the state farm system introduced all over the Soviet Union in agriculture. It was the primary institution in reindeer herding from the 1950s until the collapse of the Soviet Empire, and in some regions still functions very similar to its Soviet predecessor. The zootechnician used to be a very important figure in Soviet reindeer herding, having agricultural and veterinary education, being responsible for the health and migration routes of the herds. After the Soviet Union, the importance of these people was largely ignored, which resulted in their dissatisfaction. Therefore, many searched for alternative occupations.

and cash economy; in most cases cash enters between shipping the *panty* from the village to a processing facility and selling it to the next middleman. These traders are locals not outsiders, because specific knowledge of the local production market is needed from the outset. This knowledge is the key to entering this specific niche market. Firstly, one must know the 'lords of the transport', for example the regional administration, the boss of the air cargo company, some leading pilots, and the owners of the kerosene supply, in order to access the transport infrastructure. But having obtained a helicopter or other means of transport is not sufficient if the trader does not know the migration routes of the herders, or the location of the slaughtering corrals. In Anderson's (2000) words, the entrepreneur working with reindeer herders has to 'know the land', otherwise they will not be able to find particular herders and their herds in the wide arctic tundra. This is why the trading pattern described by Humphrey ends at the gates to the tundra, in the villages. The expertise needed for the *panty* business cannot be bought, but is obtained through building up friendships and 'trust networks'. I remember a trader from Moscow coming to a reindeer herding village where I stayed, who wanted to make money in the *panty* business. With an arrogant approach he wanted to pay natives for access to people who know the land, and the location of herders in *panty*-time. He was defeated, because nobody would agree to work with this capitalist, even for good money. Both Russians and Nentsy were proud of themselves, stressing their solidarity in the North, where, in certain spheres, networks of solidarity are more important than money.

The Yamal example shows that barter networks do not, as Humphrey (1999) pointed out, undermine trading, but are in fact necessary for starting it. Three major local reindeer entrepreneurs have developed from beginnings in the mid-1990s into vertically integrated reindeer herding enterprises. They engage in all aspects of the business from owning the animals, sawing the fresh antlers, collection by helicopter, trading goods for antlers with reindeer herders, freezing, drying, and cutting them into chips, and taking antlers to Moscow to sell to Korean traders. They also organise the autumn slaughtering of animals, processing the meat, and selling it on the market. Conceptually, their asset is that they know both sides of the tundra/city border, as well as that of the barter/cash border. In an evolving market economy, characterised by a high number of middlemen in the economic chain from the producer to the consumer, such a vertical integration is highly exceptional. It allows the local enterprises to control the access to the tundra, and act as gatekeepers in the exchange relations between the tundra economy and the capitalist economy of twenty-first century Russia.

The Position of Herders in the Business

Most models of world system or dependency theory assume that once the periphery begins to be integrated into the global economy, the population experiences marginalisation and exploitation (Frank 1978, Meillassoux 1981, Wallerstein

1989). Indeed exchanging one kilo of *panty* for five dollars, when it goes on to sell for US$ 30-60/kg seems rather exploitative. However, the herders themselves choose their trading partners. In a survey of 25 reindeer herding households in Yamal in 2000-2001, almost all reported feeling no obligation around deciding on trading partners. Most act pragmatically, diversifying risks and working with all enterprises coming to collect *panty*. If they feel that they have not been offered favourable goods, they will complain or not work with those traders the following year. Reindeer herders show surprisingly little interest in obtaining cash for their *panty* or meat. They say in the tundra, there are no stores and shops anyway, and they feel more comfortable ordering things from knowledgeable people whom they trust, rather than leaving the tundra and their herds too often. Therefore the local vertically integrated enterprises exchange the cash that they receive for the *panty* for goods, which they then bring to the herders in helicopters. So cash is needed for an enterprise to purchase better and fresher goods, but is not needed to start doing business with the reindeer herders. This challenges the assumption that barter is a major obstacle for the development of a market. On the contrary, many primary producers in this sphere (herders) will not even consider selling their products for money.

In keeping with this cashless exchange, it is my observation that herders are inclined to work with relatives or neighbours engaged in this business. Although enterprises in the villages act as gatekeepers to the tundra, a 'social boundary defence' is also operating around the herders (Cashdan 1983, Casimir 1992). Only those who belong to the group are permitted to do business with reindeer herders, despite gatekeepers providing access. Gudeman (1998, 2001) has conceptualised this twofold notion of exchange as the 'community' sphere and the 'market' sphere. In both spheres goods are exchanged, but in the tundra, the 'community' sphere is dominant, and pure market relations do not work, let alone a cash economy. This does not mean that the herders do not gain wealth from this business. It only means that there is a clear notion of socially meaningful behaviour to be performed in order to get access to the community. Because of this difference between community and market, the border of which is the village, the tundra dwellers have not become the marginalised producers of a neoliberal world economy that is assumed by world system analysis. In Evers' words (1996: 169), the *panty* economy among Yamal reindeer herders shows a 'hybridisation' of subsistence and the global economy. In this connection between the 'global' and the 'local', we see the proof against a widespread evolutionary assumption that we have subsistence economies in 'pre-modern' societies, whereas commodity exchange in the global economy is the brutal face of capitalism that threatens subsistence practices (Nash 1994). Evers (1996), and Russel (2004) note that both can occur in one society, and at the same time.

The income from this new commodity changes the economic worldview of the reindeer herders. From a market point of view, producing *panty* during the decade following the Soviet Union's collapse was more profitable than producing meat, but despite this, *panty* is neither the sole nor even the major economic activity of the reindeer herders. One reason for this is the reputation of *panty* cutting among the Nentsy. Only a tiny minority of the herders interviewed claimed that cutting *panty* does not affect the health of the reindeer, as is stated in earlier literature (Iudin 1993). Although this did not prevent many from cutting, they were aware that the animals might become weaker from the cutting, which affects their transport capacity, they become more vulnerable to mosquitoes, and less competitive in fresh pastures. This then leads to a certain loss of the physical quality of the herd. This perceived negative effect on the reindeer is one reason for the bad reputation the *panty* production has among herders. Moreover, some herders also felt the beauty of their herd would be diminished if they cut *panty*, and said they would do so only in cases of exceptional need for particular commodities. But the most important argument against *panty* is its perception as highly risky and an unreliable source of income. Reindeer herders have experienced the bankruptcy of early enterprises (one run by an Armenian and another by a Khanty businessman), failing to receive goods in exchange for their product. Many herders believe it is better not to rely too heavily on *panty* income. Given this experience and the extreme price fluctuations, they would prefer to diversify and produce both meat and *panty* simultaneously. Cutting *panty*, year by year, enables them to obtain goods previously unseen in the tundra. These are mostly luxury goods such as fresh bread, spices from Caucasus, fruits, and leisure items, music players, music and the like, as well as expensive equipment such as snowmobiles. Income from *panty* enables the herders to increase their standard of living above subsistence level. As we saw in Anniko's example, a household with a normal size herd can cut more than 200 kg of *panty* per year, and earn half the price of a snowmobile. Some families buy a snowmobile every second year, for example to equip their boys, just as some school children get a car for their high school graduation or their eighteenth birthday. This clear cut difference between basic and non-essential needs allows us to conceptualise *panty* income as a sort of special purpose money, but it is not stigmatised as 'dirty money' as Hutchinson (1996) suggests, where a low prestige business such as carrying the excrements of the rich to the waste deposit generates only low prestige income that cannot be spent for 'noble' purposes. It is more that despite *panty* income being seen as unhealthy for the reindeer economy, it is used to improve the standard of living. This compares with the role of stock exchange incomes in some western settings. Like the stock exchange, the *panty* market is seen as hectic, unstable, and full of dishonest people. Relying too much on *panty* income would mean investing in an unpredictable branch of the economy and therefore people do not count on it for their everyday life. This is why the backbone of the reindeer economy in Yamal is still meat production, as it has been for the last 100 years. Despite its bad reputation, cutting *panty* is seen as a legitimate source of income, as Pine (2002: 77) noted in a Polish example 'almost any means of obtaining it [money] is legitimate'. Perhaps *panty* expresses best the herders' quest to incorporate their nomadic economy into world capitalism without becoming detached from their communities' social processes.

Complex Dynamics of Economic Spaces

Nowadays, many reindeer nomads feel their scepticism towards *panty* has been justified. The broad introduction of Viagra and other developments in the international male potency market has affected *panty's* price, and production costs have also increased, mainly due to air transportation costs. The *panty* business has therefore almost completely collapsed or become insignificant for the reindeer herding economies of Russia. In Yamal, the local air company has a monopoly on the official *panty* market, and renting helicopters solely to collect *panty* is now unprofitable. As is the case for any reindeer product, transport costs from the site of production to the customers are the bottleneck of the entire economy. This is not likely to change as the natural setting for reindeer herding, is remote areas with low human population density. So, although they hold a sceptical view, business has not decreased because of herder opinion, but rather due to market perturbations that are beyond their influence.

Since 2005 *panty* entrepreneurs have solved the transportation problem in two main ways: (a) Using a new innovative means of ground transport: the 'Trekol'. The Trekol is a six-wheel all terrain vehicle with very big soft tyres that put minimal pressure on the fragile tundra and still offer excellent off road capacity close to that of a tank. Trekols have normal car engines which are comparably easy to maintain. Their cargo capacity is up to 500 kg. Some owners equip their vehicles with an additional trailor. *Panty* entrepreneurs now use this means of ground transport to collect *panty*. This decreases their radius of operation in comparison to the helicopter, but is still profitable. (b) Sharing helicopters with the gas company. Helicopters fly ever more frequently over the Yamal Peninsula, where Russia's biggest on-shore gas field is being developed in Bovanenkovo (Forbes and Stammler et al. 2009). During the development phase, workers and material are brought to construction sites and base camps on the tundra by helicopter. Entrepreneurial gas industry workers with local standing have been using these helicopters for *panty* collection. In such cases the basic cost of the flight is covered by the gas company, only leaving the extra time and fuel to be paid for. In 2005 during fieldwork the following practices were noted: Helicopters bringing a new shift of subcontractors to the gas deposit arrive just before the weekend. The helicopter crew also stays for the weekend, ostensibly as a business trip to the gas deposit. A locally knowledgeable Russian gas worker and *panty* businessman would arrange a flight to the reindeer nomad camps. This requires long flying hours and even though there is 24 hour daylight in June and July, flying at night and for too many hours without interruption is forbidden. To circumvent this, the crew flies to a camp where they officially stay, before removing the helicopter's 'black box', and refuelling independently. Making sure that these changes occur in the tundra means that the whole operation is untraceable by outsiders. With this specially prepared helicopter, they fly to all the camps and collect as much *panty* as possible. To avoid flight tracking radar, the pilots reveal their incredible skill, never flying more than 50 m above ground. I was deeply impressed by the lead pilot's talent in handling the massive MI 8 helicopter and effectively tracing the relief of the Yamal tundra by flying so low. During the collection process, herders were told they could come to the deposit at any time and get their goods in exchange from the warehouse. At the end of the weekend, the black box would be reinstalled and the helicopter refuelled. Next morning, some workers take the flight back to the city, but the rest of the helicopter is loaded with the *panty*. In the city, the *panty* is met by the main entrepreneur, who would dry them, or sell them on to China raw.

Both trekol and helicopter collection lower transportation costs significantly and work within a margin that makes the *panty* business continue to be worthwhile for all sides.

The gradual introduction of mobile communication into the tundra is another significant innovation, with far reaching consequences for nomadic societies in general, as I have outlined elsewhere (Stammler 2009). With the influx of oil and gas workers to the tundra, the human population density reaches a critical mass in particular places, making it profitable for mobile phone providers to install reception and transmission facilities. Reindeer herders benefit from this by rapidly entering into a world of global mobile communication, whereas most of them did not have any other communication except physical travelling as recently as seven years ago. Mobile phones as well as GPS make the logistics of *panty* trade significantly easier. It is no longer necessary to undertake initial exploratory trips to herder camps to ascertain what they want to order or to find out where exactly they are located. This information is now just a call away, provided there is electricity to charge the phone batteries. Intensive fieldwork focusing on this technological change is needed to capture the whole dimension of this dynamic. In particular, the advent of mobile communication might undermine the role of gatekeepers. Outsider businessmen could theoretically directly contact herders by phone and agree on locations, dates, terms, and conditions. However, in my current understanding, this is less likely, as herders often switch their phones on only to make a call. Leaving a phone on constant standby would consume too much electricity and require frequent recharging.

Entrepreneurs and gatekeepers have thus reacted to the new situation and are continuously searching for optimal possibilities to engage in this business, using new developments for their own advantage. Once the gas deposits are developed and helicopters fly less often, the option of helicopter hitchhiking with *panty* will change again. A railway connecting the Bovanenkovo gas deposit with the regional capital is almost completed. Once power relations and decision making on that railway are set, it is most likely that *panty* will continue to be collected by Trekol-ground-transport from the camps to the deposit, and from there by train southwards to customers.

Such developments reveal the continued creativity with which entrepreneurs deal with the remoteness of the tundra as a production site, while at the same time using the distance from efficient controls as an asset for overcoming the economic shortcomings connected to remoteness. It is also remarkable that such entrepreneurship bridges ethnic boundaries, through friendships between Nenets brokers and pilots, Nenets businessmen collecting *panty* on ground transport and gas industry staff who have decision making power over flights and trains. As long as the space for such creative approaches on the tundra is not eliminated, we will see the volatile side-business of reindeer herding continue.

Conclusion

This contribution has illustrated the process of economic integration of a remote arctic community into an international trading network. Drawing on ethnography of the developing velvet reindeer antler business between Korean/Chinese customers, Russian traders and Nentsy reindeer herders, I have argued that market integration and commoditisation does not always have to be accompanied by a marginalisation of nomads or a change of social strata among them. The *panty* business in this case is not, as mentioned, the brutal face of capitalist globalisation that threatens a marginalised livelihood as shown by Nash (1994). The continued practice of *panty* business throughout two decades of ever-changing economic conditions argues for a highly adaptive nomadic community. They respond to transformations in the world around them by meeting the demands of a newly developed market that offers fast returns today, but that might collapse just as quickly tomorrow. Against the background of recent world system studies I also show that the integration in the world economic system is not mutually exclusive with subsistence. I argue that the Nentsy have succeeded in organising their engagement with new commodities post 1990 by influencing the conditions for business with outsiders. This is achieved by controlling access to the place of production, the reindeer pastures of the West Siberian Tundra. Access is only gained through cooperation with the knowledgeable locals involved, who organise the production, cutting, packing, storing, shipping, and drying of 'velvet' reindeer antlers, before selling it to Russian traders, who sell it on to Chinese or Korean customers. In exchange for the *panty*, Nentsy traders receive hard currency income, which enables them to provide a range of new imported products to the tundra. This is how reindeer herders access the consumer goods of the global economy, such as Chinese music players, American instant soups, ketchup, and Japanese snowmobiles or power stations.

I argue that in spite of the significant flow of commodities between the 'nomadic' and the 'sedentary' space, the borders between these two 'worlds' are still significant, and the reindeer herders still see their engagement with the commodity economy as supplementary to their subsistence and production of meat for local markets. The split between two items of economic exchange – antlers and meat – is mirrored by a split of expenditures. Income from meat and fur production is spent to satisfy basic needs, whereas income from the antler business is used for 'luxury' goods not necessary for survival. This finding links to recent anthropological works about the meaning of money as an item of generalised exchange versus 'special purpose money' (see Parry and Bloch 1989, Pine 2002). It is still seen however as risky money, not to be relied on.

Ethnographic material from northwest Siberia allows us to understand the process of commoditisation in a nomadic community, since its beginning in the 1990s. People in a seemingly remote community can engage in global markets without sacrificing their distinctive culture. In this way the Nenets approach to market economy follows the same pattern as their response to most other innovations brought from outside into the tundra: I have argued elsewhere (Stammler 2005) that the distinct characteristic of their reaction is a very flexible openness to everything new, but as usefully supplementing already existing set of practices of a distinctly nomadic livelihood, rather than replacing their own Nenets cultural practices. So here we see market capitalism practiced according to the rules of tundra nomads, rather than nomadic tundra economy practiced according to the rules of capitalism. Russell (2004: 2) argues that we should 'move subsistence and informal economies from the fringes and portray them instead as belonging to the plural, unruly, and ungovernable set of economic practices that make up "capitalism"'. The case of *panty* illustrated in this chapter suggests an interpretation the other way round: Taking on the view of reindeer nomads, we can perceive market capitalist economic activity as being on the fringes of a diverse, plural, and community-internally regulated set of practices that make up the mixed economy of these nomads today.

Acknowledgements

Parts of this chapter have been published in earlier publications as Stammler, F. 2007. Domestic Economy and Commodity Trade Among West Siberian Reindeer Herders. *Arctic & Antarctic Journal of Circumpolar Sociocultural Issues*, 1(1), 227-253; and Stammler, F. 2004. The Commoditisation of Reindeer Herding in post-Soviet Russia: Herders, Antlers and Traders in Yamal, in *Segmentation und Komplementarität. Organisatorische, ökonomische und kulturelle Aspekte der Interaktion von Nomaden und Sesshaften* (Orientwissenschaftliche Hefte Nr. 14, Mitteilungen des SFB 'Differenz und Integration'), Halle, 105-122.

References

Anderson, D. 2000. *Identity and Ecology in Arctic Siberia: The Number One Reindeer Brigade*. Oxford: Oxford University Press.

Cashdan, E. 1983. Territoriality among human foragers: Ecological models and an application to four Bushman Groups. *Current Anthropology*, 24, 47-66.

Casimir, M. 1992. The dimensions of territoriality: An introduction, in *Mobility and Territoriality: Social and Spatial Boundaries Among Foragers, Fishers, Pastoralists and Peripatetics*, edited by M. Casimir and A. Rao. Oxford: Berg, 1-26.

Evers, H.-D. 1996. Globale Märkte und Transformation, in *Weltsystem und kulturelles Erbe. Gliederung und Dynamik der Entwicklungsländer aus ethnologischer und soziologischer Sicht*, edited by H.-P. Müller. Berlin: Reimer, 165-173.

Frank, A.G. 1978. *Dependent Accumulation and Underdevelopment*. London: Macmillan.

Forbes, B., Stammler, F. et al. 2009. High resilience in the Yamal-Nenets social-ecological system, West Siberian Arctic, Russia. *PNAS feature article* (29 December), 106(52), 22041-22048.

Gudeman, S. 1998. Introduction, in *Economic Anthropology*, edited by S. Gudeman. Cheltenham, UK/Northhampton, USA: Edward Elgar (The International Library of Critical Writings in Economics), xi–xviii.

Gudeman, S. 2001. *The Anthropology of Economy*. London: Blackwell.

Humphrey, C. 1999. Traders, 'disorder' and citizenship regimes in provincial Russia, in *Uncertain Transition: Ethnographies of Change in the Postsocialist World*, edited by M. Burawoy and K. Verdery. Oxford/Lanham: Rowman & Littlefield, 19-52.

Hutchinson, S. 1996. *Nuer Dilemmas: Coping with Money, War and the State*. Berkeley: University of California Press.

Iudin. 1993. *Panty i Antlery*. Novosibirsk: Nauka.

Meillassoux, C. 1981. *Maidens, Meals and Money: Capitalism and the Domestic Economy*. Cambridge/New York: Cambridge University Press.

Nash, J.C. 1994. Global integration and subsistence insecurity. *American Anthropologist*, 96(1), 7-30.

Parry, J. and Bloch, M. 1989. *Money and the Morality of Exchange*. Cambridge: Cambridge University Press.

Pine, F. 2002. Dealing with money: Zlotys, dollars and other currencies in the Polish Highlands, in *Markets and Moralities. Ethnographies of Postsocialism*, edited by R. Mandel and C. Humphrey. Oxford/New York: Berg, 75-97.

Russell, W. 2004. The people had discovered their own approach to life. Politicising development discourse, in *In the Way of Development: Indigenous Peoples, Life Projects and Globalisation*, edited by M. Blaser, H. Feit and G. McRae. Available at: www.idrc.ca/en/ev-64528-201-1-DO_TOPIC.html [accessed: 18 June 2010].

Stammler, F. 2005. *Reindeer Nomads Meet the Market: Culture, Property and Globalisation at the End of the Land. Halle Studies in the Anthropology of Eurasia 6*. Münster: Litverlag.

Stammler, F. 2009. Mobile phone revolution in the tundra? Technological change among Russian reindeer nomads, in *Generation P in the Tundra*. (Folklore 41), edited by A. Ventsel. Talinn: Estonian Literary Museum, 47-78. Available at: http://www.folklore.ee/folklore/vol41/stammler.pdf [accessed: 02 March 2010].

Wallerstein, E. 1989. *The Modern World System 3: The Second Era of Great Expansion of the Capitalist World Economy: 1730-1840s*. New York: Academic Press.

PART IV
From Commercialised Production to Integrated Markets

Chapter 13

Livestock Markets and Drought in Sub-Saharan Africa

Nikola Rass

Introduction

If policy makers or development practitioners want to support pastoralists[1] in Africa, it is essential to have a solid understanding of the vulnerability of pastoral livelihoods and their resilience strategies in response to risks. Furthermore it needs to be considered how commercialisation of the pastoral production system and existing market and nonmarket systems serve to regulate herd sizes, preserve scarce assets and enable recovery from adverse shocks.

This chapter begins with a short outline of pastoral livelihoods and continues with an analysis of the impact of drought on markets, and the associated response of pastoralists to both markets and drought risk. It sets out to explain the different income generating strategies and production objectives of pastoralists in Sub-Saharan Africa (hereafter referred to as pastoralists), that have evolved with increasing commercialisation, in order to understand the pastoralist's response to markets. It shows the manifold reasons why pastoralists might opt not to follow market incentives to sell livestock, but rather follow a strategy of herd expansion. These reasons cannot be reduced to their exposure to risk.

In a second step the impact of drought on markets is described and the detrimental effects of commercialisation on traditional strategies (such as managing size and composition of herds in consideration of subsistence needs and prevailing risk exposure) are discussed. During a drought high numbers of livestock are lost due to mortality or the need to sell at unfavorable market conditions. When combined with the process of commercialisation, this has meant that with every drought the number of absentee herd owners is increasing.

The adaptation of traditional strategies to current contexts needs to be acknowledged and supported, thus the next section debates the opportunities of technologies such as Early Warning Systems and gives examples of market focused interventions.

---

1 In this chapter, pastoralism describes production systems based on the use of extensive grazing on communal rangelands for livestock production.

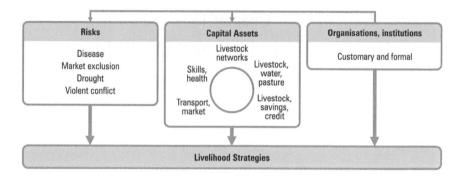

Figure 13.1   Framework of pastoralist livelihoods
Source: Rass 2007.

Pastoralist Livelihoods

The livelihood approach aims to provide a framework for holistic analyses of people's strategies within a dynamic context. Chambers and Conway (1992), describe livelihood as 'the capabilities, assets (stores, resources, claims and access) and activities required for a means of living'. The livelihood framework thus views the livelihoods of people as determined by five 'capitals', namely human capital (e.g. education), natural capital (e.g. land), financial capital (e.g. access to credit), social capital (e.g. community networks) and physical capital (e.g. livestock) and the institutional environment within which people operate (see Bohle, this volume). Within this framework, livelihoods are not static but exposed to external risks such as seasonal variability, long-term trends and shocks. Although access to assets is the most prominent component of the livelihood approach, the framework takes into account the capability to make profitable use of these assets and to combine them to improve existing livelihood strategies. Thereby the livelihood framework is flexible and aims to guide policy makers to not only consider peoples' access to assets but also the economic and institutional setting within which they can or cannot profitably transform them.

An attempt to draw a generalised livelihood framework of pastoral people (see Fig. 13.1) suggests that that the overall livelihood frame of pastoral people depends on both access to assets, such as pasture, water, animal health services, markets, credit and education, as well as the environment where these assets are combined for production and consumption purposes, namely political, organisational and institutional infrastructure. Furthermore the livelihood framework sets the welfare of pastoralists in the dynamic context of trends, seasonal variations and risks, which affect assets and livelihood strategies and determine the level of vulnerability (Rass 2007). In this understanding the livelihood framework is applied to analyse pastoralists' livelihood strategies in response to the external risk of drought and their given economic and institutional setting. Here a special focus is placed on markets as important transforming institutions.

Pastoralist Response to Markets

The prevailing pattern of pastoral livelihoods in Sub-Saharan Africa is to slaughter livestock for meat consumption only for specific occasions while the main subsistence product obtained from livestock is milk.[2] However, although much milk is drunk, cereals form the staple diet. It has been argued that with the introduction of taxes and alternative currencies by colonial administration and traders, there was a shift from more embedded subsistence-oriented pastoral production systems towards commercialisation[3] (Bohannan 1967) and that with increasing commercialisation pastoralists have become more dependent on the market for food. Currently, pastoral livelihoods have a high dependency on food purchases. As shown in Figure 13.2, in 2000-2003 pastoralists in Djibouti, Mauritania, Niger and Chad sourced 70 per cent or more of their food from crop purchase, both in poorer and better off households. Despite the high dependency on monetary income for food purchase, it was observed in the 1970s that pastoralists do not follow price incentives by selling more livestock when prices are high. Based on this observation the 'store of wealth' concept was devised as a model explaining that pastoralists use livestock as their principal store of wealth rather than as income-generating capital (Doran, Low and Kemp 1979, Goldschmidt 1975). The 'store of wealth' concept has been criticised since it does not consider the multiple and rational strategies pastoralists pursue to secure their livelihoods.

This concept was modified to become to the 'target income' concept (Dahl and Hjort 1976), which argues that if there is a lack of response price incentives to sell livestock, this is probably not because of an irrational attachment of pastoralists to their stock, but for combinations of reasons, such as the anticipation of livestock losses due to the recurrent risks of epidemics and droughts. Taking into consideration that restocking after drought will be difficult, pastoralists follow a risk minimising strategy by only selling as many animals as needed for a target income for identified needs. It is argued that pastoralists are following an opportunistic stocking strategy, accumulating livestock numbers that exceed subsistence demands during good years so that they can be assured of enough heifers and cows surviving for reestablishment after the bad years (Coppock 1994). As pastoralists aim to increase their herd they try to prevent selling female stock. Therefore typical herd structures include more than 60 per cent female animals (not including calves up to one year old), and only four per cent bulls, ten per cent immature bulls (over one year old), and seven per cent steers (Little 1985). In addition to the strategy of herd expansion, traditional strategies such as herd dispersion, formation of stock alliances and stock patronages used to function as social insurance systems.

---

2   Apart from milk, live animals can also yield blood and this has been historically exploited in eastern Africa and the horn of Africa, although it is a practice looked on with distaste by pastoralists elsewhere.

3   Commercialisation involves a partial or total shift in the goals of production from meeting subsistence needs to producing for a market.

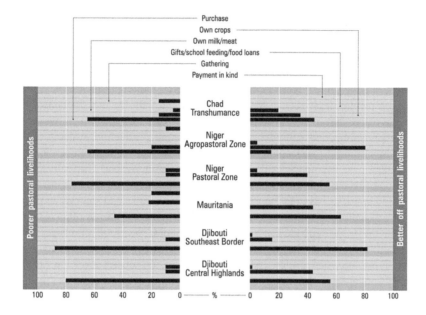

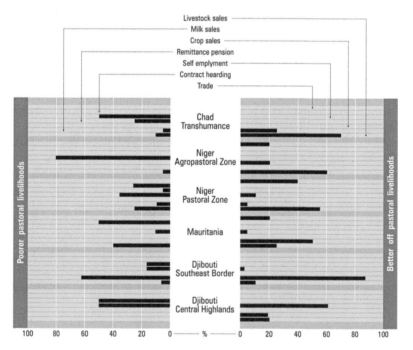

Figure 13.2   Sources of food and cash in pastoralist livelihoods: Top: Sources of food; Bottom: Sources of cash

Source: Based on livelihood profiles published on FEWSNet (2004/5).

The target income concept, with its focus on risks, is complemented by the 'capital assets' model which sets out to explain that income from livestock assets in pastoral Africa is mainly derived in the form of products produced from the livestock themselves rather than in cash obtained from livestock sales (Sikana et al. 1993). Livestock owners regard their animals as capital assets, which produce a stream of valuable products while held and have a capital value when sold or slaughtered. Stockowners determine the optimal age of sale or slaughter by comparing the expected net present value of the future stream of products with the expected net capital value of the animal if slaughtered or sold. Calculations of the net present value of live animals are least complicated for production systems where meat is the only product and more complicated where there is a complex of valuable flow and stock products. In their study on commercialisation and dairy production of pastoralists in Africa, Sikana et al. (1993) explore the factors influencing herders' decisions to consume or sell milk, versus using milk to enhance the growth of calves for market. They conclude that the determining factors traded off against each other are the use value of milk (defined by the volume of milk output, number of consumers and available female labour for dairying), the exchange value of milk (defined by prices, demand and market access) and the value of milk as an input to intensify meat production (this in turn is defined by the trade off between use value and the exchange value of live animals).

There are various factors influencing pastoralists' decisions to sell or breed their livestock, as well as to sell or consume milk. By comparing two pastoralist livelihoods in different zones of Djibouti (see Figure 13.2), it is evident that access to markets has a significant impact on the income generation strategies of pastoral livelihoods: The pastoral livelihoods who are residing in a zone with better market access (southeast border zone) rely on milk sales to generate cash income and prefer to consume less milk products, whereas the pastoralists who have constrained access to markets supplement this income with self-employment activities like sale of firewood, salt and so on. Additionally they depend upon remittances sent by family members living in the city or upon a pension received by a household member re-settled in the countryside upon retirement.

The Impact of Droughts on Markets and Pastoralist's Strategies

Pastoralists in Sub-Saharan Africa have a particularly high exposure to the risk of droughts[4] as they make a living through mobile pastoralism in drylands.[5] Droughts,

---

4   Following the definition of Pratt et al. (1997) a drought occurs when rainfall in a year is below half the long-term average or when rainfall in two or more successive years falls 75 per cent below the long-term average.

5   Drylands are defined by the UNCCD as arid, semi-arid and dry sub-humid areas in which the ratio of mean annual precipitation to mean annual potential evapotranspiration ranges from 0.05-0.65.

or periods of unusually low rainfall, are part of the expected pattern of precipitation in semi-arid Africa. A total of 46 countries in Africa suffered 236 drought incidents in the 59 year period spanning 1950 to 2009. Seven major droughts have occurred on the African continent in the last five decades: 1965/66, 1972/74, 1981/84, 1986/87, 1991/92, 1994/95, 1999/2004. The three worst time periods were 1980-1985 (38 countries affected); 2001-2004 (27 countries affected); 1988-1995 (25 countries affected).[6]

Pastoralists are particularly vulnerable to droughts. In response to this high exposure and the specific vulnerability, they have developed many livelihood strategies to increase their resilience against drought, the most prominent strategy being the mobility of livestock to track fodder in response to the disequilibrium environment (Scoones 1994). The lack of rainfall reduces water and forage availability on the rangelands, which creates an imbalance between the number of livestock and available fodder. If animals cannot move to greener pastures, they become emaciated and die. Systematic selling of stock has been assumed as another way pastoralists respond to or 'track' rainfall (Scoones 1994, Toulmin 1994).

The price effects of market gluts, however remains somewhat controversial. Analyses of price movements show that the livestock/grain terms of trade rapidly turn against livestock during drought.[7] However, some careful large-scale surveys in East Africa suggest that livestock sales fail to correlate with drought cycles (McPeak and Barrett 2001). The picture in the Sahel is clearer: Livestock prices in Niger plummeted during the drought of 2005, reaching 10 per cent or less of pre-crisis levels, even for healthy animals, and cereal purchasing power of livestock-dependent households dropped by 75 per cent (ILRI 2005).

It is argued that two predominant processes induce these price movements and aggravate each other. As the production of staples decreases in a period of drought, their price rises dramatically and pastoralists are thus forced to sell more animals than in normal years to satisfy their need for cereals. This process in turn causes the price of livestock to decrease. Additionally, during severe droughts livestock mortality increases dramatically and herders tend to sell their animals pre-emptively before they become emaciated. In such situations the livestock prices collapse (see Figure 13.3). Thus at the onset of a drought there are often gluts in the livestock market and market prices, even per unit live-weight, decline sharply, whereas towards the end of a drought, the stock that pastoralists do sell might also be in poor condition and thus fetch lower prices. According to this argument the detrimental effect of drought on pastoralists is exacerbated by the related reaction of livestock markets (Smith et al. 1999), since pastoralists not only lose their livestock due to increased mortality, but are also forced to sell more animals

---

6   Information compiled from The International Disaster Database, available at the Center for Research on the Epidemiology of Disasters: http://www.emdat.be/.

7   Fafchamps et al. 1998, FEWSnet 2004/5, Holtzmann and Kulibaba 1994, Toulmin 1994, Williams 2002a.

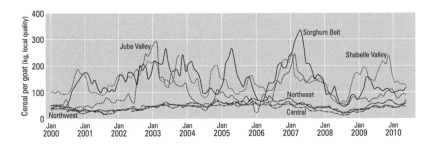

Figure 13.3   Terms of trade for kilogram cereal per local goat in Somalia
Source: FSAU (2010).

than usual for the same amount of staples because of the worsening livestock-grain terms of trade. The basis of pastoralists' livelihood is especially at risk in these situations when they are forced to sell breeding stock. This, for instance, was observed among the Borana during the 1981-84 drought: The proportion of females in livestock sales rose to 43.5 per cent from 25 per cent in normal years (Cossins and Upton 1987).

In contrast to poorer livestock keepers, wealthier herders and urban entrepreneurs are able take advantage of the market situation during drought: They buy livestock at low prices as an investment, and are able to keep stock alive through purchase of feed. Post drought they often entrust the animals to impoverished pastoralists to manage them. As a consequence, many herders are gradually transformed from herder-owners to hired herders. This shift was observed during the droughts of the 1970s and 1980s by Gass and Sumberg (1993) and White (1991). Estimates of the proportion of livestock now belonging to absentee herd owners are as high as 80 per cent in some areas, for example Mopti, Mali (Shanmugaratnam et al. 1992).

Opportunities to Support Pastoral People

Increasing integration of livestock into monetarised market systems affects traditional pastoralist coping strategies. Livestock is losing its multifunctional embedded economic and social function (Schultz 1996) as the degree of reliance on exchange mechanisms based on reciprocity, kinship ties and the genetic line of livestock diminishes. Stock alliances and stock patronages are now less common, while the number of absentee herd owners is increasing. Traditional strategies of herd dispersion and creation of social bonds through stock exchange are being partly substituted by monetary practices, such as the creation of social bonds by the donation of 'cattle without legs' (an amount of money equivalent to a cattle) (Schultz, 1996). The reduction of traditional coping strategies raises the need to support innovative strategies that are adapted to the market system and support pastoralists to maintain a viable herdsize throughout droughts or prevent the loss of breeding stock necessary to restock in recovery periods.

Satellite based and computer aided earth observation systems allow droughts to be predicted at an earlier stage, and with the help of Early Warning Systems (EWS) drought forecasts are becoming more reliable. This technological development offers a window of opportunity to support innovative strategies, adapted to the monetarised market system and enable governments and relief agencies to position themselves for more effective drought interventions (Hazell 2000).

The Development of Livestock Early Warning Systems (LEWS)

The first EWS were set up in the wake of world food crises in the early 1970s, with the aim of improving UN and other food aid institutions' management of food supply.[8] These systems were developed into livelihood focused early warning systems in response to the severe and devastating droughts of the early 1980s. Early warning systems with a specific focus on pastoral livelihoods give more attention to livestock preservation as pastoral livelihoods' most important asset. However, many international relief systems remain focused on famine prevention and food aid and are criticised for their limited perspective towards existing opportunities for sustainable interventions of the livelihoods focus (Buchanan-Smith 1992).[9]

The introduction of the Spot Satellite earth observation system (Spot image 2005) and new technological options such as the normalised vegetation index (Infocarto 2005) allow rainfall and forage production forecasts, which are important physical assets of pastoral livelihoods. However, Early Warning Systems with a livelihood perspective need to monitor all assets of pastoralists. Livelihood security and ability to cope with drought depend not only on endowments such as rainfall, crop and vegetation but much more on entitlements, which are markets, assets, rights over endowments, and opportunities to change livelihoods (Sommer 1998).

Taking this into consideration, monitoring systems with a focus on livestock (LEWS) have been enhanced by combining the satellite-based information with ground-based indicators, most notably livestock production and herd movements. These monitoring systems have a clear focus on livestock production[10] but aim to detect and react to threats on all assets of pastoral livelihoods in a timely manner. Set

---

8 The FAO-based Global Information and Early Warning System (GIEWS) is a prominent example of such an Early Warning System. It is the umbrella for ARTEMIS (Africa Real Time Environmental Information System), EMPRES (Animal Plant Pests and Diseases) ECLO (Emergency Centre for Locust Operation) and has strong links with the USAID funded FEWS (Famine Early Warning System).

9 These criticisms have been repeated by several speakers at the Workshop on Pastoral Early Warning and Response Systems in Mombasa (2001), especially referring to the late response in the drought in Ethiopia and Kenya in 1999-2001.

10 In this chapter the term 'pastoral production system' is applied, when strictly discussing the mode of mobile pastoral livestock production, whereas the term 'pastoral livelihoods' is referred to when the analysis includes all 'the capabilities, assets and activities required for a means of living' (Carney 1998).

up in such a way LEWS offers the opportunity to implement timelier movement/ destock/restock strategies allowing pastoralists to maintain their breeding assets through crises and to influence distribution of livestock.

One of the first Livestock Early Warning Systems to follow this approach is the Turkana Livestock Early Warning System in northern Kenya (see Box 1). While the system has been successful in alerting decision makers to impending stress in the local economy, there has often been a lag between warning and response, in particular when it comes to the mobilisation of emergency funds for swift intervention.

This leads to the conclusion that EWS need to not only monitor the onset and progression of droughts but also pay attention to the development of contingency and drought cycle preparedness plans. This should be combined with the decentralisation of decision-making and capacity building of local governance. A contingency fund should be provided to avoid long administrative procedures and ensure fast response.

LEWS thus now encompasses a broader development approach that includes preparing contingency plans and setting up funds and institutions. Contingency plans are developed in response to different phases of a drought, which are usually classified as the risk preparedness phase, emergency alert phase, emergency phase and post emergency phase. This classification of the drought cycle is based on the

---

Box 1: The Example of the Turkana Early Warning System

In 1984/5 OXFAM prepared a Drought Contingency Plan for Turkana District, which later formed the basis for Turkana LEWS, established in 1987. As part of the Arid Lands Resource Management Project of the World Bank, the LEWS has been extended between 1992 and 2001 to cover nine additional districts, which include Marsabit, Samburu and Isiolo.

The system relies on the monthly monitoring of the livestock economy, the environment and human welfare. LEWS collects monthly rainfall statistics and is supplied with satellite imagery and associated forage resource analyses by the USAID-funded famine early warning system (FEWS). The monitoring system further draws on monthly surveys of 30 to 40 households per district to assess livestock production and prices. Additionally, each quarter community meetings are held to review community resources and activities.

The early warning system has four phases: 1. normal, 2. alert, 3. alarm and 4. post-drought phase. Each phase is linked to a pre-programmed response as part of the drought contingency plan. Potential responses include: emergency veterinary campaigns, livestock purchase schemes, food-for-work, restocking, relief feeding, and nutritional and health support.

The LEWS and the contingency plan are managed at district level by a drought management committee, which is linked with pastoral associations and communities. Because of its comprehensive nature the information is also used for regular district planning.

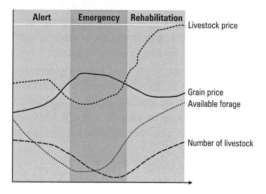

Figure 13.4   Drought phases based on forage and livestock production and livestock and grain prices

Source: Toulmin (1994).

Phase	Activity
Risk prepardness	Longer term policies for resilience: • Establishment of financial institutions • Provision of animal health services • Establishment of dry grazing reserves
Emergency alert	Mitigation activities: • Provision of access to emergency grazing • Provision of contingency feed • Livestock banking • Livestock marketing interventions (early purchase)
Emergency	Relief activities: • Emergency purchase of livestock • Provision of food and/or cash aid • Shelter
Rehabilitation	Rehabilitation activities: • Credit provision • Restocking programmes • Establishment of alternative income generating activities

Figure 13.5   Phases in the drought cycle and related activities
Source: Author.

relationship between forage production, livestock numbers, grain price and livestock price (Toulmin 1994, see Figure 13.4). Each phase provides different options for moderating the impact of drought on the pastoral sector (see Figure 13.5).

Market Interventions in the Drought Cycle

Livestock markets can play an important economic and ecological role as institutions that facilitate converting livestock to grain and adjusting livestock populations to local forage availabilities (Williams 2002b). Market interventions in East Africa follow a two-fold approach. On the one hand, the establishment of credit and savings institutions in the drought preparedness phase is meant to create another form of security, so that herders can change their risk-averse market behaviour and follow market incentives to sell livestock when prices are high instead of following target income marketing. On the other hand, in the emergency alert phase market interventions aim to stabilise livestock prices by inducing an early off-take of livestock before they become emaciated.

Together, both components should help minimise the loss herders have experienced in the past, when terms of trade worsen during droughts. The rapid removal of stock at the onset of drought and the holding of cash savings by pastoral households would not only reduce demand for forage at the early stage of drought and conserve the value of cattle, but also build up a post-drought pool of cash for restocking or investing in alternative income generating activities.

Establishment of Credit and Savings Institutions   In East Africa drought losses are usually too high for livestock insurance systems to be profitable for private providers, and public providers are affected by fraud and moral hazard. Therefore credit and savings institutions seem to be the only viable financial insurance system. The establishment of appropriate financial institutions, combined with incentives to sell livestock in good times, could encourage pastoralists to put aside some of the value stored in their livestock herds in good years into a bank account. When the inevitable downturn occurs, the money in the bank will protect them from destitution and provide capital for buying back into the system after the shock. For example Coppock (1994) estimated in Ethiopia, that (given the value of livestock during the 1980s), if animals had been sold during the inter-drought period, the cash banked, and then withdrawn to purchase grain during the drought, households would have, on average, only liquidated one third of the livestock assets they actually disposed of.

However, this concept is not yet being practised because several issues remain unsolved. For example, the ideal herd size to which the herd can be reduced in inter-drought periods and 'cash banked' still needs to be defined. Models need to be developed along the capital assets concept to help determine the ideal herd size in any given context. A more general constraint is the lack of experience in microfinance institutions that can effectively serve the scarcely populated pastoral areas. The PARIMA research on five financial services associations in northern Kenya's Marsabit district finds that the associations experience high rates of loan and share capital delinquency, low rates of savings deposits, poor profitability, and a weak level of local participation. Furthermore, the demand for loans far exceeds the supply of savings as pastoralists are not taking advantage of the opportunity to convert livestock wealth into cash savings, even during a period of drought that brought considerable herd stress and relatively high off-take (Osterloh 2001).

Ndofor (1998) finds that well functioning informal savings and credit schemes exist in northern Kenya and southern Ethiopia, where pastoralists use trusted friends/shopkeepers as savings and credit institutions. Pastoralists deposit money with a shopkeeper when they sell animals and later withdraw money in cash or goods. Shopkeepers may also offer credit in kind or cash to be repaid when the next animal or animals are sold. This example of the informal institutionalisation of a savings and credit organisation suggests that, despite the remoteness of pastoral areas, microfinance institutions located in market centers can also serve pastoral people. Nissanke and Areetey (1998) support this view and show that rural savings mobilisation is not necessarily correlated with the number of bank outlets.

Alternative microfinance models should be investigated further, with particular attention to the potential complementarity between the eligibility criteria of microfinance institutions and the traditional banking sector. Policy and legal frameworks should be similar to credit union's so that government plays the role of facilitator but does not directly provide financial services to the poor.

Inducing Early Livestock Off-Take   In the emergency alert phase, when livestock prices start to sink as forage availability falls, market access is crucial, because livestock mortality increases exponentially if pastoralists are unable to preemptively sell animals. In the droughts of 1991-92 and 1994-95, close to 90 per cent of the gross reduction in herd sizes was attributed to mortality (von Bailey et al. 1999). Coppock (1994) estimates that the 1983-84 drought in Borana, Ethiopia, reduced cattle density by 60 per cent: 42 per cent due to livestock mortality, 14 per cent due to forced sale and 4 per cent due to slaughter.

In order to support livestock prices, early purchase at the onset of a drought can be induced. Recent experiences in East Africa indicate that interventions in pastoral areas that reduce livestock marketing transaction costs can increase off-take during stress periods (Aklilu and Wekesa 2001, von Bailey et al. 1999). To date such interventions have primarily focused on part subsidisation of trader's livestock transport costs (see Box 2). However, where transport only constitutes a small proportion of livestock trading costs, as is often the case with cattle trekked to markets, additional incentives may be required, such as provision of feed and water to trekked animals. Additionally, a moratorium on livestock market taxes in times of drought might be considered.

Although access to markets is an important component to strengthen pastoralists' ability to sell their livestock pre-emptively in case of an emerging drought, pastoralists should not be encouraged to sell their breeding stock. Strategies aimed at timely livestock de-stocking should remove animals from the land sufficiently early in a drought to avoid long-term damage to vegetation and soils. However, the pivotal issue is how to sustain breeding livestock through the drought in order to ensure rapid reconstitution of the livestock economy in the post drought period. As pastoralist economies take a long time to recover after drought if breeding females are scarce, preservation of the latter is widely recognised as a key aspect of early drought management strategies. Towards the end of a drought, herders are unwilling to sell female livestock and the few breeding females on the market are poor and overpriced (Blench 2001). Strategies to preserve livestock

Box 2: Support of Marketing During Drought (Arid Lands Resource Management Program)

The Arid Lands Resource Management Program (World Bank) aims at developing a marketing system linked to LEWS technologies in order to facilitate livestock off-take during early drought phases. The investments of the Arid Lands Resource Management Project in marketing and market information include a transport subsidy. This subsidy is intended to stimulate livestock sales at the onset of a drought rather than later, when prices have collapsed. The subsidy was based on a pilot experience in Isiolo district in 1996, when a 40 per cent subsidy on the cost of transport to Nairobi was paid to traders, to induce them to buy cattle in remote parts of Isiolo. Some 3,000 cattle were purchased at KSh 6,000/head (USD 105) representing a gross injection of KSh 18 million (USD 315,000) into pastoralists' pockets, at a subsidy cost of KSh 2.5 million (USD 43,750) (Source: Reij and Steeds 2003).

Box 3: Feed Subsidies Combined with De-Stocking

As part of its de-stocking programme, the ACK-MDO Agency in Marsabit provided partial payment for livestock in the form of 22.5 kg bags of feed, destined to enable weak breeding stock to survive. The bag would sustain one small stock over a three-month period. For every three small stock a bag of feed was given, with the remaining balance paid in cash. It was estimated that the provision of supplementary feed concentrates during the drought is more cost-effective than buying animals once the drought is over (Aklilu and Wekesa 2001).

include the movement of breeding stock to less affected areas, the establishment of cattle camps, and/or subsidised transport of fodder to breeding stock.

Experiences with feed subsidies in North Africa have shown that appropriate targeting is difficult: The major share of the subsidised concentrates went to large herders and commercial farmers (Hazell 2000). Delivering feed in exchange for destocking mature male animals during the drought alert phase is a self-targeting strategy that delivers feed subsidies to poorer herders enabling them to keep their breeding stock. This has been practiced by the DFID-funded ACK-MDO project in Marsabit, Kenya (see Box 3).

Conclusion

The adaptation of pastoral livelihoods' traditional risk management strategies to an increasingly monetarised market based economy needs support. Without this support the number of poor pastoralists is likely to increase, with more livestock becoming owned by a few rich absentee herd owners. Livestock Early Warning Systems accompanied by a comprehensive risk management approach is a powerful means to support pastoralists in maintaining their core breeding stock throughout a drought. Interventions focusing on the market component include risk minimising strategies such as monetary savings and induced early livestock off-take. Early off-take, however, should only be encouraged in case of an emerging drought or when the capital assets concept of pastoral livelihoods is taken into consideration.

References

Aklilu, Y. and Wekesa, M. 2001. *Livestock and Livelihoods in Emergencies: Lessons from the 1999-2001 Emergency Response in the Pastoral Sector in Kenya.* Nairobi: OAU-IBAR.

Bailey, D. von, Barrett, C.B., Little, P.D. and Chabari, F. 1999. *Livestock Markets and Risk Management Among East African Pastoralists: A Review and Research Agenda.* (Technical Report 03/99). [Online: GL-CRSP Pastoral Risk Management Project (PRMP)]. Available at: http://www.cnr.usu.edu/research/crsp/tr399.htm [accessed: 2 September 2007].

Blench, R. 2001. *'You Can't Go Home Again'. Pastoralism in the New Millenium.* [Online]. Available at: http://www.odi.org.uk/work/projects/pdn/eps.pdf [accessed: 3 September 2007].

Bohannan, P. 1967. The impact of money on an African subsistence economy, in *Tribal and Peasant Economies*, edited by G. Dalton. New York: American Museum Sourcebook in Anthropology, 123-135.

Buchanan-Smith, M. 1992. Famine early warning systems and response: the missing link? Case study: Turkana district, north-west Kenya, 1990-91, in *Famine Early Warning Systems and Response: The Missing Link?* edited by M. Buchanan-Smith, S. Davies and C. Petty. Sussex: Institute of Development Studies.

Carney, D. 1998. *Sustainable Rural Livelihoods. What Contributions Can We Make?* London: DfID.

Chambers, R. and Conway, G. 1992. *Sustainable Rural Livelihoods: Practical Concepts for the 21st Century.* (IDS Discussion Paper 296). Brighton: Institute of Development Studies.

Coppock, D.L. 1994. *The Borana Plateau of Southern Ethiopia: Synthesis of Pastoral Research, Development and Change, 1980-91.* (Systems Study No. 5). Addis Ababa: International Livestock Center for Africa.

Cossins, N.J. and Upton M. 1987. The Borana pastoral system of southern Ethiopia. *Agricultural Systems*, 25, 199-218.

Dahl, G. and Hjort, A. 1976. *Having Herds: Pastoral Herd Growth and Household Economy.* Stockholm: Stockholm University, Department of Anthropology.

Doran, M., Low, A. and Kemp, R. 1979. Cattle as a store of wealth in Swaziland: implications for livestock development and overgrazing in east and southern Africa. *American Journal of Agricultural Economics*, 61, 41-47.

Fafchamps, M., Udry, C. and Csukas, K. 1998. Drought and savings in West Africa: Are livestock a buffer stock? *Journal of Development Economy*, 55, 273-305.

FEWSNet. 2004/5. *Famine Early Warning System. National Livelihood Profiles.* [Online]. Available at: http://www.fews.net/livelihoods/files/td/national.pdf [accessed: 6 September 2007].

FSAU. 2010. *Food Security and Nutrition Analysis Unit Somalia.* (Market Data Update, May 2010). [Online]. Available at: http://www.fsnau.org/downloads/Market_Update_May_2010.pdf [accessed: 14 June 2010].

Gass, G.M. and Sumberg, J.E. 1993. *Intensification of Livestock Production in Africa: Experience and Issues.* Norwich: University of East Anglia, School of Development Studies, Overseas Development Group.

Goldschmidt, W. 1975. A national livestock bank: An institutional device for rationalizing the economy of tribal pastoralists. *International Development Review*, 17(2), 2-6.

Hazell, P. 2000. Public policy and drought management in agropastoral systems, in *Property Rights, Risk & Livestock Development in Africa*, edited by N. McCarthy et al. Washington DC, Nairobi: IFPRI (International Food Policy Research Institute), ILRI (International Livestock Research Institute), 86-102.

Holtzmann, J.S. and Kulibaba, N.P. 1994. Livestock marketing in pastoral Africa: Policies to increase competitiveness, efficiency and flexibility, in *Living with Uncertainty. New Directions in Pastoral Development in Africa*, edited by I. Scoones. London: International Institute for Environment and Development, 79-95.

ILRI. 2005. *Top Stories from the ILRI Website, 10 August 2005.* [Online]. Available at: http://www.ilri.cgiar.org/ILRIPubAware/ShowDetail.asp?CategoryID=TS&ProductReferenceNo=TS_050810_001 [accessed: 17 January 2006].

Infocarto. 2005. *Normalised Difference Vegetation Index.* [Online]. Available at: http://www.infocarto.es [accessed:17 January 2006].

Little, P.D. 1985. Absentee herd owners and part-time pastoralists: The political economy of resource use. *Human Ecology*, 13, 131-151.

McPeak, J.G. and Barrett, C.B. 2001. Differential risk exposure and stochastic poverty traps among East African pastoralists. *American Journal of Agricultural Economics*, 83(3), 674-679.

Ndofor, A.B. 1998. *Evaluation of the Successful, Grassroots Credit Union Development in Southern Ethiopian Rangelands.* (Technical Report). SR/GL–CRSP Pastoral Risk Management Project.

Nissanke, M., and Aryeetey, E. 1988. *Financial Integration and Development Liberalization and Reform in Sub-Saharaan Africa.* London: ODI.

Osterloh, S. 2001. *Micro-Finance in Northern Kenya: The Experience of K–REP Development Agency (KDA).* (Research Brief 01). PARIMA.

Pratt, D.J., Le Gall, F. and De Haan, C. 1997. *Investing in Pastoralism: Sustainable Natural Resource Use in Arid Africa and the Middle East.* (Technical Paper 365). Washington: World Bank.

Rass, N. 2007. *Policies and Strategies to Address the Vulnerability of Pastoralists in Sub-Saharan Africa.* (PPLPI Working Paper No. 37). Rome: FAO.

Reij, C. and Steeds, D. 2003. *Success Stories in Africa's Drylands: Supporting Advocates and Answering Skeptics. Global Mechanism of the Convention to Combat Desertification.* Amsterdam: Centre for International Cooperation.

Schultz, U. 1996. *Nomadenfrauen in der Stadt. Die Überlebensökonomie der Turkanafrauen in Lodwat/Nordkenia.* Berlin: Dietrich Reimer.

Scoones, I. 1994. New directions in pastoral development in Africa, in *Living with Uncertainty: New Directions in Pastoral Development in Africa,* edited by I. Scoones. London: Intermediate Technology Publications, 1-36.

Shanmugaratnam, N., Vedeld, T., Mossige, A. and Bovin, M. 1992. *Resource Management and Pastoral Institution Building in the West African Sahel.* (Discussion Paper 175). Washington D.C: World Bank.

Sikana, P.M., Kerven, C.K. and Behnke, R.H. 1993. *From Subsistence to Specialized Commodity Production: Commercialization and Pastoral Dairying in Africa.* London: ODI, Pastoral Development Network's Research Programme on Commercial Change in Pastoral Africa.

Smith, K., Barrett, C.B. and Box, P.W. 1999. Participatory risk mapping for targeting research and assistance: An application among East African pastoralists. *World Development,* 28(11), 1945-1959.

Sommer, F. 1998. *Pastoralism, Drought Early Warning and Response.* [Online]. Available at: http://typo3.fao.org/fileadmin/user_upload/drought/docs/Pastoralism%20Drought %20and%20Early%20Response.pdf [accessed: 30 March 2011].

Spot Image. 2005. *Spot Satellite Earth Observation System.* [Online]. Available at: http://www.spotimage.fr/html/_.php [accessed: 10 January 2006].

Toulmin, C. 1994. Tracking through drought: Options for destocking and restocking, in *Living with Uncertainty. New Directions in Pastoral Development in Africa,* edited by I. Scoones. London: International Institute for Environment and Development, 95-115.

White, C. 1991. Changing animal ownership and access to land among Wodaabe (Fulani) of Central Niger, in *Property, Poverty and People: Changing Rights in Property and Problems of Pastoral Development,* edited by P.T.W. Baxter and R. Hogg. Manchester: Manchester University, 240-274.

Williams, T. 2002a. Livestock market dynamics and local vulnerabilities in the Sahel. *World Development,* 30, 683-705.

Williams, T. 2002b. *Economic, Institutional and Policy Constraints to Livestock Marketing and Trade in West Africa.* Draft version. ILRI.

Chapter 14

The UK Sheep Industry: An Introduction to its Pastoral System and Approach to Marketing and Breeding

Chris Lloyd

Introduction

The picture of a large island with green, rain-fed pastures and hills populated by occasional shepherds and their herds still comes to mind when people think of the United Kingdom. These images are no mere vestiges of the past. The UK is home to Europe's largest sheep production system, and relies on grass-based diets. In world terms our production per ewe is high but in population terms the largest flocks are in China (136 million), Australia (79 million), India (65 million) Iran (54 million), Sudan (51 million) and New Zealand (34 million) (AHDB 2010).

The UK sheep industry with its 16 million head of breeding ewes and rams focuses on sheep meat production; wool and milk production are of little significance except for a few specialist producers who focus on these commodities. Scotland accounts for 3.3 million head, Northern Ireland 0.9 million, Wales 4.4 million and England 6.9 million (Defra 2005). There has been a recent downward trend in the size of UK sheep breeding flocks, which was exacerbated in 2001, when the foot and mouth disease paralysed UK production. The numbers of UK sheep flocks are now comparable to the levels of the early 1980s, which then saw a period of steady growth from the late 1980s to the early and mid-1990s. In the peak year 1993 the number of UK breeding flock amounted almost to 21 million head.

Pastoral Production System

The UK's climate varies extremely from north to south, closely connected to the north-south topography. Thus, there are many regional microclimates in the United Kingdom, ranging from the mild maritime climate of Cornwall with mean annual temperatures of 10-11 degrees centigrade and rainfall of 1,250 mm, to the comparatively dry semi-arid conditions in East Anglia with similar temperatures but rainfall under 600 mm, to upland areas on the West Coast of Scotland which has average rainfall of over 3,000 mm and mean temperatures of 4-5 degrees centigrade. The western and northern parts of the British Isles lie close to the path of the Atlantic depressions, resulting in milder but wetter winters and cool and windy summers as the depressions travel further north. The areas of upland and hill are generally in the west and north where the upland barriers produce significant increases in rainfall. The lowlands, mainly in the south, centre and on the eastern side of the country have a drier climate and less severe winters.

There are also other general regional climactic variations: The south is warmer and has a longer growing period than the north, where the more extreme weather tends to occur in the mountainous and hilly areas, and the west is generally wetter than the east. Subsequently, UK lamb production is marked by geography and climate that lead to regionally differing management, lambing periods and resulting marketing periods. The UK lambing period lasts five months, as spring starts early in the south and wanders north. To finish a lamb for sale, it normally takes a minimum of 12-18 weeks from birth.

The early lambing period runs from November to February in southwest England and the lowlands, to provide new season lambs for the spring/Easter market. The animals are mainly born indoors to protect them from possible unfavourable weather conditions and to make any assistance and subsequent management tasks at birth easier. Lambs are weaned at 8-10 weeks and quickly finished on intensive cereal based rations. The lambs are marketed between February and June, when prices are highest due to little domestic competition. Yet in recent years, early producers have had to deal with intensified competition from New Zealand (NZ) producers selling imported chilled lamb, as opposed to frozen lamb, into the home and European market.

The main lambing period over the UK, in contrast, spans from February to April. Lambs are also predominantly born indoors, although this reduces with later lambing flocks. Ewes and lambs are turned onto grass after lambing. Lambs are often finished straight off the ewe on grass, or finished on grass post weaning, with little use of supplementing cereal rations. These lambs come onto the market from June to October, when prices are generally lower due to the abundant seasonal supply.

The latest lambing season of the year begins in April in the mountainous areas of the UK, North England, North Scotland and Wales and in some areas ends in June. The hill flocks are mainly kept outdoors on the hills with improved pastures at a premium to provide better grazing. The lambs grow more slowly due to rough grazing conditions and are sold late in the season. The lambs born in the latter lambing period are finished on conserved grass in the form or hay or silage, or specially grown forage crops of kale, stubble turnips or other root crops which they graze through the winter. Most of the female animals are kept for reproduction, while most male lambs are reserved for slaughter.

The UK pastoral system is predominantly based on grass-fed lamb production systems, in which supplementation is only used to compliment grazing systems. This explains the pastoralists' strong dependence on adequate pastures and grass availability.

Sheep Breeding – Connecting the Lowlands and Hills

The diversity of breeds found in the UK has developed historically to suit the varied climate and topography of mountains, hills and lowlands. The variety in land type and quality contributed to a stratified breeding structure. Robert Bakewell is recognised as one of the earliest pioneers of breed development dating back to 1750. This sparked a period in which British genetics found their way all over Europe and later to most parts of the world.

Crossbreeding is a central strategy to utilise each breeds' strength and increase the flock's health through hybridisation. Additionally, crossbreeding links production from the mountains to the more productive lowlands, exploiting the beneficial characteristics of the respective local breeds. One main difference between lowland and highland breeds is their body size. Hill breeds are hardy and can survive in the harsher uplands but produce fewer lambs and with comparably poorer carcasses. They are therefore crossed and their progeny (F1) utilised by farmers in the lowland areas to produce greater volumes of a more consistent product. Lowland breeds, have meatier carcasses, which can be attributed to their more favourable climate and fertile farm environment, a higher variety and availability of plants and thus the better nutritional condition of lowland farms. One typical example of the breeding stratification is the crossing of a hardy hill ewe, like the Swaledale, which originates in the hills of the north of England, with a prolific longwool ram, like the Bluefaced Leicester, to produce a recognised crossbred (F1) known as a North of England Mule. Male lambs resulting from this union are destined for slaughter. In contrast, the female lambs, the F1-hybrid ewes, are sold to lowland producers for reproduction. There they are crossed with lowland rams, like the Suffolk or Texel, with a fast growth and meaty carcasses. The resulting F2-generation comprises fast growing well-shaped lambs for slaughter. This system has created an interdependent relationship connecting the hill and lowland producers, and sharing the wealth generated by the more productive lowlands and their traditionally closer links to the food chain.

Yet, this stratified lamb production system was exposed to increasing pressure throughout the 1980s and 1990s. The influx of other European breeds into the UK over this time encouraged variations to the traditional system. UK sheep producers have always sought to utilise the best breeds available whilst adapting to market signals and government policies for food production, such as the reforms of the Common Agricultural Policy (CAP). The new European breeds brought even more variety into the UK breeding system, and an influx of better shaped, leaner animals to match consumer preference for leaner meat. Today there are in the region of 90 breeds and recognised crosses in UK breeding flock, although approximately 12 of these breeds constitute 90 per cent of production.

Changes to the Common Cultural Policy (CAP)

Sheep production fortunes are closely linked to the development of the European Community Sheep Meat Regime. In the early 1980s the sheep meat regime encouraged producers to improve efficiency and sheep numbers, and consequently the economics of sheep production flourished. However by the mid 1990s polices had begun to change, and the effects of other factors such as BSE in cattle, the subsequent knock on effects of rapidly increasing pig production and supplies of cheap poultry meats were beginning to take their toll. The boom years were over. The UK sheep industry relies heavily on exports to other European countries which in turn generates greater competition for lambs on the home market, with the presence of export buyers often driving prices up as they compete for lambs at key times of the year. However, fluctuations in interest rates and the slow reduction of live exports, in response to welfare lobbies, has meant the positive effect of the export market cannot always be relied upon.

The CAP was recently comprehensively reviewed. Payments are now decoupled from production, and Single Farm Payments introduced where land area dominates subsidy payments. These payments are on a sliding scale until 2012 with producers increasingly rewarded for delivering environmental benefits through their farming systems as opposed to production output. Consequently producers are focused more on market returns for profit. Competition from cheaper NZ imports has often driven down market prices during the later marketing season (Dec–April) when their product is readily available. Decreasing profitability has resulted in many UK producers diversifying into other systems or leaving farming altogether. Those who remain are focused on finding ways to make their systems more productive or to increase the value of their lamb product through niche marketing, organic production systems and methods to improve meat quality.

Herd Mobility, Disease and the Vulnerability of the Livestock Industry

While UK lambing was formerly concentrated in spring, the season now runs from November to June. Alternative crops are grown to extend the grazing season and sheep flocks are frequently moved to fresh grazing within farms. This is done according to the availability of pastures and management decisions about when and where the animals are put on the market. There are different ways of lamb marketing. Lambs are either sold directly to an abattoir, a live auction market or a via livestock agent. Agents act as intermediaries and have a range of clients and contacts for whom they sell sheep to slaughterhouses, or to other livestock producers for further finishing or breeding. For example, a hill farmer could wean lambs and sell them via an auction market or an agent to a lowland farmer for fattening.

The concentration of livestock markets and slaughterhouses in the last 30 years has also increased distances animals must travel. However improved road networks and lorries mean longer journeys are quicker and less stressful for animals than previously. The percentage of sheep sold through markets has remained constant at around 65-70 per cent, reflecting the dispersed nature of the industry and the beneficial role markets play of allowing the combination of smaller batches into economic numbers for purchase and transport over longer distances to the larger abattoirs. On the domestic market, breed variation leads to some variation in carcase size (typically 14-22 kg). Whilst this can be a challenge it does provide flexibility to supply a variety of markets both at home and abroad. When it comes to exports, there are a variety of desired specifications for carcase size but they need to be consistent in terms of weight and finish to the desired markets' specification. For example, smaller lambs with a leaner finish are preferred in southern Europe whereas typically Northern Europe favours a heavier lamb with good shape. Unlike cattle and pigs, sheep are rarely traded singly, usually only males and occasionally purebred females purchased for reproductive purposes are sold and traded individually. Sheep are usually sold in batches of up to 50 or more.

The number of slaughterhouses across the UK has reduced from over 1,500 in the 1980s to less than 400 today. The rationalisation has resulted from increased competition and the subsequent need for efficiency (which has required investment) and the cost of rising hygiene standards and compliance with legislation. In general it has been the smaller abattoirs which serviced local supply chains which have suffered most.

Sheep Markets and Lamb Exports

Annually the UK, one of Europe's largest sheep meat suppliers, produces approximately 302,000 t of lamb meat. Multiple retailers purchase 34 per cent of UK lamb, food services, such as restaurants and caterers, buy 27 per cent. Twenty-two per cent of the lambs are exported, ten per cent are acquired by the butchers for high street outlets and seven per cent by other retailers such as farm shops and so on.

The UK mainly exports its lamb to other member states of the European Union. Total lamb exports – 22 per cent of the entire lamb production – amount to approximately 76,000 t. France is by far the biggest importer of UK lamb taking one of every five lambs produced in the UK. In 2005, for instance, 73 per cent of the lamb exports were directed to France. Other key markets were, nine per cent to the Belgium market, six per cent for Italy, five per cent for Germany and seven per cent to other countries. Generally, the heavier lambs are headed to Northern Europe, while the lighter lambs predominantly from the hills were destined for Southern Europe.

Lamb exports are an important source of income for the United Kingdom's sheep producers, since the European prices are generally higher than the domestic prices. Yet, export quantities have been sinking in the last ten years. In 1995 UK pastoralists exported nearly 150,000 t, with large numbers exported live for slaughter in other European member states. The following years were marked by dwindling volumes. By 2000 lamb exports had fallen to less than 100,000 t. In 2001 with the outbreak of foot-and-mouth-disease these numbers dramatically crashed to 30,000 t. However, the demand for lamb in the main European markets has remained high and since 2001 lamb exports have recovered although now almost entirely exported as carcasses or selected cuts/joints.

Lamb exports are not distributed evenly throughout the year. Spring and early summer are the major lambing periods and it takes time to finish the lambs. Thus, the major export season for the spring lambs is autumn and winter, from September to December. For instance, in 2004 and 2005, lamb exports for these months averaged about 8,000 t per month, but December 2005 stands out with lamb exports reaching 12,000 t. From April to July, right within the main lambing season, exports are at their lowest point, about 5,000 t, until they rise again in August. The winter months January to March are marked by moderate exports of 6,000 to 7,000 t.

The reverse is true for the UK lamb import profile. The UK is not only an exporter of lamb meat, but also a large consumer of foreign lamb to balance seasonal supply and demand. Each year it imports approximately 134,000 t of lamb, most of it coming from New Zealand (NZ). Lamb imports are highest from March to May. Through the mid 2000s an average of 12,000 t of lamb was imported in each of these spring months. From June to July imports sink slightly as domestic production increases, and then reach their lowest point during the home lamb marketing season from August to November. The average monthly lamb meat imports from August to November in 2003-2006 amounted to about 6,000 t. In December the imports recovered to 9,000 t and remained on that level until February. NZ remains the biggest importer of lamb into the UK.

The New Zealand Factor

The lamb production relationship with NZ is uneasy. On the one hand, the UK needs imports to secure lamb supply throughout the year and provide customers with a consistent product in seasons with a low domestic production. On the other hand, as home-grown spring lambs become available, they face fierce competition from the last of the main stream NZ lamb imports.

From a UK pastoralist's perspective NZ is seen as the major competitor, with relationships heavily influenced by major retailers and food service buyers, looking for consistent supplies and competitive prices when home grown lambs reduce in number, become more expensive and due to varied finishing systems and age have more varied consistency in size and flavour. NZ sheep production is a predominantly export driven industry and works hard to secure export markets. Due to its relative proximity to the equator it has a longer grass growing season making it ideal for pastoral production. It also has a long history of exporting product and today accesses a much greater range of markets worldwide to absorb its considerable production. In contrast, the UK concentrates on the home market and exports almost solely within Europe.

Due to restructuring of the pastoral system in NZ, its products are more competitive. For example, the national flock is now smaller but the amount of lamb meat produced remains the same. This is due to increased lambing percentages and the increase in average carcass weight of NZ lambs from 13 kg in the 1980s to 19 kg today. More money is made with heavier lambs, as pastoralists make more profit on each individual head of sheep and spend less on rearing and finishing. The heavier the lamb gets, the better the returns on investment. However, there is also an upper limit to marketable lamb weight for both NZ and UK producers. The 21 kg-mark presents a ceiling to producers, as UK supermarkets do not want bigger legs or joints, because large size becomes a unit price issue. This is where the customers' demand for consistency is crucial: They want stable, consistent prices, quality and sizes of products.

Novel Programmes and Incentives in the Lamb Industry

Among other pressures, strong foreign competition has led to the creation of several initiatives and programmes aimed at improving the competitiveness of UK lamb producers on the domestic and export markets to help farmers to cope with the changing environment. UK producers have had the benefit of beef and lamb technical and promotional organisations for the past 35 years. More recently (2000-2001) devolution of some political powers to the individual countries in the Union has filtered down to agriculture and the establishment of country specific levy bodies for promotion and technical development. In the English context the English Beef and Lamb Executive (EBLEX) was founded in 2001 to promote its products and provide English producers with a competitive boost. It aims to provide technical information and advice to producers, monitor market and industry information at producer and consumer levels, and develop a marketing strategy for English beef and lamb. To accomplish this, it has developed its own quality standard mark for beef and lamb, this currently being the only UK assurance scheme regarding both the eating quality and production standards. EBLEX is financed through a levy paid on all sheep and beef animals slaughtered in England and is responsible for spending money on promotion and technical development on behalf of the English pastoralists.

Recently EBLEX launched the Better Returns Programme with the objective of improving producer efficiency and production. Its main goal is to improve efficiency at the beginning of the animal supply chains, and thus directly benefit livestock producers. In the realm of lamb production, the programme consists of different target areas of technical efficiency like ram selection using performance recorded figures, lamb management, knowledge of market requirements, managing the grazing and ewe management with a focus on maintaining the stocks and boosting lamb output. It encourages producers to pay closer attention to key management areas such as grassland, feeding regimes, pregnancy, weaning, on-farm labour and equipment to influence business productivity.

In the frame of this programme the Better Breeding Project promises to increase lamb performance by means of careful ram selection based on known genetic information for variables such as weight, muscle and fat depths. Another aspect of ram selection is the training of breeders to assess sheep performance and to precisely record their own ram's and flocks' performances. The programme encourages producers to focus on meeting market specifications in order to achieve better returns.

At the other end of the supply chain EBLEX also works with consumers such as school chefs, restaurants, caterers, butchers and the public procurement sector, since food service and catering are other key areas of activity. In the supply chain sector, EBLEX continually has a variety of marketing led projects running with various beef and lamb businesses and retailers. These projects cover areas as diverse as nutrition, retail packaging, butchery techniques and product innovation.

All programmes and projects are preceded and paralleled by knowledge transfer from the EBLEX board to farmers. In order to catch the farmer's attention and sharpen awareness to certain issues and new scientific findings, EBLEX has to display ingenuity. For example, to get livestock producers involved, it has designed computer games like the *fantasy ram selection* and frequently holds on farm events with market sorting trials, training in lamb selection and butchery demonstrations. These aim to help the pastoralists to know how to immediately assess the fat and muscle proportions of a given lamb and consequently should help them to market their lambs at the right time to meet market specifications that are in demand. This kind of knowledge helps to connect the farmers to the other end of the supply chain and help them respond to new demands and trends.

New Marketing Strategies

Lamb supply chains are increasingly consumer driven – especially those directed to the European market. Producers must adjust to market demand. Knowing what customers want has become a key to success. New health food trends have left a lasting impact on the UK sheep production system as producers must market products to fit in with shifting consumer attitudes. Lamb is often viewed as being 'too fat', which while traditionally acceptable, now no longer fits with modern consumer health driven sensibilities. This transformation means that although previously fatter lambs were considered ideal, consumers are now after a leaner lamb, which has led to changes in animal diets and more attention on the correct timing of selection for slaughter. The view of lamb as a modern meal option also varies across Europe. In Germany, the low consumption of lamb may be connected to the notion of lamb as a heavy, fatty and unhealthy meat. Changing these preconceptions and exploiting new opportunities is a major goal of lamb marketers.

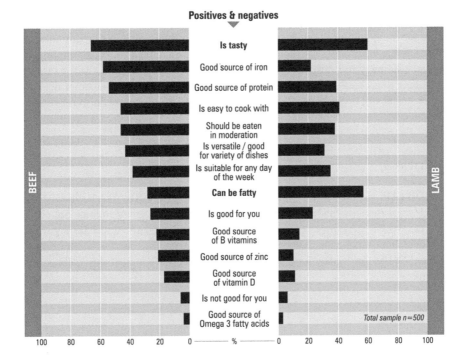

Positives & negatives

Is tasty	
Good source of iron	
Good source of protein	
Is easy to cook with	
Should be eaten in moderation	
Is versatile / good for variety of dishes	
Is suitable for any day of the week	
Can be fatty	
Is good for you	
Good source of B vitamins	
Good source of zinc	
Good source of vitamin D	
Is not good for you	
Good source of Omega 3 fatty acids	Total sample n=500

BEEF   LAMB

100 80 60 40 20 0 ——— % ——— 0 20 40 60 80 100

Figure 14.1   Image of meat
Source: EBLEX (2010).

Societal changes and their impact on eating habits have also shifted meat demand. Today people spend less time cooking than in the past and the way we eat has also changed. Many people, even families, eat at different times of the day and 60 per cent of people eat alone. For instance, in 1980 the average time spent for meal preparation was 62 minutes. Today people spend on average 13 minutes. This means that for lamb to compete with other meal options, its preparation must be more convenient. It is essential that lamb marketing responds to these societal changes and creates an image of lamb as being tasty, easy to cook, convenient, and healthy (Fig. 14.1).

The environmental friendliness of livestock production systems is also of increasing importance to consumers. Many customers care about the origins of animals, that they are raised predominantly on pastures not in stables, and fed only on natural grass and substances without hormones. They also care about different stages of the lamb supply chain, like the adequacy of animal transport, and the guarantee of food safety for every production phase. Lamb production needs to respond to consumer requirements for ethical and environmentally sound practices as well as ensuring high quality lamb.

The new and changing demands tied up with cultural changes will have lasting and profound effects on the United Kingdom's pastoral system which must change in sync in order to remain competitive. This is complicated by competition from cheaper lamb imports that are also attempting to meet consumer requirements. The UK sheep production sector is rising to these challenges, by attempting to refine efficiency to boost competitiveness, and providing programs (for example EBLEX) that creatively link sheep production, marketing and consumer desires. How successful these practices will be in an era of strong competition remains to be seen.

References

AHDB (Agriculture Horticulture Development Board). 2010. *International Meat Market Review*. ISSN no. 0263 2217. Contact: redmeatmi@ahdbms.org.uk.

Defra. 2005. *Livestock Census*. [Online]. Available at: www.defra.gov.uk/evidence/statsitics [accessed: 29 November 2010].

Eblex. 2010. *Consumer Survey*. Unpublished material. Contact: brp@eblex.org.uk.

Chapter 15

Market-Making and Livelihood Challenges in Contemporary New Zealand's Dairy and Sheep Pastoral Economies

Richard Le Heron

Introduction

New Zealand (NZ) is well known internationally for its livestock industries. Yet little attention has been paid to the market-making activities that have shaped, and livelihood prospects that stem from, NZ's leading presence in the world livestock economy. In the context of neo-liberalising economic reforms, NZ's dairy and sheep (meat and wool) economies have changed dramatically. The practices of 'doing dairy' and 'doing sheep' in NZ in a globalising era are no longer what they were in the past. Nowhere is this more so than in the forging of market relations. With this has come a new generation of issues relating to livelihood, the ramifications and implications of which are only slowly edging into the public spotlight. Globalising economic entities and globalising governmental frameworks are reworking the relationship between markets and livelihoods in New Zealand.

This chapter considers the broader processes altering the market-livelihood interface of NZ's pastoral economies in the early twenty-first century. The chapter identifies a range of old and new issues that follow from recent developments in the interface and argues that agri-food inquiry should more explicitly recognise recent globalising developments and their inevitable but unpredictable politics.

The chapter centres market-making by drawing on conceptual, theoretical, methodological and empirical developments in the social sciences literatures over the past quarter century, especially those emphasising the constituted and constitutive nature of human activities.[1] In particular the chapter uses post-structural political economy framings (Larner and Le Heron 2002a, Le Heron 2009, Lewis 2009) to explore connections between investment and institutional processes that are implicated in market-making and livelihood arrangements. To say market-making rather than markets is to shift the focus from a preoccupation with market conditions, such as prices of commodities, to a concern about revealing the breadth and nature of agency that go into creating market relations, and to open up

---

1   Carolan (2009), Larner and Le Heron (2002b), Larner et al. (2007, 2009), Lewis et al. (2008), E. Le Heron et al. (2011), Le Heron and Lewis (2009).

consideration of the translation of market arrangements into territorial livelihood possibilities.[2] I am therefore thinking about the emerging pastoral economy very differently from much orthodox (Emerson and Rowarth 2009, Oram 2007) and new political economy of agriculture (Friedland et al. 1991) writing.

I seek to make more visible dimensions of the dairy and sheep worlds in the form of relationships, actors, places, ecologies and possibilities. These aspects and their interconnections are integral to understanding the globalising dimensions of the circulation and exchange of pastoral economy products and services (Campbell et al. 2009, Holloway et al. 2009). This is vital when concern includes getting a better sense of who is gaining or losing, where, and, what is being gained or lost where, as sectors continue to emerge.

Specifically the chapter poses the question: 'How might agri-food researchers and people in NZ talk in a globalising world economy about the nature and consequences of contemporary engagement of NZ's pastoral economy actors, in the making and remaking of markets, and livelihoods?' Why this question?

Foremost the question reflects the performative turn in the agri-food, economic geography and rural geography literatures.[3] This involves a growing commitment to producing new styles of knowledge and narratives about socially significant phenomena.[4] Setting out how different models of organisation confer recognised or unrecognised choices, about socio-economic futures, in and beyond the pastoral economy, opens up discussion over processes needed to review choices and their wider ramifications and implications.

NZ is used twice in the question. This is to indicate that how we think about and position NZ may be important to where we take any discussion of markets and livelihoods. As the mix of actors in NZ's pastoral economy has altered, it has become increasingly important to develop understanding of both the nesting of these actors in wider networks and how NZ is then understood by actors whose world of operation is far more than NZ (Gray and Le Heron 2010, Roche 2011).

The question confronts the nexus of production and consumption relations, which is at the heart of developing greater consciousness of links between market relations and livelihood experiences. This nexus is often inadvertently downplayed in the literature, at least in NZ. There the focus has usually been on economy-market links per se, a particular truncation of the nexus that reflects a NZ interest in exports, and not taking the next step of linking market relations to possible livelihood outcomes. Knowledge about the actual and potential links should assist with enlarging the decision space regarding organisational choices and the futures they imply. The question thus opens up wider issues that precede specific analyses of livelihood outcomes. Additionally, the NZ experience, where the 'stretched out around the world' materiality of its production-consumption

---

2   Freidberg (2004), Gertel (2007), Maye et al. (2007), Moran et al. (2006).

3   Le Heron (2003), Maxey (2007), Panelli (2007), Woods (2010).

4   Anderson and Harrison (2010), Carolan (2008, 2009), Gibson-Graham (2008), Gibson et al. (2010), Thrift (2007).

relations is an enduring feature, shows how the rise of globalising processes has forced new knowledge production strategies to deal with these changes. This then is a new conceptual space in which the chapter's explorations about the NZ pastoral economy are placed.

The question also emphasises the relationality and mutuality of markets and livelihoods. The word engagement is therefore deliberately used because it positions actors and their circumstances, constraints and contributions, in varied, fluid and geographically specific contexts. It privileges neither consumption nor production activities. Instead it accents as very important the production-consumption interconnections enacted by actors. Connection has to be achieved, worked on, and maintained. The question is thus suggestive of different directions of understanding, deliberation and negotiation over livelihood. In approaching market relations in this manner I argue for a situated framing of economy-market understandings, and a very different exploration of how markets and livelihoods are co-constituted.

The chapter attempts to answer the question in four ways. First, it opens by framing the contemporary 'contrast in fortunes' of the dairy and sheep sectors. This strategy sets the scene for learning through the experiences of contrast. The contrast creates conceptual space to link market-making efforts and livelihood outcomes to historically specific commitments to particular models of individual and social or collective organisation in the pastoral economy. Second, dairy and sheep experiences are outlined in trajectory terms. The discussion provides the background to the contrast in fortunes established in the first section. This depiction immediately enables the relations of markets and the relations *with* and *in* markets to be seen both in detail and in their direction. But more usefully the trajectory-based approach reveals key moments in each trajectory where the basis and nature of market relations have been challenged, whether from external pressures or developments internal to the dairy or sheep economies, and where, consequentially, shifts in investment patterns may arise. Simplified trajectories show how successive sets of differently empowered actors relating to each sector keep meeting dilemmas over appropriate forms of organisation in and across spheres of activity.[5] Reappraisals, contests and politics over 'best' coordination and control arrangements shape how actors in the trajectories behave, when faced with multi-sourced pressures. Third, the trajectory framework is then used to interrogate the nature and implications of possible organisational choices of the very different contemporary moments in the dairy and sheep trajectories. The contrast in fortunes is used as a point of entry into an investigation of livelihood issues (though not the detail of actual outcomes) that flow from the emerging dynamics in each. By following both what the key actors in each trajectory are doing with respect to market relations and theorising what new structural dynamics are possible in each of the dairy and sheep economies, a very different understanding of the geography

---

5   Begley et al. (2009), Bryson and Taylor (2010), Hayter et al. (2005), Le Heron (1988), Le Heron, Roche and Anderson (1989), Taylor and Thrift (1982a, 1982b).

of livelihood possibilities in NZ is reached. Fourth, the new understandings about the complexities of the very different contemporary moments in the dairy and sheep trajectories are used to consider how different 'interests' as well as obvious 'stakeholders' could constructively re-approach the debate about organisational issues.

Contrasting Pastoral Economies – Dairy and Sheep

Dairy and sheep remain integral to NZ's economic fortunes. However, their special significance for the purposes of the chapter lies in their distinctive forms of individual and wider organisation, within NZ and overseas, that each trajectory features. The long sweep of politicised and problematic engagement, of first sheep, then dairying, in overseas markets, initially under preferential arrangements of supply to Britain and more latterly, through participation in an increasingly globalised world economy, offers a unique view upon the always in-the-making aspects of market relations. It is through the lens of difference that the range and character of contemporary issues relating to markets and livelihoods are brought into sharper relief.

The dairy and sheep situations of the 2010s highlight the breadth of issues springing from continuing encounters of actors, with dilemmas over the nature of agency and structure relating to organisation. Recent developments mark a dramatically changed field of actors, efforts to find 'better' organisational models at every level in the supply chain, and unprecedented lines of differentiation in the two trajectories.

Commentators generally agree that the dairy economy has been a strong performer over the past decade. This positive assessment, though not without its sceptics (Fox 2009), results principally from high returns to farmer-shareholders in Fonterra, NZ's global dairy corporate, which processes over 95 per cent of NZ's milk production and has strategically expanded internationally as a major ingredients processor (Gray et al. 2008). In response to dairy's perceived strength as an income stream from land use (but see Sheather (2009) for a counter view), large scale conversions from sheep into dairying have occurred in many parts of the country (FFNZ 2008).

In contrast the sheep economy has been viewed as a weaker, if not weak, performer, with wildly fluctuating returns (Morgan 2011). Several factors, in meat and wool respectively, have contributed to this situation. Although NZ has sizeable sheep and beef meat processors in their own right, they individually lack the international presence of dairy's Fonterra. Intense competition amongst the larger private and co-operative meat processors for livestock has been longstanding. Arguably, this rivalry has regularly stymied efforts to develop a collective approach to coordination of economy-market relations (Roche 2011). A 2010 merger proposal, one of a number in a decade, had no more success than its predecessors. The recently announced Meat Industry Strategy initiative involving the Primary Growth Partnership, a central government science funding arrangement (Adams 2010a), may suggest renewed interest in collective solutions. Nonetheless, a specialist lean meat segment has grown steadily. This consists of smaller specialist companies, who are creating new and often shorter supply chains into markets.

The wool segment has not fared well, suffering declining demand for many decades, and much erosion of industry infrastructure, including dismantling of the New Zealand Wool Board and the closure of the Wool Research Organisation of New Zealand. Responding to this structural crisis has been uncoordinated and intensely competitive. Two largely independent consortia of coarse wool producers sought in 2010 to re-launch an international organisational presence, the first significant development since the demise of the NZ Wool Board in 2003 (Bradford 2009). This move had two parts. The Australian agri-food company, Elders, and the Primary Wool Cooperative launched their Just Shorn brand, aimed at retailers. Wool Partners International with PGG Wrightson, a New Zealand agri-food company, along with Wool Grower Holdings, launched the Laneve brand, targeting carpet manufacturers (Slade 2009, 2010a, 2010b, 2010c). There has, however, been recent success. In the fine wools segment Merino wool producers re-invented themselves during the 2000s by shortening the gap between producers and consumers through new generations of Merino products (Saporito 2010).

The chapter now goes on to argue that organisational choices are deeply implicated in the stark 'contrast in fortunes' displayed by the two pastoral economies, by outlining the trajectories of organisational emergence. Given NZ's prominence in the internationally traded sphere of pastoral products and services, an examination of the organisational dimensions of the dairy and sheep economies should provide insight into what might be possible or not possible in each, with the contemporary mix of actors and the pressures and conditions in which they are operating.

Trajectories of Emergence in NZ's Pastoral Economies

This section presents NZ's pastoral economies as trajectories. Each trajectory is discussed in terms of composites of actors mobilising and investing in activities contributing to the creation, maintenance and expansion of value and values centred on livestock and livestock products and services. The argument is that trajectories of market relationships frame into visibility the circulation and exchange or non-circulation and non-exchange of economic value *and* the circulation and exchange of other values. This section focuses on dairy and sheep emergence under the national development project (McMichael 2000). Subsequent sections explore their emergence as part of globalisation (McMichael 2005).

While the sheer dollar contribution of dairy and sheep to NZ GNP (Gross National Product), economic impacts in rural and urban areas, volume of production on and off farms, employment dependence and so on are unassailable facts, this emphasis limits how immediate and future oriented discussion might proceed. It strangles discussion over what is happening to investment streams in each pastoral economy and their attendant organisational, technological and territorial characteristics. Further it deflects discussion of what is changing and who is benefitting or losing.

At the outset it must be said that NZ agri-food researchers for the most part find it hard to recognise biases that creep into their explanatory accounts, stemming from NZ's position geographically, in relation to markets, and in relation to key dynamics in the globalising world economy. It is often assumed that distance from centres of population and economic activity influence the ability to build market relations, that the ability to produce is more important than abilities to secure and maintain market relations, and that producers in the NZ economy have a special place in that economy. The biases have led to a privileging of producer views over other views, which has led to accounts of developments and issues that restrict the scope of most discussion to producer interests. An important exception is Brooking and Pawson (2010) who show how grass, grassland and grassland farming were co-constitutive of colonial dependencies in the nineteenth century.

For much of the twentieth century the broad aim, as articulated in government policies, advocated through farmer political support and affirmed electorally, was to position NZ as an agricultural export nation, into primarily the UK and European markets. This was accomplished not through a grand design that was directly imposed, but from the gradual adoption of organisational models and principles of organisation at the farm, processing and marketing levels. The models and principles developed were attempts to reap the best from and contain the worst of capitalism into which pastoralism in NZ was being inserted. This occurred over many decades. Central to this vision and its supporting narratives are efficient units of farming, processing and marketing. But the pathways followed by and the experiences of the dairy and sheep industries to achieve efficiencies at every level in their commodity chains and in the system as a whole have diverged. This has meant that at any time the stresses and successes met by farmers, processors or marketers were never the same. Before providing a genealogy of wider organisational developments a short summary of formative influences in each trajectory is outlined.

The establishment of NZ's pastoral economies by European settlers reflected strong class origins. The sheep economy, starting in the 1840s both led and developed with inland exploration of NZ. It was extensive in nature, quickly consolidated into large estates in the high country of the South Island and the Hawkes Bay and Wairarapa in the North Island, initially grew wool and then meat once refrigerated shipping was available, and was closely integrated into colonial and then NZ politics. Sheep grazed on indigenous then exotic grasses. Run holders, because of their capital base and size, typically organised the shipment of wool and meat to Britain, retaining ownership until sale in Britain, hoping to secure both higher prices for the product and a higher percentage of the total value created (Roche 1999). By contrast, dairying started half a century later, especially in

Taranaki and the Waikato, involved intensive farming of small holdings owned by under-capitalised farmers. Again, a grass-based ecosystem using supplementary feeds was adopted. The organisational solution to the pooling from many small producers of a perishable product, milk, was the formation of dairy cooperatives to manufacture butter and cheese. These cooperatives struggled to sell dairy produce in Britain at prices that gave adequate returns to NZ farmers. Success in this early period came from NZ's unique relationship with Britain. Described as 'Britain's farm', much access to the British market was guaranteed through colonial and settler bonds. Nevertheless it should be noted that Britain was also the world's largest importer of food supplies and it imported from many countries. The open and enduring question, however, was 'Could farmer-settlers achieve an adequate livelihood by farming within this distant geo-economic and geo-political relationship?'

Pricing difficulties of the 1920s and 1930s prompted major legislative response that laid down new regulatory frameworks and set new conditions for each sector (Belshaw et al. 1937). The main model of the reorganisation was the Producer Marketing Board. This moment of organisational inspiration was significantly shaped by pastoral economy politics. In dairying the cooperative ethos was deeply enshrined. Accordingly dairy farmers favoured a model of single desk selling to represent their collective interests abroad, combined with a collective and stable mechanism to distribute returns so as to help farming viability. Dairy farmers thus effectively sold their milk and rights at the farm gate, trusting that the cooperative dairy company they sold to would be an efficient producer and that the industry's Producer Marketing Board would be able to successfully stand in the market because of quality and quantity advantages that it could build up. The sheep industry preference diverged, favouring instead the more atomised approach that it had grown up with. The marketing structure that resulted nevertheless had notable differences. For wool marketing responsibilities remained with farmers. Two pathways were available – sell locally to competing private companies or retain ownership of product into overseas markets. By comparison, meat was processed by overseas freezing companies (Roche 2011) who had established networks in the UK and US. These options exacerbated competition, amongst farmers, processors and marketers, up and down, and across, the value chain built up around sheep production.

In the post-World War II long boom NZ's pastoral economies underwent further change (Johnson 1992). This was a period when the dairy and sheep economies expanded to satisfy external demand (such as unexpected demand for wool occasioned by the Korean War or meeting emerging demand for dairy products in Asia and the Middle East) and to domestic pressures (such as the expansion of agriculture as part of national development, bringing more land under agriculture).

Efficient farming was aided by farm system innovations that flowed from state-funded research and farmers and farm labour who had been university trained in diploma and degree courses and was actively assisted by agricultural extension services. Efficient processing too rested on state-led research, but it also rested on economies of scale that continued to accrue from a productive farming sector. Government policy was ideal for the dairy sector where synergies accrued from more milk at the farm gate flowing through to processing and on to available product for sale. It was less ideal for the sheep sector. Although grassland expansion and grassland productivity was incentivised by government (Le Heron 1989a, 1989b, Winder 2009), the processing sector was dominated by large scale foreign owned meat freezing works (Le Heron 1991). This became a major bottle neck. Spirited struggles took place to create farmer owned meat processing cooperatives to reveal the 'excessive' profits of foreign owned operations.

In dairying, the cooperative resolution meant farmers were shareholders of NZ dairy manufacturing cooperatives and this conferred vertical integration, with the New Zealand Dairy Board (NZDB) the marketing solution, for the industry, and an allocation framework, to dairy farmers in particular. Legislation gave the NZDB single seller status, access to low cost credit, stabilised and to a degree guaranteed farmers income. In this regulatory context farmers could push their farm frontiers of efficiency. They could also expand their farm size through purchase outright or of portions of 'inefficient farms' as they came on the market. Share milkers (contracted farm labour/managers who owned their own herd but not land) could expect a future in dairying. They could climb the ladder from small herd owner in a share milking contract with an established dairy farm owner to a farm owner in their own right.

The social and infrastructural dividends from this state-led 'whole system' organisational configuration of the pastoral economy probably peaked in the 1960s (Britton et al. 1992). Specialised agriculture in the form of dairying and sheep/beef farming was the mainstay of the rural economy in many parts of the country (Britton et al. 1992, Le Heron and Pawson 1996, Moran et al. 1996). This was when the number of farm households, households linked to processing, rural suppliers and rural contractors servicing each industry, was at a maximum. This national territorial configuration could be said to be organised for volume production and for maximum export revenue.

This was, however, volume production of undifferentiated commodities. What was sent 'to the market' in the UK was further processed by others into additional products. Whether it was the sale of carcasses to butchers who made the cuts for consumers, or bulk butter that was packed into small pats for distribution, or the slicing up of cheese rounds by grocers, NZ farmers (and NZ) did forego much value obtained at point of final sale. This was not a situation that could be sustained indefinitely.

How might the ongoing developments discussed above be assessed in more conceptual and theoretical terms? Figure 15.1 provides a stylised and highly simplified trace of dairy and sheep developments as they have been played out. The figure re-maps dairy and sheep as involving a number of interlinked dimensions and key market-making moments. The organisation of the figure requires comment. Its main argument is that shifts in regimes of regulation and accumulation possibilities have had major implications for NZ's pastoral systems. This is represented through

	Period	Market Making Moments	Governmental Framework		Key Coordinating Actors		
			Values framing	Values feedback	Dairy	Sheep meat	Wool
National Development Project	1930s	Depression	Colonial relations	Price/volume	Cooperatives Produce Marketing Boards	Foreign meat freezing companies	Independent farmers .
National Development Project	1940s	Post-World War II	Commonwealth relations	Price/volume	Cooperatives Produce Marketing Boards	Foreign meat companies	Indepentent farmers Wool Board
National Development Project	1970s	EEC/EU	Nation state relations	Volume/access	New Zealand Dairy Board	New Zealand Meat Board	Wool Board
Globalisation Project	1980s	Neoliberalising restructuring	Company-company relations	Quality/price	New Zealand Dairy Board	NZ companies NZ cooperatives	Wool Board
Globalisation Project	2000s	World Trade Organisation, peak oil, peak food	Big global actors	Multiple/ indeterminate	Fonterra	NZ cooperatives Landcorp	Independent farmers

Figure 15.1 Simplified genealogy of New Zealand dairy and sheep economies
Source: Author.

the 'market-making moments' column. In the political economy of agriculture and food literature, moments are regarded as periods when established patterns of investment, investor behaviours and governmental frameworks are strained by unexpected issues, events and developments. The idea of moments is a conceptual device to assemble dimensions of change to enable reflective discussion. Moments are thus a way to expose or problematise, through their conceptual assemblage work, the basis and nature of existing relationships amongst actors, the difficulties actors have in maintaining such relationships, experimentation that actors engage in to find workable arrangements in a new context and hints of relationships that seem to be successful. The succession of moments shown indicates an emerging set of engagements with contextual relationships of organisation. The next two columns bring wider governmental influences on actors into the picture. For each successive moment the basis of resolution over the next generation of market-relations has differed. This point has been neglected. Its use in Figure 15.1 highlights what has been at stake; explicit and implicit value and values signals that must be understood in terms of both their presence and their absence. The extent of feedback around value and values shapes how producer-consumer links are institutionalised, and this in turn, is important in guiding governments and pastoral economy investors. To complete the figure the main groups of actors in the dairy and sheep economies are also shown. These actors have been responsible for the coordination that ensured that market relation expectations were met. Such coordination has never been easy or simple. Rather coordination is better regarded as an ever incomplete, contested, negotiable and negotiated commitment. The figure thus treats market-making as something that is both ongoing and is also characterised by periods of intensified activity. What particular actors do under conditions of uncertainty is especially crucial to the definition and settlement of market relationships and the nature of feedback or the lack of feedback that helps hold together market relationships. Finally what should also be obvious from Figure 15.1 is that success and failure, performance and competitiveness, cannot be separated from the market relations that have been made and that actors use to reference some or all of their activities, strategies and behaviours.

Restructuring and Ongoing Reorganisation – The Emergence of 'Big' Investors Amidst Competing Governing Mentalities

NZ's economic reforms in the mid 1980s impacted immediately, directly, deeply but also differentially in NZ's pastoral economies. Rather than discuss in detail the early impacts, which have been previously examined in Le Heron (1988) and Le Heron et al. (1989), the section provides an overview of the cumulative impacts, especially as they relate to changes in organisations and organisational features. The dimensions of change that resulted can be understood as arising from the investment decisions of producers, who in the particular dynamics of the pastoral economies, adjusted rapidly to globalisation pressures, largely without supportive institutions of the state (Le Heron and Pawson 1996).

Figure 15.2 points to key developments. When state-led restructuring commenced in the 1980s, the very different actors of dairy, sheep meat and wool faced vastly different business conditions. With strong signals that the producer marketing board framework would be eventually terminated, the overarching concern of all actors became coordination from production to consumption. This amounted to a re-invention of supply chain relationships, though what this meant was fundamentally changing. Hitherto the term supply had been understood in NZ as referring to the placement of NZ product through NZ producer marketing boards into preferential (for example UK) overseas markets. Restructuring in NZ's pastoral economies, however, coincided with the rapid rise of private sector coordinated supply chains in the global economy. Two problems immediately became apparent. These were stabilisation of farmer-processor relations that had been fractured with closures and competition (Le Heron et al. 2001) and relations between the farmer-processor nexus and supermarkets and food manufacturers abroad as these entities became more and more dominant (Campbell and Le Heron 2007). By the mid 2000s supply was understood by actors in NZ to mean contract arrangements set by big overseas supermarkets, food manufacturers and institutional buyers who had chosen temporarily, from a global menu of supply sources, NZ sourced products. Maintaining a credible market presence in many parts of the world had become increasingly difficult.

Over a quarter century the mix of actors at every level changed inexorably and fundamentally. This has resulted from the disappearance of many established actors and the entry of new (usually small and niche oriented) entrants. All entities were either rapidly globalising or engaging more and more as globalising entities. They have moved from trade dominated activities to deeper integration into global production and links into global finance circuits. These players were meeting other globalising entities and the globalising trajectories of these entities. Especially

	Dairy	Sheep Meat	Wool
Long Term Conditions	Expanded demand for dairy products and internationally traded dairy products	Narrow world traded market for sheep meat	Collapsing demand for wool internationally
Dimensions	• intensification of dairy farming • continuing amalgamation of dairy cooperatives • larger scale processing facilities • continuing shift to milk powder • deepening of ingredients focus • entry into high value products (functional food, nutriceuticals) • reorganisation of NZ Dairy Board, NZ Dairy Group and Kiwi Dairy into Fonterra • Fonterra's emergence as a global dairy corporate • rise of Fonterra's global dairy supply chain • new domestic entrants • foreign entrants • increasing capitalisation of dairy farming	• removal of supplementary minimum prices • decline of sheep farming triggered by declining sheep numbers • rise of private NZ meat companies • exit of foreign meat companies • entry of new small meat processes • realignment of sheep meat supply chain relationships • emergence of large sheep farming entities • experimentation to link farmer groups and dedicated processors into special marketing channels • reduced role of Meat Board • consolidation of meat companies	• gradual collapse of wool auction system • collapse of wool scouring in NZ • collapse of woollen manufacturing in NZ • decline of carpet manufacturing in NZ • shrinkage of WRONZ • phasing out of Wool Board in 2003 • farmers efforts to develop new marketing channels • merino revival driven by merino farmers forging new market relations
Resulting Actors by 2000s	Global corporate: Fonterra Small dairy cooperatives: Tatua, Westland Small proprietary companies: Open Country New entrants: Synlait	Three big meat companies: Silver Fern Farms, Affco, Alliance Diversity of small specialist meat companies with contracted farmers: Rissington Meats, Lean Meats	Diversity of farmers and small specialist wool companies

Figure 15.2   Recent transformation in New Zealand dairy and sheep economies
Source: Author.

important have been the supermarkets, food and fibre manufacturers, institutional purchasers and science/research and development providers. These developments have culminated in an overall structure of connections into private global supply chain that centre on a few big actors and a range of small actors (Le Heron 2009b).

But restructuring has led to two other interlinked trends. An increasing proportion of income of many NZ domiciled pastoral economy actors is coming from activities outside NZ, this means the very word 'NZ' must be increasingly qualified. Actors from 'elsewhere' are actively engaging in NZ – this is not new but it is under very different conditions – and their increasing prominence is further blurring the interpretation of NZ produced product. The productionist world view of the national development era, with its NZ-proclaiming and value-for-NZ rhetoric, is being challenged by structural realities and complexities in the pastoral economies.

The second source of competitive pressure has come from cross cutting supply chain interactions. Open to globalisation processes, the re-established supply chain relations of dairy and sheep have been 'scrutinised' by an increasing range of actors who have questioned the very practices upon which re-forged production relationships are based.

The trajectories thus show how the present actors have come into existence, what they are facing and the very different national and global arrangements and dynamics in which they are set and participating. There is a gradual realisation that meeting supply chain demands requires more than marketing product, since challenges are springing from wide ranging claims about how value and values should be understood.

Economic actors (for example processors, farmers) have been forced to publicly defend or try to explain why the outcomes of their individual actions (for example dairy effluent on farms) led to off farm environmental impacts (for example declining stream qualities). Under the neo-liberalising state, facilitation by the state has been heavily dependent on private sector initiatives, to the point where the facilitative state has become supportive of private sector governance. Significant initiatives have appeared. For example, dairy stakeholders formed the Clean Streams Accord (Blackett and Le Heron 2008). The dietary, health and nutritional merits and demerits of dairy and sheep meat products have been re-scripted (Dixon 2009).

What are the wider ramifications and implications of these developments for the organisation of pastoral economies in NZ? And, what livelihood possibilities are being dismantled or built in this context? In the section the concern is to gain a sense of the bigger picture from broad based information and not delve into specific measurements or calculations of livelihood impacts.

Linking Market-Making and Livelihood Issues in NZ-Located Pastoralism

This section argues that NZ is *one* of many sites in the global landscape of pastoralism – hence the wording NZ-located in the section heading. It is moreover, a site of contestation as existing and would-be investors, from potentially anywhere, attempt to maximise what they can do in the opportunity space of NZ and secure conditions for their expanded and profitable investment. Discussion in the section is confined to examination of *potential* pastoral sector dynamics and *potential* livelihood consequences in NZ, given already observable changes in the mix and characteristics of major actors, and, the breadth of external and internal influences and adjustments affecting their supply chains discussed earlier. Such is the likely scale and location of change in NZ as a result of evident developments that established and assumed notions of the pastoral economy may be increasingly tenuous.

NZ's pastoral economy is now structured around a handful of dominant NZ-originated actors (Fig. 15.2). These actors have emerged from the restructuring and reorganisation over the last 25 years. They are supported by a new assemblage of infrastructure, industry suppliers, marketing agencies, logistics specialists, scientific entities, lobby groups and supportive frameworks. The main business focus of the actors is to be globally competitive. But while these big NZ originating actors have sprung from NZ pastoral economy dynamics they are embedded in the structural possibilities of both national and global dynamics. What are some of the wider possibilities that are opening in the globalising economic and institutional world that these actors inhabit, and how might the dairy and sheep trajectories alter as a result? Where will competitive pressures felt in NZ come from and where might NZ-generated competitive pressures impact elsewhere? What does market-making mean for NZ-based actors operating in the open horizon of the global economy and actors from elsewhere who are assessing NZ as a location for pastoral economy investment? And what kinds of livelihood questions might be anticipated in this ever changing context?

The most recent moment is one of historically unparalleled circumstances. It now no longer strains credibility to suggest that the NZ pastoral economy is open to any investor, from anywhere, any sector, any production system, any business model, or any labour arrangement. Few actors can be said to have a NZ mandated focus, in contrast to early 'whole system' dedication to the national interest that characterised the national development project. While it is difficult to shed the inclination to think *only from NZ* purely NZ-centric questions are insufficient in this moment.

Figure 15.3 uses three questions that centre on investment: What interests are attempting to make advantageous investments in the pastoral economy? What have these interests been responding to or seeking to initiate? What sites are available to work on existing and new organisational dimensions?

The questions create a conceptual space in which to reflect on the increasingly contested nature of pastoral economy experience in NZ. While obviously an incomplete summary of contemporary interactions, Figure 15.3 helps with positioning the multiplicity of normally un-associated developments and draws on recent press coverage in NZ, to indicate a variety of issues and concerns. It should be added that NZ's Resource Management Act framework and other legal processes mean advance warning of major investments occurs. This allows proposed developments to be scrutinised for their effects. The major drawback to this, however, is that effects are effectively examined on a case-by-case basis, with minimal attention to how they might begin a new pattern of investment, or add to, or deviate from, existing investment patterns.

Figure 15.3 highlights three matters. First, it exposes a range of positions from which interests and specific actors might voice views about actual, proposed and possible investment. Voice may take the form of challenges, claims, promises, announcements, apologies and so on, all of which bear upon investment decisions and decision makers. Figure 15.3 makes it much easier to see what is and is not being spoken about when particular developments are examined. Second, it speaks of deeply challenging re-evaluations of every facet of the pastoral economies as generally understood in NZ. Partiality and patchiness abound in the understandings brought by actors to discussion over the worthiness and credibility of different investment directions. Third, the organisational dimensions need to be seen for their potential to facilitate the total scale of investment. Larger scale investment proposals are often attributed to pressures to achieve economies of scale and economies of scope. By asking what is happening, in which part of NZ,

Domain of chains / places	Interests	Market-making sensitivities / matters of concern
Sites where investors can potentially alter existing or create new organisational features and configurations	Actors / interests positioned in chains/places with capacities / capabilities to mobilise political projects for change / gain	Recent examples drawn from NZ media
• Farming • Processing • Food manufacturing • Transport / logistics • Supermarkets / retailing • Institutional purchasers • Regions • Nations • Trading blocs / World Trade Organisation	• Farmers (e.g. family, corporate, Maori Incorporations, Landcorp) • Processors / marketers (e.g. dairy, meat, wool companies) • Central government regulators (e.g. Ministry of Agriculture and Forestry, Department of Conservation, Ministry of Science and Innovation) • Knowledge founders (e.g. Ministry of Science and Innovation, Marsden Fund, AgMardt) • Knowledge providers (e.g. Universities, Crown Research Institutes, think tanks) • Consultants (e.g. sector, accountancy, legal) • Certifiers (e.g. GLOBALGAP, HACCP, AsureQuality) • Regional government (e.g. Regional Councils, District Councils) • NGOs (e.g. Environmental Defence Society, Greenpeace, Fish and Game, Forest and Bird) • Multi-interest alliances (e.g. Sustainable Land Use Forum, Global Agricultural Research Alliance, Global Sustainable Dairying Alliance, Meat Industry Strategy Initiative, Wool Industry Taskforce)	• Effluent (Balme 2010, Fonterra 2010) • Human health and nutrition (Barton 2010) • Animal welfare (Kerridge 2010) • Intensification (Fallow 2010) • Farm inputs (Davison 2009) • Land use conversion (FFNZ 2008, Save the Farms 2010) • Foreign ownership (Adams 2010b, Anon 2009a, 2011b,c, Gould 2010) • Returns on investment (Palmer 2011, Sheather 2009) • Farming systems (Booker 2010, Perley 2007) • Farming productivity • Labour relations (Hembry 2010) • Rural infrastructure – water (Oram 2010, Taylor 2010) • National interest (O'Sullivan 2009, 2010) • Land prices (Gawith 2010) • Cooperative structure (Gaynor 2010) • Career ladder in farming (Fox 2009, 2011, Anon 2009b) • Farm debt (Scherer 2011) • Marketing strategy (Anon 2009c, Slade 2010a,b,c)

Figure 15.3   Range of issues challenging New Zealand dairy and sheep actors
Source: Author.

the potential dimensions and extent transformation and impacts of investments can be explored. That the main actors are big actors, suggests major changes could be pending. Once cumulative effects of up scaled investment have become a visible part of the economic landscape, new organisational arrangements are probably well developed, and the basis of livelihood changes already in place.

In the open, and facilitative (as distinct from interventionist) and market-serving regulatory environment of NZ's pastoral economies there are several crucial system 'features'. These are often downplayed or missed, but are hugely important in obtaining any sense of what might emerge from the interactions amongst present and new actors. These are the degree of independence of any investor, the responsibilities of different sorts of investor entities, and the assumption that whatever government does, it should be improving investor certainty (O'Sullivan 2010). Independence comes from the fact that in the end it is investors, not regulators, who actually make investment decisions, and it is the configurations that stem from these decisions that impact most on the nature of jobs available, where, who and with what skills and expertise are likely to be attracted and so on. The question of responsibilities in the environment of economic freedom (NZ is ranked fourth in the world on this indicator (Anon 2011)), on the other hand, is often ignored. Yet directors of corporations, for example, have a fiduciary responsibility to act to preserve the entity (Gaynor 2010), ahead of other more social responsibilities such as livelihood stability or environmental quality. These system features mean that advocacy becomes a very central part of the pastoral economy. By voicing issues and concerns (Fig. 15.3) different interests can be supportive or reactive, and narrate here views to reveal or hide crucial information, interpretations or activities. In this context, then, what are some of the main possibilities inherent in the present structural realities of the dairy and sheep economies? The remainder of the section explores several scenarios.

Dairying is highly illustrative because it still appears in NZ to be primarily a single entity arrangement, in the form of Fonterra, with perhaps limited scope for fundamental change. Several scenarios, however, disturb this thinking. First, what might a 'business as usual' scenario bring? Here, the threads of change in NZ include significant gains from NZ competitors such as Open Country and Tatua and major growth of overseas processors such as Singaporean Synlait in dairying or investors in rural services such as the Chinese corporate, Agria, in PGG Wrightson (Hembry 2010). What labour relations might accompany these ventures? Will 'promised' market access actually translate into net positive jobs in processing? Chinese owned Natural Dairy (2010), for example, in its bid to buy failed Crafar Farms, took a full page advertisement in the national daily, the *NZ Herald*, in which it asserted major generation of employment and market access in China. Second, attention should also focus on NZ dairy or sheep investors who are also investing abroad. Through farm sales, for instance, Fonterra farmer shareholders could increasingly become business entities involving arrangements between NZ and foreign investors. NZ dairy farmers have invested in Tasmania and Victoria in Australia, Chile, the mid west of the United States and Russia. The case of NZ Farming Systems Uruguay (NZFSU), championed by PGG Wrightson, epitomises the challenges of developing a successful business model from NZ for another country. NZFSU, according to Gaynor (2010) made the mistake of purchasing land at the food price peak in 2008 and exhausted development monies. But more troubling would be public statements on NZ issues, such as those noted in Figure 15.3, by directors, who are by definition NZ located farmers, but who might also be investing abroad though not disclosing their interests in farming and related businesses in other countries. Third, the difficulties that Fonterra seems to be having in raising on-farm productivity to obtain further processing economies may lead to a complete rethink of the nature-production model that has underpinned NZ dairying for many decades. Should dairying move from an outdoor to indoor pasture model? What are the implications of the import and use of palm oil feedstock in dairying (Davison 2009)? Will dairy and other interests lobby successfully for expanded irrigation of pasture, in regions such as

Canterbury and the Waikato? What are the effluent implications of large scale and proximate investments? This could be advantageous to both existing and different farming models. Which regions will be seen has having major advantages or disadvantages for which nature-production model? The Manawatu River's (Booker 2010) international notoriety as the most polluted river in the world exposes the contestation between Horizon, the regional government regulator, and Fonterra, over responsibilities and appropriate responses. The furore over proposed dairy barns in the Mackenzie Basin, illustrates another edge to land use and production system conflicts in NZ dairying. Fourth, a very different scenario could spring from any fundamental change to the dairy herd. Opting, for instance, to supply reputedly health beneficial A2 milk, rather than A1 milk (Woodford 2007), could prompt changes in farm and processing configurations.

In contrast, the sheep economy has many different ownership structures operating (Adams 2010b) and the present scene is far more complex than how the dairy sector operated in the pre-Fonterra days. Whereas NZ dairying, through Fonterra, is accessing over 120 national markets, NZ sheep meat continues to be oriented to the UK, which is NZ's major market, accounting for 25 per cent of lamb and lamb leg, and 60 per cent of lamb export value respectively (Champion 2010). Germany is the second major market. In the case of the UK, sheep numbers have dropped, and British lamb is being exported to the continent because of a favourable exchange rate. The organisational arrangements that have enabled the retention of the UK market may now be inappropriate.

Conclusion

The chapter has argued that any attempt to understand NZ's pastoral economies must examine dimensions of NZ and global capitalist political economy. This scrutiny reveals two key points. First, both the dairy and sheep economies in NZ are now dominated by big actors who grew and invested in an open NZ regulatory context following restructuring that encouraged expansion by existing entities. Second, and simultaneously, the very regulatory context that enabled growth and dominance is now being re-appraised itself, for hitherto unrecognised investment opportunities, or opportunities previously perceived as unrealisable. In looking at the capitalist political economy, the trajectories approach adopted in the chapter helped greatly with identifying and summarising the main actors, what they are doing and how their activities interact with those of others, and who is entering and exiting the pastoral economies as a result of actor circumstances and capabilities. Indeed an important part of the argument of the chapter is that such trajectories relating to investment might enable actors to more readily engage in wider debate, because they can position themselves and their own actions in the landscape of the wider pastoral economy.

The NZ structural realities, as outlined, bear immediately on current and prospective livelihood scenarios. The NZ situation throws much light on whether market integration, the single most prominent strategy round the world, is a simple and dependable strategy to improve and diversify livelihood outcomes. As it should be obvious from the chapters actor-centred framing of NZ's pastoral economies, a whole series of issues and concerns are actively extending the imaginaries of all directly and indirectly connected with these economies. For the most part recognition of the connections between potentially different market-making scenarios and their directions and dynamics of organisation and livelihood arrangements is limited. The trajectory based approach shows the inherent tensions that persist, even during relatively stable periods, as well as during intensified restructuring phases. By revealing just how complex and complicated the pastoral economies are, any idea of straightforward solutions to the development dilemmas of the current moment would seem to be baseless and naïve, in theoretical, empirical and day-to-day terms. Like the past when investors faced perplexing challenges, more collaborative and collective organisational arrangements are attracting attention, perhaps even gaining ground. But, as in the past, the enduring difficulty is about finding some commonality of interest around what sort of nature-production model should be widely promoted, what goals would guide behaviours and what roles and necessary commitments would ensure behaviour that was both responsible and accountable. The trajectory approach provides one easy way to highlight where issues and concerns come from, and also who brought them into existence and who will be doing the immediate acting.

At this time, in NZ and internationally in the 2010s, a new generation of social science is available to aid in shaping and consolidating, from a basis of some understanding, collective and collaborative organisational arrangements. A burning question nonetheless remains – is there any realisation that a vital issue is that of developing capacities and capabilities to reconfigure social components so as to give improved outcomes for livelihoods (for all plants and animals) from emergent market-relations? Adopting *this* focus would be a monumental shift in how the problems of NZ's pastoral economies might be discussed – a shift from the narrow focus of what needs to be done, to the more demanding and ultimately crucial focus of how dairy and sheep might be *practiced*, in, into and from NZ, fairly for all life, and for the longer term.

Acknowledgements

The chapter is an outgrowth of the presentation originally given at the 2006 University of Leipzig conference on 'Pastoralists and the world market. Problems and Perspectives'. Recent research funded by a Marsden Award, 'Biological Economies: Knowing and Making New Rural Value Relations', UoA09024, has greatly contributed to the development of the argument in the chapter. This support and the interactions of the project team are gratefully acknowledged.

References

Adams, C. 2010a. Chinese to buy Synlait majority stake. *New Zealand Herald*, 20 July, B1.

Adams, C. 2010b. Sheep and beef farms 'on way to extinction'. *New Zealand Herald*, 4 August, B1.

Anderson, B. and Harrison, P. 2010. *Taking-Place: Non-Representational Theories and Geography*. Farnham: Ashgate.

Anon. 2009a. Dairy farmers eye Russian opportunities. *New Zealand Herald*, 9 January, A10.

Anon. 2009b. Dairy industry too important to be stifled. *New Zealand Herald*, 11 September, A10.

Anon. 2009c. Big buyout of farms raises suspicions. *New Zealand Herald*, 19 December, A7.

Anon. 2011a. NZ rates fourth in world for economic freedom. *New Zealand Herald*, 14 January, B3.

Anon. 2011b. Germans cleared to buy South Island dairy farms. *New Zealand Herald*, 2 February, B3.

Anon. 2011c. We did not modify expiry date: Synlait. *New Zealand Herald*, 3 February, B3.

Balme, I. 2010. Fonterra's approach tardy on dairy-farm pollution. *New Zealand Herald*, 30 March, A10.

Barton, C. 2010. The fat of the land. *New Zealand Herald*, 2 October, A9.

Begley, S., Taylor, M. and Bryson, J. 2009. Firms as connected, temporary coalitions: Organisational forms and the exploitation of intellectual property. *Journal of Knowledge Management*, 7(1), 11-20.

Belshaw, H., Williams, D., Stephens, F., Fawcett, E. and Rudwell, H. 1937. *Agricultural Organisation in New Zealand: A Survey of Land Utilisation, Farm Organisation, Finance and Marketing*. Melbourne: University of Melbourne, with Oxford University Press.

Blackett, P. and Le Heron, R. 2008. Maintaining the 'clean green image': Challenges for governance of on-farm environmental practices in the New Zealand dairy industry, in *Agri-Food Commodity Chains and Globalising Networks*, edited by C. Stringer and R. Le Heron. Aldershot: Ashgate, 75-87.

Booker, J. 2010. Abandon giant dairy plans, urges watch dog. *New Zealand Herald*, 28 January, A6.

Bradford, L. 2009. *A Complicated Chain of Circumstances: Decision Making in the New Zealand Wool Supply Chains*. Unpublished PhD thesis, Lincoln University.

Britton, S., Le Heron, R. and Pawson, E. 1992. *Changing Places in New Zealand: Geographies of Restructuring*. Christchurch: New Zealand Geographical Society.

Brooking, T. and Pawson, E. 2010. *Seeds of Empire: Environmental Transformation of New Zealand*. London: IB Taurus.

Bryson, J. and Taylor, M. 2010. Mutual dependency, diversity and alterity in production: Cooperatives, group contracting and factories, in *Interrogating Alterity*, edited by D. Fuller, A. Jonas and R. Lee. Farnham: Ashgate, 75-93.

Campbell, H., Burton, R., Cooper, M. et al. 2009. Forum: From agricultural science to biological economies? *New Zealand Journal of Agricultural Research*, 52, 91-97.

Campbell, H. and Le Heron, R. 2007. Big supermarkets, big producers and audit technologies: The constitutive micro-politics of food legitimacy food and food system governance, in *Supermarkets and Agri-Food Supply Chains*, edited by G. Lawrence and D. Burch. Cheltenham: Edward Elgar, 131-153.

Carolan, M. 2008. More-than-representational knowledge/s of the countryside: How we think as bodies. *Sociologia Ruralis*, 48(4), 408-422.

Carolan, M. 2009. 'I do therefore there is': Enlivening socio-environmental theory. *Environmental Politics*, 18(1), 1-17.

Champion, S. 2010. Knoweldge is key for our sheep farmers. *New Zealand Herald*, 24 May, B13.

Davison, I. 2009. Fonterra says 'no' to ship protest plea. *New Zealand Herald*, 17 September, A1.

Dixon, J. 2009. From the imperial to the empty calorie: How nutrition relations underpin food regime transitions. *Agriculture and Human Values*, 26, 321-333.

Emerson, A. and Rowarth, J. 2009. *Future Food Farming: New Zealand Inc. Meeting Tomorrow's Markets*. Wellington: NZX Ltd.

Fallow, B. 2010. Political challenge over water allocation. *New Zealand Herald*, 23 September, B2.

FFNZ (Federated Farmers New Zealand). 2008. *Factsheet T150*.

Fonterra Co-operative Ltd. 2010. An open letter to all New Zealanders. *New Zealand Herald*, 6 November, A29.

Fox, A. 2009. Fonterra still needs to face big questions. *New Zealand Herald*, 21 September, B18.

Fox, A. 2011. Talks with iwi aim to clear way for milk plant. *The Dominion Post*, 25 January, C2.

Freidberg, S. 2004. *French Beans and Food Scares*. Oxford: Oxford University Press.

Friedland, W., Busch, L., Buttel, F. and Rudy, A. 1991. *Towards a New Political Economy of Agriculture*. Boulder, CO: Westview Press.

Gawith, A. 2010. Restricting ownership comes at a cost. *New Zealand Herald*, 3 August, B2.

Gaynor, B. 2010. Farm firms fail to grow on sharemarket. Most agricultural companies delist after struggling. *New Zealand Herald*, 25 September, C2.

Gertel, J. 2007. Mobility and insecurity: The significance of resources, in *Pastoral Morocco. Globalizing Scapes of Mobility and Insecurity*, edited by J. Gertel and I. Breuer. Wiesbaden: Reichert, 11-30.

Gibson, K., Cahill, J. and McKay, D. 2010. Rethinking the dynamics of rural transformation: Performing different development. *Transactions of the Institute of British Geographers*, 35(2), 237-255.

Gibson-Graham, J.K. 2008. Diverse economies: Performative practices for 'other worlds'. *Progress in Human Geography*, 32(5), 613-632.

Gould, B. 2010. Dairy industry depends on keeping land in NZ hands. *New Zealand Herald*, 28 April, A13.

Gray, S. and Le Heron, R. 2010. Globalising New Zealand – Fonterra and shaping the future. *New Zealand Geographer*, 66(1), 1-13.

Gray, S., Le Heron, R., Stringer, C., and Tamasy, C. 2008. Does geography matter? Growing a global company from New Zealand, in *Agri-Food Commodity Chains and Globalising Networks*, edited by C. Stringer and R. Le Heron. Aldershot: Ashgate, 189-199.

Hayter, R., Rees, R. and Patchell, J. 2005. High tech 'large firms' in greater Vancouver, British Columbia: Congregation without clustering?, in *New Economic Spaces. New Economic Geographies*, edited by R. Le Heron and J.-W. Harrington. Aldershot: Ashgate, 15-28.

Hembry, O. 2010. German firm among several nations investing in NZ dairy industry. *New Zealand Herald*, 10 July, C5.

Holloway, L., Morris, C., Gilna, B. and Gibbs, D. 2009. Biopower, genetics and livestock breeding: (Re)constituting animal populations and heterogeneous biosocial collectivities. *Transactions Institute of British Geographers*, NS 34, 394-407.

Johnson, R. 1992. Farming change, in *Changing Places in New Zealand*, edited by S. Britton, R. Le Heron and E. Pawson. Christchurch: New Zealand Geographical Society.

Kerridge, B. 2010. South Island tragedy highlights animals' plight. *New Zealand Herald*, 30 September, A13.

Larner, W. and Le Heron, R. 2002a. From economic globalisation to globalising economic processes: Towards post-structural political economies. *Geoforum*, 33(4), 415-419.

Larner, W. and Le Heron, R. 2002b. The spaces and subjects of a globalising economy: A situated exploration of method. *Environment and Planning D: Society and Space*, 20(6), 753-774.

Larner, W., Le Heron, R. and Lewis, N. 2007. Co-constituting neoliberalism: Globalising governmentalities and political projects in Aotearoa New Zealand, in *Neoliberalisation: States, Networks and People*, edited by K. England and K. Ward. Routledge: London, 233-247.

Larner, W., Lewis, N. and Le Heron, R. 2009. The state spaces of 'After Neoliberalism': Co-constituting the New Zealand designer fashion industry, in *Leviathan Undone? Towards a Political Economy of Scale*, edited by R. Keil and R. Mahon. Vancouver: University of British Columbia Press.

Le Heron, E., Le Heron, R., Lewis, N. 2011. Performing research capability building in New Zealand's social sciences: Capacity-capability insights from exploring the work of BRCSS's 'Sustainability' theme, 2004-2009. *Environment and Planning A*, forthcoming.

Le Heron, R. 1988. State, economy and crisis in New Zealand in the 1980s: Implications for land-based production of a new mode of regulation. *Applied Geography*, 8(4), 273-290.

Le Heron, R. 1989a. A political economy perspective on the expansion of New Zealand livestock farming, 1960-1984. Part 1: Agricultural policy. *Journal of Rural Studies*, 5(1), 17-32.

Le Heron, R. 1989b. A political economy perspective on the expansion of New Zealand livestock farming, 1960-1984. Part II: Farmer responses – Aggregate evidence and implications. *Journal of Rural Studies*, 5(1), 33-41.

Le Heron, R. 1991. Reorganisation of the New Zealand export meat freezing industry: Political dilemmas and spatial impacts, in *The State and the Spatial Management of Industrial Change*, edited by D. Rich and G.J.R. Linge. London: Routledge, 108-127.

Le Heron, R. 2003. Cr(eat)ing food futures: Reflections on food governance issues in New Zealand's agri-food sector. *Journal of Rural Studies*, 19(1), 111-125.

Le Heron, R. 2009. 'Rooms and moments' in neo-liberalising policy trajectories of metropolitan Auckland, New Zealand: Towards constituting progressive spaces through Post Structural Political Economy (PSPE). *Asia Pacific Viewpoint*, 50(2), 135-153.

Le Heron, R. and Lewis, N. 2004-2009. Performing research capability building in New Zealand's social sciences: Capacity-capability insights from exploring the work of BRCSS's 'Sustainability' theme. *Environment and Planning A* (forthcoming).

Le Heron, R. and Lewis, N. 2009. Theorising food regimes: Intervention as politics. *Agriculture and Human Values*, 26, 345-349.

Le Heron, R. and Lewis, N. 2011. New value from asking 'Is geography what geographers do?' *Geoforum*, 42(1), 1-5.

Le Heron, R. and Pawson, E. 1996. *Changing Places. New Zealand in Nineties.* Auckland: Longman Paul.

Le Heron, R., Penny, G., Paine, M. et al. 2001. Global supply chains and networking: A critical perspective on learning challenges in the New Zealand dairy and meat commodity chains. *Journal of Economic Geography*, 1, 439-456.

Le Heron, R. Roche, M. and Anderson, G. 1989. Reglobalisation of New Zealand's food and fibre system: Organisational dimensions. *Journal of Rural Studies*, 5(4), 395-404.

Lewis, N. 2009. Progressive spaces of neoliberalism? *Asia Pacific Viewpoint*, 50(2), 113-119.

Lewis, N., Larner, W. and Le Heron, R. 2008. The New Zealand designer fashion industry: Making industries and co-constituting political projects. *Transactions of the Institute of British Geographers*, NS 33, 42-59.

Maxey, L. 2007. From 'alternative' to 'sustainable' food, in *Alternative Food Geographies. Representation and Practice*, edited by D. Maye, L. Holloway and M. Kneafsey. Amsterdam: Elsevier, 55-76.

Maye, D., Holloway, L. and Kneafsey, M. 2007. *Alternative Food Geographies. Representation and Practice*. Amsterdam: Elsevier Science.

McMichael, P. 2000. *Development and Social Change: A Global Perspective*. Thousand Oaks: Pine Forge Press.

McMichael, P. 2005. Corporate development and the corporate food regime, in *New Directions in the Sociology of Global Development*. (Research in Rural Sociology and Development, Volume 11), edited by F. Buttel and P. McMichael. Amsterdam: Elsevier, 265-299.

Moran, W. Blunden, G. Workman, M. and Bradly, A. 1996. Agro-commodity chains, family farms and food regimes. *Journal of Rural Studies*, 12, 245-258.

Morgan, J. 2011. Wool and meat returns: Less painful now, but still not cured. *The Dominion Post*, 25 January, C2.

Natural Dairy (NZ) Holdings Ltd. 2010. Why we want to buy the 16 Crafar farms. *New Zealand Herald*, 17 August, A9.

Oram, R. 2007. *Reinventing Paradise: How New Zealand Is Starting to Earn a Bigger, Sustainable Living in the World Economy*. Auckland: Penguin.

Oram, R. 2010. Sacked ECan board no damp squib. *Sunday Star Times*, 4 April, D2.

O'Sullivan, F. 2009. What's good for Fonterra is good for NZ. *New Zealand Herald*, 9 September, B2.

O'Sullivan, F. 2010. Fears grow we're selling the golden goose. *New Zealand Herald*, 28 July, B2.

Palmer, T. 2011. NZ super starts buying up farms. *New Zealand Herald*, 3 February, B1.

Panelli, R. 2007. Time-space geometries of activism and the case of mis/placing gender in Australian agriculture. *Transactions of the Institute of British Geographers*, 32(1), 46-65.

Perley, C. 2007. Achieving multiple landscape benefits in the farmscape. *Primary Industry Management*, 10(2), 40-41.

Roche, M. 1999. International food regimes: New Zealand's place in the international frozen meat trade, 1870-1935. *Historical Geography*, 27, 129-151.

Roche, M. 2011. *Disruptive Moments Past and Present: Pastoralism in New Zealand with Specific Reference to the Meat Industry*. (Proceedings of the International Colloquium: Disputed Territories New Spaces of Agriculture and Social Reproduction Systems, 11 & 12 December 2008). Leipzig: Leipzig University.

Saporito, B. 2010. Icebreaker is a natural. *Time*, 22 March, 37-38.

Save the Farms. 2010. Going going gone. *New Zealand Herald*, 2 October, A27.

Scherer, K. 2011. The $47 billion hangover. *The Business Herald*, 4 February, 12-15.

Sheather, B. 2009. Farm values divorced from reality. *New Zealand Herald*, 7 November, C8.

Slade, M. 2009. Industry body struggling to knit competing wool factions together. *New Zealand Herald*, 13 August, B3.

Slade, M. 2010a. Revival plan for ailing wool industry. *New Zealand Herald*, 4 January, B16.

Slade, M. 2010b. Rival wool brands battle it out in Vegas. *New Zealand Herald*, 13 February, C5.

Slade, M. 2010c. Growers look to best strategy to sell country's finest fibre. *New Zealand Herald*, 13 February, C5.

Taylor, G. 2010. Environment Canterbury report puts all regional councils on notice. *Environmental Defence Society, Media Release*, 19 February.

Taylor, M. and Thrift, N. 1982a. Industrial linkage and the segmented economy: 1. Some theoretical proposals. *Environment and Planning A*, 14(12), 1601-1613.

Taylor, M. and Thrift, N. 1982b. Industrial linkage and the segmented economy: 2. An empirical reinterpretation. *Environment and Planning A*, 14(12), 1615-1632.

Thrift, N. 2007. *Non-Representational Theory: Space, Politics, Affect*. London: Routledge.

Winder, G. 2009. Grassland revolutions in New Zealand: Disaggregating a national story. *New Zealand Geographer*, 65(3), 187-200.

Woodford, K. 2007. *Devil in the milk: Illness, Health and Politics: A1 and A2 Milk*. Nelson: Craig Potton.

Chapter 16

Contested Market-Relations Around Value and Values: Live Sheep Exports from Western Australia

Jörg Gertel, Erena Le Heron and Richard Le Heron

Introduction

In most parts of the world when sheep are traded live across borders they barely attract attention, other than from border authorities. When live sheep are exported from Australia all hell seems to break loose. It is almost as though the bodies of sheep have suddenly been placed differently, and as animals they have gained new respect. They journey as live animals in the cultural domain of Australian pastoral norms. Particular values define the complicated and contested qualities of *life* of the sheep. When they travel from Western Australia and disembark at their destinations in the Arab Gulf countries, they are repositioned, entering the cultural realms of the Islamic domain. There, differing values are mobilised to sanction and prescribe morally correct procedures for their journey to *death*.

Our approach in the chapter is informed by work that considers phenomena like agri-food commodity chains as a series of ongoing performances and enactments (Busch 2007, Caliskan 2010, Law and Urry 2004, Le Heron et al. 2011, Thrift 2008). We are especially interested in establishing how imaginings of commodity chains, by academics, actors connected to the chains, and wider publics, are altered through successive moments of event-full performance. In this work we have been inspired by agri-food research around following (Cook 2004) and animal politics and welfare (Buller and Cesar 2007, Holloway et al. 2009). These strands of research helped clarify the methodology to enquire into the political and ethical mobilisations that are made visible through a performative approach.

The chapter explores what has happened to live sheep in Western Australia as they have come into the spotlight of multiple and different valuing systems. We argue that the Western Australian experience is an extraordinary window on the constitution of market exchange relations. In the chapter we begin by establishing the presence and magnitude of Australia's live sheep exports and the nature of the live sheep chain from production in Australia to consumption in the Middle East. Because of the particular cultural dynamics and interfaces associated with live sheep export we outline dimensions of the trading and consumption context into which live sheep move. We then focus on two very different moments in the

2000s when the whole live sheep chain has been intensely scrutinised. These were disruptions to the expected normality of chain. The first moment discussed is the voyage of the *MV Cormo Express*, a voyage that became emblematic of failing efforts to complete exchange relations involving the cargo of exported but still living sheep. The second moment reviewed is the contest over whether there will be live sheep exports from Australia or not. One still contentious development in this second moment was the recent decision of the city of Fremantle to stop live sheep exports from its port (see Fig. 16.1). By examining the displacement of live sheep bodies in moments of valuing, by different actors living the contested and fluid norms of different cultures, we are able to learn more about dimensions of market relations that usually go unnoticed.

Setting the Scene

Australia's sheep industry has historically produced mainly wool. However, live sheep have been exported from Western Australia to the Middle East in large numbers since the early 1980s. Over the past 3 decades some 250 million live sheep have been sent to the Middle East from Australia. This trade started in the early 1960s. Sheep were loaded in Fremantle and Adelaide into small ships and sent to ports in the Persian Gulf, principally to Kuwait. Over the 1970s converted oil tankers carried up to 125,000 sheep a vessel between Western Australia and the Middle East. The increasing scale of the trade, largely driven by increasing oil revenues in the Arab Gulf, triggered confrontations between producers and the meat processing industry. Meat industry workers blamed the live export for the closure of abattoirs, saying that jobs were being exported to the Middle East. With the *Farid Fares* disaster, a livestock carrier that sank on its way to Iran with over 40,000 sheep on board on 27 March 1980, the animal welfare movement entered the scene. Despite criticisms and the loss of the Iranian market, due to the Iraqi-Iranian war, the trade grew. In 1982/83 about 7 million sheep were exported, representing a monetary value of AUD$ 190 million. In March 1983, a second disaster occurred: About 15,000 sheep died in the Portland (Victoria, Australia) feedlots as a result of cold weather, stress and exposure. Both animal welfare organisations and the government authorities were largely concerned about the cruelty of the trade and the level of mortalities. The then Australian Minister of Primary Industry stated: 'The current levels of mortalities cannot be explained, understood nor justified. The industry seems intent on ignoring these dying sheep and the pleas of the concerned public' (PCA 1985: 7). From this time on live sheep trade was no longer pure economics, as value and values competed to sustain, change or eradicate highly contested market relations.

Despite a shrinking sheep flock (73 million head in 2009), Australia is still a major sheep meat exporting country, dominating this world trade alongside New Zealand. Excluding intra-EU trade, both countries hold a joint world market share of 88 per cent (see Fig. 16.2). In 2007-08 the cross value of lamb and mutton

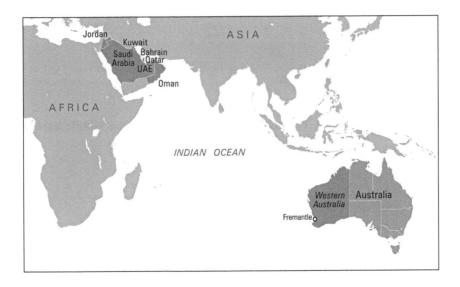

Figure 16.1 Middle East countries and Western Australia
Source: Authors.

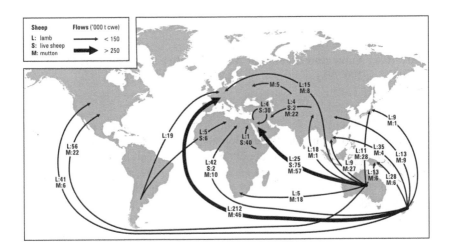

Figure 16.2 Global sheep meat and live sheep trade (2007)
Source: ACIL Tasman (2009).

production in Australia reached AUD$ 2.2 billion. Of this 45 per cent of lamb (AUD$ 800 million) and 82 per cent of mutton (AUD$ 400 million) were exported. In the same year about 4.1 million live sheep, valued AUD$ 280 million, and accounting for 11 per cent of total sheep turnoff in Australia, were also exported (ACIL Tasman 2009). In this trade too, Australia is dominant. Almost all live sheep

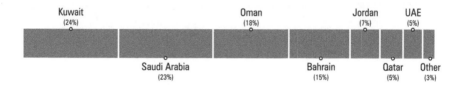

Figure 16.3   Shares of Australia's live sheep exports to the Middle East by country (2007/08)

Source: ACIL Tasman (2009).

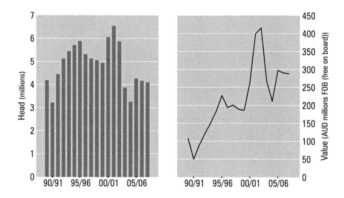

Figure 16.4   Value and volume of Australian live sheep exports (1989/90 to 2007/08)

Source: ABARE (2009).

exported, 97 per cent, are destined for markets in the Middle East, including Saudi Arabia, Kuwait, Bahrain, Oman, and Jordan (see Fig. 16.3). Western Australia is the largest supplier, provisioning about three quarters of all live sheep exports (for live sheep exports from Sudan see El Dirani et al. 2009, for the Horn of Africa see Behnke 2008). However, the market is volatile (see Fig. 16.4), due to disease outbreaks, trade restrictions, climate variability in production regions and size and status of the international sheep flock.

Dimensions of the Live Sheep Trade Chain

Markets are being continuously constructed and re-constructed, confronting value and values, and delineating exchange relations. In order to understand the constitution of the whole live sheep commodity chain, its spatial and societal dimensions, and the relations between producer and consumer, the following questions are raised: What kind of political regulations and social settings are shaping the exchange relations? What kind of properties are negotiated, what kind of contracts are involved and where do they stop? How does the transfer of ownership contrast with the (spatial) limits of responsibility? Who is involved in the discursive framing of these market relations? How are power structures distributed, and who finally, has the right to narrate the fate of the sheep body?

We proceed by briefly introducing the range of actors at each stage of the chain, from the farm through trade and exchange, to consumers and consumption, together with the regulators and interested parties who attempt to mediate and modify the conduct of affairs in, and nature of, the chain. This simplified framing and positioning of actors reveals the emerging character of the live sheep chain, the power relations being exercised by various groups of actors, and how and why the chain, and trade, in live sheep has persisted, in spite of mounting opposition. This discussion also can be read in two steps. First, it is a summary of the physical links through which sheep (as bodies) move, covering producers, traders and consumers who are directly involved in the chain. Second, it then considers the rise of the assemblage of visible, vocal and violent interpreters, government and non-government institutions, who focus on the bodily meanings of sheep in movement from Australia to the Middle East.

Producers

A farm survey of broad acre farms in Western Australia (with more than 300 sheep) reveals that the number of properties is decreasing, from 6,250 (2000-01) to 5,790 (2005-06) (Drum and Gunning-Trant 2008: 7). In 2005/6 the average farm was 4,576 ha and held 3,882 sheep. Compared to the previous five years average flock sizes (3,766 head) slightly increased. Turnover of sheep sold, however, experienced a large growth. In 2005/6 the number of animals sold per property increased to 44.9 per cent (1,744) against a five year average of 37.1 per cent (1,399). In particular, the share of animals sent to the live export market more than doubled and increased from the previous five year average of 21.7 per cent (304 sheep) to 39.6 per cent (691 sheep). Despite this growth, the farms still only earned 7 per cent of their total cash receipts from live sheep exports. Most cash income on these farms has and continues to come from cropping receipts.

What kinds of animals are traded? In general, Middle East consumers have a preference for 'fat tail' varieties of sheep such as the *awassi* breed. Despite this preference, the majority of the sheep shipped to the Middle East are, however, Merinos, further signalling the relatively marginal relationship of the live sheep trade segment to the Western Australian sheep industry. Already in 1985 about 90 to 95 per cent of export sheep were Merino wethers. The main non-Merino breeds are the Polwarth and Corriedale (PCA 1985). More than a decade later, in 2001, only 64,000 fat tailed sheep were estimated to be in Australia. Moreover, Merinos typically fall into the lower price categories for live export. The majority (79.1 per cent) of sheep shipped are weathers (comprising 61.3 per cent adults, 6.3 per cent hoggets, and 11.5 per cent lambs), followed by rams (12.2 per cent) and ewes (8.7 per cent) (Norris and Norman 2009: 6).

Trade

Trade has relatively few agents. Export companies, pastoral houses or private livestock agents purchase the sheep from the farmer. By the mid 1980s market control was concentrated in the hands of four vertically integrated Middle East companies, which were involved in the ownership and operation of some or all aspects of sheep export trade: from feedlots and feed mills in Australia, separate buying organisations in Australia, livestock carriers, feedlots and other facilities in the Middle East, to wholesale and retail outlets in the Middle East. These four entities owned about 50 per cent of feedlots in Australia, while most of the others were Australian owned (PCA 1985). Two of the companies also owned Australian feed mills. The big four exported around 6 of the 7 million live sheep in 1985, although through their vertical integration these importers were not 'tied to Australian supply but can, and do, import sheep from other countries' (PCA 1985: 16).

In 1984, for example, 16 exporters were licensed and 24 ships were approved by the Federal Department of Transport to export live sheep to the Middle East, the ships ranging in capacity from 12,000 to 125,000 sheep. None of the carriers was Australian owned. Companies from Kuwait and Saudi Arabia owned 12 ships, with an annual capacity of 4.24 million sheep. The remaining ships formed a charter segment (PCA 1985: 13). In the range of Middle East destinations, unloading facilities, feedlots, transport and slaughter of the sheep are largely controlled by the integrated companies, while different government and commercial organisations in each country are responsible for regulating aspects of the flow.

What then are the main features of the value chain in the 2000s? A 2004/05 analysis of the live sheep marketing chain, starting from the farm gate price in Western Australia until unloading in Saudi Arabia, assesses total costs per head at AUD$ 99.13 and total cross value at AUD$ 103.16. This translates into an added value of AUD$ 4 per head (Hassall & Associates 2006: 15). The distribution of total costs incurred in the chain are as follows: sheep producers (53 per cent), livestock agents (2 per cent), road transport providers (4 per cent), shearing (1 per cent), fodder suppliers (6 per cent), others (6 per cent), ship owners (22 per cent), and exporters (4 per cent). The study notes: 'The value added has decreased from [AUD] $ 8.75 in 1999/00 and the proportion of the values to the producer has increased from 33% to 55%. This highlights the variability in the value chain as market conditions and prices received alter' (Hassall & Associates 2006: 15).

Not surprisingly, the Hassall report was widely challenged. It was commissioned by LiveCorp and Meat and Livestock Australia and undertaken by a company whose director was a former Chairman of LiveCorp. A counterview contended, for example, that 'the Hassall Report is flawed', as it was

> counting as live export industry employees, those that would be employed similarly if the animals were killed in Australia (truck drivers, farmers, shearers etc), failing to consider the significantly greater GDP and job creation in the Australian meat processing sector if live export ceased and sheep meat exports increased to the

Middle East; and failing to take into account the historical instability and unreliability of the live animal export trade and its effect on Australia's trade reputation (http://www.liveexport-indefensible.com/facts/ halal.php; access 18.4.2011).

Consumers

The relations of trade over live sheep between Western Australia and the Middle East are built around particular cultural practices in Islamic societies. These practices need to be understood as both enduring and dynamic, with implications for producers wishing to connect to the live sheep market, and for traders who are attempting to negotiate completed sales, at every point in the live sheep value chain.

The global *halal* market amounts to US$ 2.1 trillion and is growing at an annual rate of US$ 500 million (ACIL Tasman 2009). Live sheep are part of this. Demand for live sheep in Muslim societies is year round, with two seasonal pikes, which are largely driven by two of the five essential practices for all (Sunni) Muslim, relating to the pillars of Islam: namely, fasting during the month of Ramadan, and the Pilgrimage to the holy sites of Mecca and Medina in Saudi Arabia. While during Ramadan Muslims must (among other things) not eat or drink from dawn to dusk, food consumption in the night is often much higher, and the demand for meat increases during Ramadan, sometimes up to 30 per cent, as in Jordan (Obeidat 2010). At the end of Ramadan meat consumption further increases with the celebration of *Eid al-Fitr.*

The Festival of the Sacrifice *(Eid al-Adha)*, related to the pilgrimage to Mecca, is celebrated in all Islamic countries. Many families slaughter live sheep on this occasion. The festival causes also a special local demand in Saudi Arabia as it attracts up to two million pilgrims annually. One of the rituals during pilgrimage requires sacrificing an animal – most commonly a sheep. Traditionally the pilgrims would slaughter the animal themselves or witness the slaughtering by someone else. Today centralised butcher houses established in Mecca will sacrifice a single sheep for each pilgrim, then package the meat and donate it to charity. The sacrifice feast is celebrated annually on the 10th day of the month of *Dhu al Hijja.* Its date is approximately 70 days after the end of Ramadan.

These two religiously driven demand spikes for live sheep are, however, not related to agricultural seasons, they shift during the year, as the Islamic calendar is determined by the movement of the moon, and therefore shorter. The festivals move about 10-12 days earlier each year. Demand structures for live sheep in the Middle East are hence disconnected from agrarian production cycles, shaping provisioning structures via a-seasonal drafting, and ultimately translate into production and flock patterns of and pressures on farmers in Western Australia. For instance, in 2009 Australia lost the contract to supply sheep for pilgrims due to falling stock numbers. Instead, live sheep were sourced from Somalia.

In Islam, the word *halal* extends beyond a specific notion of food; it represents more generally things or actions that are lawful and permissible. Its meaning is situated between *haram*, prohibited, and *fard*, required action to fulfil the obligations of Islam (such as to fast during Ramadan, and to perform the pilgrimage). Between *haram* and *halal* a further differentiation is delineated with the meaning of *makruh*, representing things and action that are frowned upon and undesirable. They are not explicitly prohibited, but they tend to be close to *haram*. A careful Muslim will try to avoid the latter to be sure in the sphere of *halal*.

This reference system translates into dietary laws that are deducted from *Quran* and *Sunnah* (i.e. the words, actions and practices used by the prophet Muhammad), the two most important sources of Islamic law. In general all foods that are not forbidden are allowed. Prohibited foods are alcohol, blood, pork meat or meat from animals that have been strangled, beaten to death, killed by fall or savaged by beast of prey. The Sure 5.3, among others, is important in this respect:

> Prohibited to you are dead animals, blood, the flesh of swine, and that which has been dedicated to other than Allah, and [those animals] killed by strangling or by a violent blow or by a head-long fall or by the goring of horns, and those from which a wild animal has eaten, except what you [are able to] slaughter [before its death], and those which are sacrificed on stone altars, and [prohibited is] that you seek decision through divining arrows. That is grave disobedience. This day those who disbelieve have despaired of [defeating] your religion; so fear them not, but fear Me. This day I have perfected for you your religion and completed My favor upon you and have approved for you Islam as religion. But whoever is forced by severe hunger with no inclination to sin – then indeed, Allah is Forgiving and Merciful (http://quran.com/5).

Live sheep, once landed in the Middle East, are slaughtered in four settings: in slaughterhouses, in retail markets, in butcher shops and in private households predominantly during religious festivals. To add to the complexity, practices of slaughtering are potentially controversial within the Islamic domain. Finally, the Middle East market is changing: demographic change (youth), and purchasing and consumption patterns (growth of supermarkets) foster an increasing demand for chilled and frozen meat. Supermarket chains account for 70 per cent of the sizeable total sheep meat imports from Australia. Between 2004 and 2008 supermarket sales increased by 77 per cent and hypermarkets sales by 105 per cent, while meat products (fish, meat and poultry) comprised over 30 per cent of products purchased by food service operations in the Middle East (ACIL Tasman 2009: 17). There has been a significant increase in logistics investment and cold storage facilities in the Middle East. Between 2004 and 2006 public-cold-storage facilities increased by 650 per cent, due in part, to the Iraq war and the investments the US Government has made to supply fresh produce to its armed forces in the region.

Governance

In the next sub-sections we consider regulation and governance of live sheep as an assemblage of mediating strategies enacted by entities that are in themselves evolving and which may be aligned or not, around particular, but often changing issues. We argue that the trajectories of government and animal welfare entities have steadily intersected. Thirty years ago, under the interventionist Australian state, the government was expected to resolve any chain issues that arose, and usually did so. With neo-liberalising and globalising developments a new mix of actors has appeared. This comprises the animal welfare sector. That co-constitutive effects should arise from government organisations mandated to oversee live sheep exports, and non-government organisations concerned with governing those working with live sheep if exported, was probably not foreseen. In the 2010s live sheep bodies are objects of governmental glare.

Government Regulations   Unquestionably the Australian government responded to the incidents and publicity of the 1980s by strengthening the regulatory and governance framework at the Commonwealth and State levels for live sheep exports. The compliance component has two main foundations. First, the issuance of a Livestock Export Licence by the Australian Quarantine and Inspection Service that is required to initiate a shipment. Second, a set of standards, The Australian Standards for the Export of Livestock (ASEL), which cover all aspects of the live export process from property of origin, road transport, pre-export assembly, loading and shipment. Australian protection ceases once sheep arrive in a foreign country.

Currently responsibility for governance, in contrast, is delegated to the industry. This industry role is supported by several prominent entities, the most visible of which is LiveCorp, an industry sponsored entity dedicated to providing

> research and development, marketing, training and communication services to the Australian livestock export industry. It works closely with stakeholders to help the industry continuously improve by enhancing animal welfare outcomes and improving environmental performance, maintaining existing and developing new markets; facilitating the ongoing support of Government and the broader community, and enhancing the skills (LiveCorp 2011: 1).

LiveCorp has a supportive infrastructure. The Australian Livestock Export Animal Welfare Group (LEAWG) was established, to provide factual information about livestock exports for consumers and key stakeholder groups. The LEAWG, with Australian Government support and a secretariat, responds quickly to emerging issues. The LEAWG website intones, 'Australia is proud to supply livestock exports to the world and ... is equally proud of its commitment to the welfare of animals.'[1] Significantly, the top link when 'live sheep exports Australia' is googled, is an industry sponsored link entitled 'We Care'[2] a sign of pressure the industry faces.

---

1   Available   at: http://www.liveexportcare.com.au/HeaderFooterNav/ContactUs/ [accessed: 18 April 2011].

2   Available at: http://www.liveexportcare.com.au/ [accessed: 18 April 2011].

Animal Welfare Groups   The extent of national and international concern about live sheep welfare should not be underestimated. Whether in the past sheep on ships were seen in a different light to sheep being trucked or penned in Australia is hard to assess. Though the sheep industry has been affronted by a steady succession of disasters, reactions have tended to convey the idea that disasters are inevitable mishaps of big shipments. This easy to accept accommodation was toppled in the 2000s by internet activity. Animal welfare organisations around the world started to systematically question, monitor, and document, on an increasingly real time basis, the journeys actually lived by sheep and the actual deaths that sheep died. Sheep shipments, and sheep, are being performed and re-performed in new ways.

A significant development took place in 2007 when ten animal protection charities internationally formed a coalition, Handle With Care, which seeks to give a global voice to the view that 'animals matter'. People for the Ethical Treatment of Animals claims, for instance, that with two million members it is the largest animal rights organisation in the world.[3] The strategy of these groups is twofold: stopping live exports and improving conditions for animals on ships. This initiative is motivated by the assumption that the industry is incapable of self regulation. A year later RSPCA Australia, in conjunction with its partner organisation the World Society for the Protection of Animals (WSPA), commissioned ACIL Tasman to look into the adjustments required if live sheep exports from Western Australia (WA) were to cease. The Report concludes that adjustment costs and the impact on farmers will be modest if the trade is phased out over five years and a transferable quota system is implemented to manage the gradual reduction in the number of sheep available for live export over that period. Although the ACIL report raised political consciousness, and opened the space for an economic shift away from shipping sheep, it unavoidably limited discussion of animal welfare per se. Nonetheless, it galvanised debate, which we use to highlight interpretive dimensions that feed into understandings of live sheep as a governable object.

Recent Moments of Market-Making and Un-Making

The market relations forged between Western Australia and Middle East countries through live sheep movement are always potentially problematic. The relations are embedded in assumptions on the part of different actors about the behaviour, motivation and norms of others. They are embedded in different views about the nature, place and purpose of sheep, sheep meat, how sheep meat might be eaten, by whom and under what conditions and so on. Islamic norms, procedures, conventions and practices obviously profoundly influence how market relations are allowed, or are made to operate, as do Australian norms and practises. They also penetrate back into the realm of sheep production in Australia and handling of sheep en route to the Middle East.

---

3   Available at: http://www.peta.org/about/default.aspx [accessed: 18 April 2011].

By 2010 animal welfare issues had moved to centre stage in the scrutiny of Western Australian live sheep trade. The earlier *Farid Fares* and the Portland disasters of the 1980s had been compounded by the *MV Cormo Express* saga in 2003. This section focuses on two very different moments of the live sheep export trajectory. First, a short review of the unfolding events of the *MV Cormo Express* situation, from the point of view of how various valuing considerations enter into deliberations about action, provides a way to examine the difficulties of securing realisation and non-realisation of value of the shipment of live sheep from Fremantle to Jeddah. The discussion relies on two main sources about the shipment, (Wright 2003, Wright and Muzzatti 2007) and uses insights from Caliskan (2010) to probe some of the unexpected exchange dimensions encountered during the voyage. The stretching out over time and space of the movement of the ship and its live cargo meant a trace is available of some of the modes of relating that are required to connect producers to consumers. Second, a summary of excerpted passages from internet blog sites *after* the release in 2009 of ACIL Tasman's report on Western Australian live sheep exports, allows detailed scrutiny of lines of argument and opinion that shape the imaginaries of decision maker in different settings. We present this pro and contra trade debate because it reveals the geographic breadth of engagement in Western Australian sheep production and trade and the breadth of comment brought to bear upon the live sheep export chain. No issue is sacrosanct, no issue is unimportant, all are embodied articulations of views. In the internet era the mediation of production, exchange and consumption can be rapidly re-constituted and re-directed. On 27 May 2010 the City of Fremantle banned the export of live sheep through its port. This unprecedented act is a dramatic development, moving far beyond industry regulation and governance strategies, to removal of an established link in the live sheep export chain.

Thin Threads of Market Exchange Relations

Wright and Muzzatti (2007: 138) argue that the distinguishing feature of the voyage of the *MV Cormo Express* was the 'international politicization of the cargo and ensuing aftermath'. The voyage had been preceded by 45 other shipments, so when the ship prepared to dock:

> It became obvious to the ship's captain and crew that something was seriously wrong ... There were no stevedores, no truck and no activity that suggested any arrangements had been made to unload the sheep. Jeddah is only about 100 km from Mecca, and normally there would be frantic activity on the wharf in preparation for the movement of the animals to feedlots where they would be kept until the Haj early next year (Wright 2003).

The following synopsis lightly sketches some noteworthy dimensions to the struggles over actual and potential exchange. First, the owner of the shipment, an experienced Saudi livestock importer was embedded in wider platforms of power in the western Saudi Arabian livestock market, and may have offended rival traders. The details matter little, the effects that ensued, however, are illustrative of the difficulties of forging relationships that lead to a market. Second, the assumed qualities of the sheep, what Caliskan (2010) refers to as part of the documentary route of circulation, was put in doubt as the sheep presented for landing were inspected by Saudi veterinarians. They initially assessed the one deck as having 30 per cent of livestock suffering from scabby mouth viral infection. While this was adjusted downward to 6 per cent for the whole ship, it still breached the official threshold of 5 per cent for trade between Australia and Saudi Arabia. Attempts by the Australian veterinarian on board to arrange a re-inspection were refused. Third, the Saudi shipment owner who had charted the ship then forced the Australian government's hand by directing the ship's captain to sail for Australia. Urging of animal welfare groups who voiced concern about conditions on the ship resulted in the Australian government purchasing the sheep, thus realising the value of the cargo for the shipment owner. Fourth, this then set in motion a flurry of activity by the Australian government to find a buyer. Approaches to Jordan, the United Arab Emirates, Pakistan, the UK, the US Army on behalf of Iraq, and Afghanistan to buy the shipment, and a number of countries to accept the cargo as food aid, were either rejected or collapsed after initial interest. This attests to the fact that international relations of exchange are always derived locally, and that once contracted relations have been broken, making new connections can be incredibly hard. Fifth, the 'floating feedlot' – with its potential for disease, fire, cruelty, drowning, and death – was being discussed internationally as yet another example of being an 'inherent and long standing feature' of the trade. Sixth, when Prime Minister John Howard suggested at the 55 day mark that the sheep be brought home, farm organisations and the livestock industry distanced themselves because of alleged threats of disease. Eventually the sheep were offloaded in Eritrea, the Australian government donating the cargo, paying expenses for transport, feed, feeding and slaughter.

Public Discussion and Framing with Consequences

The ACIL report (2009) was released in September 2009. Shortly afterwards Clover Moore (CM), the Lord Mayor of Sydney, triggered a lively and controversial forum debate with her intervention: *Why we must ban export of live sheep*. Clover Moore states:

> Australia … sends 4 million live sheep to the Middle East every year. There is no question that this practice is cruel. During the harrowing four-week journey animals crammed on boats commonly suffer heat stress and diseases such as scabby mouth, pink eye and salmonellosis. Each year thousands die during the journey; more than 40,000 died last year. Those animals that do survive the trip are cruelly dealt with in countries that either lack animal welfare laws, or else do not enforce

them. Livestock are handled and slaughtered in ways that not only are illegal in Australia, but would shock and appal most people. Meanwhile, the live exports industry continues to maintain that it is working to ensure animal protection and argues that its existence prevents worse treatment of animals by other potential suppliers. Yet live exports do little for the Australian economy. A report by ACIL Tasman in October showed that each sheep processed domestically for meat is worth 20 per cent more to the Australian economy than a live export sheep, due to the potential to add value from processing in Australia (Moore 2009).

Over the next three days (15 to 17 December 2009) this elicited 65 comments (see Moore 2009). Forty people (61.5 per cent) backed CM's position and were against live sheep exports, 13 people (20 per cent) supported live exports, and 12 people (18.5 per cent) commented on other issues. This was almost entirely an Australian debate. Most participants hailed from Sydney, some from Newcastle and Perth, others from Melbourne, Canberra, Campbelltown, Bathurst, Winston Hills, Thornbury. Only one claimed to be speaking from the Middle East. Text extracts are quoted, according to sequence of publication on the blog.[4]

Pro Trade

> The live sheep export market is there for exactly that reason, its a market! Why should drought ravaged struggling farmers have to lose money because some inner city vegetarian feels sorry for the poor sheep! (Rodney Johnson | Sydney, 1)

> Clover, don't you advocate – in all it's manifest PC glory – the cultural diversity of Sydney? And isn't it the right of people in other countries to practice their own religion and perpetuate their own culture? Is not the ritual killing of animals for consumption not part of this? (Colin | Sydney, 4)

> I am a sheep farmer ... The live export is integral to WA markets and if stopped would cost WA dearly. The exporters have done a great job at transporting sheep by ship and the animals are in better condition when they arrive at the destinations than when they started, they know what they are doing! (Bernie, 10)

> If Australia doesn't export it, they will simply get live sheep from other countries. (Anthony | Paramatta, 16)

> It's pathetic to see so many people taking the moral high ground about this issue, simply because they are so remote from the whole process. It would be very interesting to see how many of these opinions would change if these people had a direct economic interest in this procedure, as the sheep farmers do. I would like

---

4   The number indicates location of comment, i.e. 1 = first comment; ... indicates abbreviation of comment text.

to know how many of those that support Clover Moore's stance refuse to wear clothes made in China, bearing in mind the sweatshop conditions that many of these clothes are produced in and the cruelty that exploited workers are put through. It's easy for anyone with no knowledge of the industry to sit back and say 'you'd make more money by exporting chilled meat'. At the end of the day, the sheep farmer's livelihoods are at stake, for many of them, it would be economic suicide to attempt the change over from live export to chilled meat export. (Alistair | Bligh Park, 30)

Instead of worrying about the poor little sheep that feeds someone, why don't you concern yourself with solving the injustices committed against human beings in your area! I know for a fact that there are people living in worse conditions than the sheep in the centre of sydney right now. (truthseeker | Newcastle, 62)

Contra Trade

Good on you Clover – regardless of whether she's a vegetarian or not – there is something called 'conscience'. We humans seem to be so proud of how we have evolved but yet we still find ourselves putting our furry friends pain and suffering below profits. Also, Rodney if you read the facts Australians drought stricken farmers are losing their jobs because of the small amount of large live export producers. (Justin | Sydney, 6)

No country should export live animals. It is inhumane and barbaric … 'The greatness of a nation and its moral progress can be judged by the way its animals are treated' Mahatma Gandhi. (Lauren | Sydney, 7)

There used to be a very profitable 'export market' in live slaves too. (Mystie. | Australia, 8)

The reality is that rather than create jobs, the live export industry actually takes jobs away from Australian meat workers, which is why their own Union has spoken out against it. It's all well and good to support other people's culture and their right to freely practice their religion but we also have a moral obligation to have the courage of our own convictions, of the convictions of OUR nation. If we wouldn't stand for this treatment of animals in Australia, then why are we sending our sheep into this situation in another country? Profit is not a good enough reason, people once profited from slavery but it doesn't make it right. (Lainie80 | Campbelltown, 11)

How do you know animals dont suffer the same as humans, after all they eat, sleep, have sex, experience sadness and joy just like us? Have you ever stopped to think? IN SUFFERING THEY ARE OUR EQUAL! (Sylvia | the planet, 14)

This is a cruel and unnecessary industry. Humanely halal slaughter the sheep here – farmers still sell their product and our sheep don't suffer. (Patti | Sydney,17)

Once they arrive, the (sic) cruel handling and slaughter methods in the Middle East have been well documented over many years and continues despite thousands of dollars being put into improving standards. (Dean, 21)

The Australian Federation of Islamic Councils has issued a press release saying that they do not support live exports. Halal meat is routinely prepared in Australia in a humane way, and in much larger quantities than live exports. So accusation that Clover Moore is a sort of cultural imperialist are flakey ... The Australian Meat Industry Employee Union are one of the strongest opponents of live exports – since the 1970s some 25 Australian meat processing plants have closed down due to live exports (figure from the AMIEU). (Bryan, 27)

This trade needs to stop, its cruel and so 30 years ago. If New Zealand, one of the largest sheep exporting countries can ban it, so can we. (Angela | Perth, 36)

Actually Bernie, it is impossible for you to claim that animals arrive at any port in Australia healthy and fit to export, because there are NO inspections of every animal by any government that is federal or state authority. The claims you make are totally unsubstanciated. The live export trade is just about self regulating because the cursory inspections by AQIS are just that cursory. They do not inspect at feedlot, inspect the trucks or at portside. The latter is left to the exporters vet. (Compassion | Perth, 37)

Kanzi – Let market forces reign? Do you think the same about the heroin trade? There are poor farmers involved in that too. (Bryan McLeod, 51)

(T)he United Nations put out a report in 2006 'Livestocks Long Shadow' that quoted that methane is 18 times more potent than all the cars trucks and all the transport in the world put together. Also Nitrous oxide from animal wate is 296 times more potent than all the transport in the world put together. Also that meat eating is the # 1 killer on the Planet ... The Worldwatch Institute also put out a report just recently that stated that 51% of the GHG's are related to animal agriculture. If we want to save this planet we have to eat more plant based food. (Penelope | Thornbury, 63)

Discussion: Values and Value

The Lord Mayor of Sydney's provocation 'ban exports', is especially interesting for two reasons. First, the wording used is 'ban', not 'stop'. This is simultaneously an invitation for governmental intervention, and an implicit recognition that a call to the main economic actors to stop, would be insufficient to bring the shipments to an end. Second, this call hails from a large city. Sheep are being read not from the rural, but from the urban. Clover Moore identifies the protagonists, names their claims and exposes the central sensitivity, sheep.

The extracts suggest heartfelt convictions and geographic imaginings. Normally unexplored transactions linking the Middle East consumers and Western Australia producers are blown up into something of enormous concern within Australia, and a little beyond. The pro-trade responses largely leave the live sheep out of view. The contra-trade remarks write the sheep back in. And although New Zealand's decision to ban live export seems to be a moral triumph, a close inspection of that move suggests it is hardly a demonstration of conviction. Live sheep exports were a very small part of New Zealand's export economy, and when combined with New Zealand's self-promotion and self-valuing as a clean and green country, arguments of moral economy were able to trump economics.

Six months later, on 27 May 2010, the Mayor of Fremantle Brad Pettitt announced that the time had come for the 'cruel and unnecessary live sheep trade to be phased out and replaced with a trade that supports local jobs' (Tromp 2009, Wilkins 2010). At this time, a Galaxy Opinion Poll[5] reported that 79 per cent of Australians believed live sheep exports were cruel, and 86 per cent thought government should phase them out if there is an alternative that saves Australian jobs.

But the Controversy isn't Settled!

An Australian Broadcasting Corporation (ABC) 7.30 Report on 2 December 2010 screened graphic footage of the cruel treatment of sheep during an Islamic festival in Kuwait (ABC 2010). We draw on several lengthy passages from the broadcast because they place argument and object of argument into the same frame, using live footage, and assemble into the same documentary the contradictory dimensions of any discussion about sheep, bodies. The transcript begins, 'One industry body, Meat and Livestock Australia, has conceded the pictures are appalling. They show sheep bound with wire and dragged into car boots, live animals dumped on top of the dead and dying and clumsy attempts to slaughter the animals' (ABC 2010). The Animals Australia advocate said there was:

> No sense from those mistreating the animals that there was anything to hide ... They have completely assumed these practices are acceptable to us because we've (Australia) been supplying them with animals, millions of animals for decades ... From the moment that morning prayers were over, animals started to be dragged to slaughter. They are thrown onto their sides and their legs, right throughout this footage, are trussed tightly with wire. The method of

---

5   Released on 13 October 2010. Available at: http://www.wspa.org.au/latestnews/2010/United_call_for_an_end_to_live_sheep_exports.aspx [accessed: 18 April 2011].

transportation routinely is in a car boot. At one stage we documented three sheep going into a small car boot. We were seeing animals trussed up, lying on top of dead and dying animals, awaiting their throats cut. Rarely were we seeing throats cut properly. It was just mass animal cruelty. There's no other way to describe it. And from knowing Islamic principles, as I do, there was no compliance with Halal guidelines whatsoever. ... Animals are on-sold from feedlots all year round for home slaughter, but this is definitely the worst time of animal suffering in the region, which is exactly why we put a proposal to the (Kuwaiti) industry that they enter into agreements with importers for animals not to be on-sold so that they could be protected from this. Now that proposal even had the support of the Western Australia Pastoral and Graziers Association, the farmers themselves, and still they didn't act on it. So much of what we documented would be prevented if the industry stopped the on-selling in the region (ABC 2010).

The voice of Princess Alia Bint Al Hussein (no details provided) questions assumptions about Middle East behaviours.

It has nothing to do with Arab or Islamic culture and I very much appreciate your asking that, because we keep being told it's Adal Althein, it's to do with Islamic culture, and it's not. It's sadly a sign of the times, a sign of modern times, where everything is too quick and people are becoming extremely insensitive to all other creation. And they think that also people from your part of the world approve of it, and so we really need everyone – we need Arabs and Muslims to reassert and underline what our true teachings are and really focus on that and we also need you to make a fuss about it, please (ABC 2010).

The Meat and Livestock Australia spokesperson had the final word

I think what I can say is I'm confident that we (Australia) are making a massive improvement in animal welfare through the Middle East. I think we need time to make sure we get across the region. We can't change everything overnight, but we are making change (ABC 2010).

Is the covenant of protection over the livesheep chain edging into a convenant of obligation that is now being institutionalised by Australian actors in Middle East markets? Are there grounds to the view that appeared a fortnight later on the industry website, www.liveexportcare.com, that 'we all care about animal welfare'? A later statement, 'Australian livestock export industry and animal rights groups all want to improve the welfare of Australian sheep in overseas market places. The difference between us is how we go about making these improvements',[6] suggests

---

6   Released on 13 December 2010. Available at: http://www.liveexportcare.com.au/LatestNews/We+all+care+about+animal+welfare.htm [accessed: 18 April 2011].

however that despite the interwoven dynamic of officialdom and activism, shared understandings of the object of governance, sheep bodies, are in their infancy.

Conclusion

When we began this chapter we had not realised how much a focus on the sheep body would affect how we would understand market and livelihood questions. Once we looked into the genealogy of the whole live sheep chain in Western Australia we found the chain to be fluid and emerging. Many new features grew from the mutability of sheep bodies, and the possibilities that changed understandings and terms, meant doing live sheep exports differently. We had originally thought too, that highlighting the distinction between value in a simple economic sense and the range of other values named by actors would form the main part of the chapter. Instead we noticed we were considering aspects of *valuing*, that is we were identifying 'the various ways in which individuals, processes and places matter, our various modes of relating to them, and the various considerations that enter into our deliberations about them' (O'Neill et al. 2008: 20). This line of thinking we suggest is more in keeping with the realities of the life and death issues of live sheep exporting.

Despite moving through the chapter, from detached analysis of statistics and the chain model of the live sheep journey, to historical reminders of breakdowns in an assumed smooth passage of sheep from Australian producer to Middle Eastern consumer, to statements about alleged best and worst practice with respect to live sheep export, to a resolution of debate when Fremantle banned shipments, we still had not made the ontological leap that would make the sheep body indisputably visible. We have the ABC to thank for this. Their live footage and accompanying commentary introduced new actors and connections to the politics of live sheep exports.

We draw several conclusions. First, at the obvious level the chapter is a story of the resilience of live sheep export links between the Western Australian sheep industry and consumers in distant lands. This link has been almost an escape valve for the industry, part of a portfolio of revenue options. What makes it special are the sheep, the live sheep. Commodities, such as live sheep, are not only valued in relation to a monetary price, but are also situated social and moral norms. These intersect, but are far from being identical from context to context. This leads to the second conclusion. In a more complex way the chapter is a narration of emergence, of how Australian values have been projected onto a little understood Middle East, and of Australian understanding about its own values. We did not have a lot of information on Middle East specifics, nor does this matter. What we have succeeded in revealing is that many actors have been narrating on behalf of someone else. Australian (urban) consumers made claims on behalf of Middle East consumers. Governments, welfare groups, farming groups and so forth regularly and forcefully have had their say. In various ways live sheep have gradually gained a narrative presence. This has been extended from detached comment over counted sheep, to heartfelt anxieties about how sheep are living and dying. Third, the chapter shows that contests over value can have direct consequences. That this valuing, more accurately a re-valuing, led to turmoil in Australia and in other places, in the Middle East, indicates the in-the-making character of all livelihood-market linkages. By closing its port, the City of Fremantle has altered the basis of how at least part of Australia engages in live sheep exports.

Acknowledgements

We wish to thank Dr Steffen Wetzstein and Professor Matthew Tonts, Department of Geography, University of Western Australia for providing many valuable leads on the Western Australian live sheep meat exports. Igor Drecki, School of Environment, University of Auckland, very ably drew the figures. A New Zealand Marsden Fund award, 'Biological Economies: knowing and making new rural value relations', UoA09024, 2010-2013, made possible the wider exploration of ideas relating to the knowing and creating new rural value relations, some of which have informed this chapter.

References

ABARE (Australian Bureau of Agricultural and Resource Economics). 2009. *Australian Commodity Statistics*. [Online]. Available at: http://www.abare.gov. au/publications_html/acs/acs_09/acs_09.pdf [accessed: 19 April 2011].

ABC (Australian Broadcasting Corporation). 2010. *Animal Cruelty Throws Spotlight on Live Sheep Trade*, Broadcast: 02/12/2010, 7.30 Report. [Online]. Available at: http://www.abc.net.au/7.30/content/2010/s3083255.htm [accessed: 18 April 2011].

ACIL Tasman. 2009. *Australian Live Sheep Exports*. [Online]. Available at: http://www.rspca.org.au/assets/files/Campaigns/ACILTasman/WSPAlive exportsreport071009_FINAL.pdf [accessed: 18 April 2011].

Behnke, R. H. 2008. The economic contribution of pastoralism: Case studies from the Horn of Africa and southern Africa. *Nomadic Peoples*, 12(1), 45-79.

Buller, H. and Cesar, C. 2007. Eating well, eating fare: Farm animal welfare in France. *International Journal of Sociology Food and Agriculture*, 15(3), 45-58.

Busch, L. 2007. Performing the economy, performing science: From neoclassical to supply chain models in the agrifood sector. *Economy and Society*, 36(3), 437- 466.

Caliskan, K. 2010. *Market Threads. How Cotton Farmers and Traders Create a Global Commodity*. Princeton: Princeton University Press.

Cook, I. 2004. Following the thing: Papaya. *Antipode*, 36(4), 642-664.

El Dirani, O.H., Jabbar, M.A. and Babiker, I.B. 2009. *Constraints in the Market Chains for Export of Sudanese Sheep and Sheep Meat to the Middle East.* Nairobi: ILRI.

Drum, F. and Gunning-Trant, C. 2008. *Live Animal Exports: A Profile of the Australian Industry.* (ABARE Research Report 08.1 for the Australian Government Department of Agriculture, Fisheries and Forestry). Canberra: ABARE.

Hassall & Associates. 2006. *The Live Export Industry: Value, Outlook and Contribution to the Economy.* Sydney: MLA (Meat and Livestock Australia).

Holloway, L. Morris, C. Gilna, B. and Gibbs, D. 2009. Biopower, genetics and livestock breeding: (re) constituting animal populations and heterogenous biosocial collectivities. *Transactions Institute of British Geographers*, 34, 394-407.

Law, J. and Urry, J. 2004. Enacting the social. *Economy and Society*, 33(3), 390-410.

Le Heron, E., Le Heron, R., and Lewis, N. 2011. Performing research capability building in New Zealand's social sciences: Capacity-capability insights from exploring the work of BRCSS's 'Sustainability' theme, 2004-2009. *Environment and Planning A*, 43, 6, 1400-1420.

LiveCorp. 2011. *OH&S Policy.* [Online]. Available at: http://www.livecorp.com.au/About/OHS_Policy.aspx [accessed: 28 February 2011].

Moore, C. 2009. *Why We Must Ban Export of Live Sheep (online forum 15-17 December 2009).* [Online]. Available at: http://www.smh.com.au/opinion/society-and-culture/why-we-must-ban-export-of-live-sheep-20091214-ks6l.html?comments=65#comments [accessed: 4 April 2011].

Norris, R.T. and Norman, G.J. 2009. *National Livestock Export Industry Shipboard Performance Report 2008.* Sydney: MLA (Meat and Livestock Australia).

Obeidat, O. 2010. Meat prices expected to drop ahead of Ramadan. *Jordan Times* [Online], 3 June 2010. Available at: http://www.jorday.net/index.php?option=com_content&view=article&id=121:meat-prices-expected-to-drop-ahead-of-ramadan&catid=45:englishjorday&Itemid=59 [accessed: 4 April 2011].

O'Neill, J., Holland, A. and Light, A. 2008. *Environmental Values.* London: Routledge.

PCA (The Parliament of the Commonwealth of Australia). 1985. *Export of Live Sheep from Australia, Report by the Senate Select Committee on Animal Welfare, Canberra.* [Online]. Available at: http://www.aph.gov.au/senate/Committee/history/animalwelfare_ctte/export_live_sheep [accessed: 10 June 2010].

Thrift, N.J. 2008. *Non-Representational Theory: Space, Politics, Affect.* London: Routledge.

Tromp, B. 2009. Australia's live sheep exports. *ABC Rural Bush Telegraph* [Online], 12 August 2009. Available at: http://www.abc.net.au/rural/telegraph/content/2009/s2653580.htm?site=riverina [accessed: 18 April 2011].

Wilkins, D. 2010. Fremantle council votes to end live sheep exports (including 183 online comments). *Perth Now* [Online], 27 May 2010. Available at: http://www.perthnow.com.au/business/fremantle-council-votes-to-end-live-sheep-exports/story-e6frg2qc-1225871875456 [accessed: 4 April 2011].

Wright, T. 2003. Sheep of fools. *The Bulletin* [Online], 22 October 2003. Available at: http://liveexportshame.com/news2/index.php/topic,68.0.html [accessed: 4 April 2011].

Wright, W. and Muzzatti, S. 2007. Not in my port: The "death ship" of sheep and crimes of agri-food globalization. *Agriculture and Human Values*, 24, 133-145.

Chapter 17

Conclusion: Embodied Risks of Exchange Relations

Richard Le Heron and Jörg Gertel

Introduction

Economic Spaces of Pastoral Production and Commodity Systems: Markets and Livelihoods has been a co-learning project, and related to it, a politics of knowledge project. The collective learning project has been to probe deeply and disturb, especially through empirics, the idea in much contemporary thought that markets and livelihoods can ontologically be separated. The partitioning breaks down when these categories are theorised from a perspective of situated knowledge production. As Barbara Harriss-White argues, the performance of markets is affected by and continually changes relations of authority, and these contested realities always have implications and ramifications for livelihoods. Hans-Georg Bohle, considering livelihood as a category, points to how livelihoods are deeply politicised, realities too that spring from power relations which make, and in which, exchange takes place. The partitioning falls down further when brought into the empirical specifics of the spaces of pastoralism and pastoral economies. The book, hence, reveals the embedded, embodied and constitutive market-livelihood interdependencies of exchange relations.

The introduction laid out three questions for the book that sharpened up the spirit of this theoretical and empirical co-learning in the realm of pastoralism and pastoral economies. These questions gave focus to how the contributors were grounding their analyses and interpretation. Now, having resourced ourselves with the co-learned insights of the book's project, we can speak more about our wider horizons and ambitions.

The politics of knowledge project is still tentative, but it can be broadly expressed as a motivation to do much better in our academic attempts to pose questions that demand novel and different categories and enactments. This leads to the immediacy of what we are doing in our knowledge project. We conclude the book with some reflections about how we might improve our *capacities and capabilities* to produce different knowledge systems. The enquiry of the book into the economic spaces of pastoralism and pastoral economies tables once again, the urgency of taking seriously knowledge approaches that assemble situated framings that are hospitable to a wider range of inhabiting people, animals, plants and things.

In the twenty-first century extensive grazing of animals continues to be practised in many forms, in many parts of the world. But pastoralism and pastoral economies, as constituted spaces of nature-society intra-actions, should not be seen as simply instances of third world or peripheral economy case studies or isolated regional geographies. It could be argued, as Bennett (2010: 37) encourages us, that pastoralism and pastoral economies be viewed as a 'vitality (of materiality) distributed along a continuum of ontological types', with 'the human-non human assemblage a locus of agency'. This is a very challenging conceptualisation, but it is one where the book might be seen as a stepping stone towards such thinking. It is our premise that the multiplicity of pastoral assemblages brought into visibility by the book is a starting point in this direction.

An Overview of Themes

We identify a number of themes about the nature of the emerging economic spaces of contemporary pastoralism and pastoral economies that emerge from the book. These themes highlight insights into market-livelihood-relations.

First, the book reveals the exposure and opening of pastoralism and pastoral economies to globalising and regionalising processes. Without exception these processes are the day-to-day realities directly or indirectly and consciously and unconsciously encountered by pastoralists. The authors' probing and documentation ranged across the experiences and making of integration and social differentiation, shifting and emerging market spaces, how best to organise different sorts of engagement, border relations that opened up or closed down flows of livestock or products, and so on. This work firmly establishes the existence of pastoralism and pastoral economies in the variegated landscape of life on the planet. The situated research engagements over lengthy periods meant we are able to convey the makings of market relations and the challenges and opportunities for livelihoods in these emerging realities.

Second, the book shows pastoralism and pastoral economies as very relevant and meaningful ways of making a living, and living, in the twenty-first century. Moreover, connections (or lack of connections) with global markets are being achieved in innovative and diverse ways. The different points of entry of each chapter in the book's sections reveal great awareness about and deliberation over the strategies of engagement and non-engagement from different contextual positions.

Third, the continued presence of colonial and other restrictive relationships of organisation is an inescapable part of the discussion. That these dimensions continue to actively shape responses is an important argument for sketching historical and spatial trajectories of experience in any investigation of pastoralism and pastoral economies.

Fourth, governments are deeply implicated in the mediation, and sometimes, the actual making, of market relations and in defining the range of livelihood possibilities that are open to communities, industries, households and families associated with pastoralism and pastoral economies. This reality disturbs any notion that markets and livelihoods can be considered independent of nation-states, local-states, pan-regional configurations or free trade blocs under the WTO framework.

Fifth, pastoralists in whatever shape or form they might be, come to market making and livelihood making with a mixed bag of social and economic resources, motivations, experiences of prior engagement, knowledge of the wider contextual influences in which they are embedded, strong senses of self, household and family awareness, and sophisticated understandings of their intimate relationships with the flora and fauna upon which they depend and by which they identify themselves. This is not a trivial conclusion, but one that speaks to insights gained from globally and regionally sensitised local knowledge.

Sixth, trust is a resource that is known to perform, in that it provides predictable and stable outcomes of particular kinds, when called upon in exchange relations. Accordingly trust is mobilised and re-established wherever possible, because of what it is known to enable. The reliability or dependability of person-to-person or group-to-group relationships has saliency, whether in terms of face-to-face or in at-a-distance relationships.

Seventh, re-imagining alternative ways of doing things in the worlds of pastoralism and pastoral economies is sometimes hard, especially when trust and communal and collective structures erode away or collapse and when resources become depleted. Indeed, much of the discussion chapter by chapter is about flux and innovations, but also about deliberate or accidental experimentation to find new ways of performing pastoralism and pastoral economies.

Eighth, while market relations are understood as being part of (pastoral) livelihoods, the human body represents the last locus of social struggles, and ultimately is inscribed by forces that are, to a certain extent, beyond the local control of pastoralists. This is seen (a) in the enclosure of pastoral spaces in Siberia or Morocco, where the subsequent loss of access to land and fodder threatens livelihoods; (b) in the exposure to state-induced conflicts as in Sudan; or (c) in the breakdown of market chains in Mongolia. At times, pastoralists are bodily exposed – facing malnutrition and hunger, or being killed during theft or warfare.

The themes link, as answers, to the book's opening questions. They show that at the everyday level it is hard to make a livelihood from pastoral activities, whatever the context and situation of the pastoralists. In today's world, pastoralists, pastoralism and pastoral economies can neither fully escape markets as an arena of exchange, nor can they fully transcend more traditional exchange forms. Further, efforts by the state, international agencies, community groups, cooperatives and such like to engage constructively and productively in market relations are governed by (vested) interests and do not necessarily represent sustainable long-term solutions to livelihood issues.

We would go further however. Our questions have begun to nurture new research imaginaries into life. There is no doubt that pastoralism and pastoral economies have been understudied, and when they have been examined, the approaches deployed have de-emphasised assemblage-like features in favour of bounded, human-centric and passive analyses and narratives. What will happen to pastoralists? What will be the fate of particular livestock industries? Rather, as the book highlights, we believe we should be able to ask knowledge production questions such as: 'Should we be exploring the significance of multiple expressions of pastoralism and pastoral economies, and putting new categories in place in order to explore how such worlds are in-the-making?' Alternative questioning about now known presences, practices and patterns of pastoralism and pastoral economies seems a next and urgent step.

A hallmark of the empirics that distinguish the book is the longitudinal and latitudinal nature of the research underpinning the chapters. We have followed pastoralists in many and changing places. The last few decades have upturned the worlds of the pastoralist and the book examines many transformations. This has given an unparalleled set of windows to identify some key aspects. The contributors highlight how trajectories of contestation can be fuelled, resolved or dissipate and also expose the short and long run manifestations of effects. We are further able to comment on the playing out of longstanding obligations forged in market relations. In some instances detailed study of entities reveal the consolidation and aggregation of bigger and bigger actors, with new positioning and powers. The often very long run processes of concentration tendencies, with their new livelihood relations, are shown to be co-constituted. Static and comparative static analyses are replaced with trajectory and performative accounts, which represent origins as distinct from causes, and the successes and failures of experiments in human intentionality. Longitudinal ethnography has meant rare insight into the spatio-temporal dynamics – the spaced times and timed spaces – of pastoralism and pastoral economies.

Very few research programmes are conceived and funded to yield insights in the cumulative manner identified here (Gertel and Breuer 2011). In coming together we have gradually begun to sense where pastoral and pastoral economies have come from, and also how they are actively in-the-making. While few chapters speak of assemblage, the book will join agri-food, economic geography and development literatures that are increasingly exploring the idea of open wholes or assemblages (Li 2007). What is portrayed in the book is nonetheless human centred, which Bennett (2010) sees as a severe limitation because of its anthropocentrism. We would counter, however, with the suggestion that our knowledge production practices have created a significant resource to inform future research that is committed to de-centring humans in pastoral and pastoral economies understandings.

From Book to Literatures

We posit that the book's chapters can be read mindful of at least four distinctive and potentially generative research emphasises in the international literature. These are the knowledge production projects of the *diverse economy* (Gibson-Graham 2006), *following* (Cook 2004, Cook et al. 2007), *spacings and timings*

(Murdoch 2006, Thrift 2008) and *human and more than human materiality* (Bingham 2006, Bingham and Hinchliffe 2008, Whatmore 2006). We offer a number of suggestions aimed at aiding dialogue with these research directions in order to craft new categories that will segue into different ways of knowing, and making, worlds. This preliminary juxtaposing or rethinking of examples in the book to other theories, gives a taste of what is possible.

Approaching Diverse Economies

We begin with Gibson-Graham's (2006) diverse economies project for two reasons. First, an important raison d'être of their project is to provide guidance and knowledge production technologies that can be used to build community economies. They reject structural imperatives that are typically incorporated in development discourses, specifically the claims that

> communities and regions 'need' some sort of engagement with the global capitalist economy in order to develop, industrialisation is the singular pathway towards economic well-being, promoting capitalist enterprise will bring social and economic dividends to the whole community, in the current era of neoliberal global capitalism regions must develop their export base to be economically viable [and that] growing the output of the capitalist commodity-producing sector is good for all (Gibson-Graham 2006: 167).

Second, this rebuttal of widely adopted conventional wisdom resonates closely with this book's interrogation of markets and livelihoods. Gibson-Graham rethink the economy using a conceptual matrix as a technology of thought. They take existing but separated categories and re-imagine the content and inherent possibilities of economy, where economy is known as an assemblage of much more than is ordinarily dealt with in orthodox accounts of economy. The generative effects of the matrix are increasingly traceable (see the August 2009 Special Issue of *Asia Pacific Viewpoint* and the website www.communityeconomies.org).

What differentiates the Gibson-Graham work, and gives it formative qualities, is their concern with developing practices that will kick-start initiatives spawned from ethical debates and decisions. To this end they propose a four step community-based process (Gibson-Graham 2006). First, create through community processes an inventory of the diverse economy, and then explore how this now known community economy might build on the strengths and assets already present. Second, look at ways of generating, marshalling and distributing surplus. Third, examine decisions communities have made to consume resources in ways that have set them on unpredictable development pathways. Fourth, explore ways of replenishing and enlarging the commons that is a shared base of material, social and spiritual sustenance for communities.

Strong elements of the Gibson-Graham principles of community building are found in the chapters on Africa, Asia and post-socialist countries. But there is at least one important difference that rapidly becomes apparent. All the African studies, for example, detail ecological intra-connections. That is, they acknowledge and try to situate and sketch these constitutive components as a priority, because this vital materiality cannot be prised away from the content of community understandings. Thus to illustrate, Chapter 4 by Wiese on Chad, delineates pasture possibilities as broadly threefold – dry, green and mixed – depending on what water can be accessed using local technology. It is the pasture-water nexus that meets (and vice versa) the mobile pastoralists and their attendant animals, things, as well as distant plants. What arrives in the Chad basin are relational bundles. These mobile bundles can be summarised in relational notation, Sahara-camel-date, Sahel-cattle-millet and Sudan-cattle-sorghum. The chapter goes on to probe livelihood achievements. Two matters are prioritised – the diversity of the economic means available and the impact of decisions made in conditions of unexpected contingency – a subset of the principles advocated by Gibson-Graham. In sum these aspects reach forward to Gibson-Graham, in a manner that affirms their claim, that 'the specifics of place and pathway' is 'deeply disturbing to economic certainty ... in development discourse' (Gibson-Graham 2006: 169). Stammler emphasises for Siberia: 'The case of *panty* illustrated in this chapter suggests an interpretation the other way round: Taking on the view of reindeer nomads, we can perceive market capitalist economic activity as being on the fringes of a diverse, plural and community-internally regulated set of practices that make up the mixed economy of these nomads today' (Stammler, this volume). In this line of argument for performing and diverse markets, Berndt and Boekler (2009: 544) comprehend markets as 'economic quasi-entities ever only stabilised temporarily by a double process of framing and overflowing'. While framing – according to Callon (2007: 140) – means 'to select, to sever links and finally to make trajectories (at least temporarily) irreversible' overflowing expresses that framing is a delicate process 'that easily gets out of control and is never completed' (Berndt and Boekler 2009: 544). This insight opens up to quite some situations depicted by in the volume. The notions of 'peace' and 'smuggling' markets elaborated by Komey in his Sudan study about market exchange in situations of war, reveals not only the shifting emergence of market places both in space and time, but also the constant search for framing exchange situations, aiming to elaborate a local code of conduct in an extreme volatile and violent environment. This again reads as complementary to Mahmoud's chapter on re-aligning exchange systems in risky environments where place of exchange and consumption (in Nairobi) are fixed, while the routes of transport and the ways of capital transfer have to be diverse, hidden, and contingent to buffer insecurities.

Facilitating Following

In examining pastoralism we are acutely conscious of the importance of near and far connectivity, for the politics of community and individual survival and achievement. Whereas Gibson-Graham are especially concerned with mobilising understandings of the wider asset-base of communities, the methodology of following offers a way to heighten understandings about wider human and non-human dependencies. A premise of following is the rejection of the notion that knowledge is simply 'out there', waiting to be discovered. Instead as Longhurst (2009: 429) outlines, knowledge is both 'embodied and situated, that is, it is made by individuals who are situated in particular contexts'. The focus on agency then, emphasises links and associations rather than nodes, and the durability of what are commonly referred to as actor-networks.

In the agri-food literature the methodological intervention of following is being innovatively trialled (Cook 2004, Cook et al. 2006, 2007) beyond ethnographies of the social life of things (Appadurai 1986). Developments in approach have been inspired from two mainsprings, Actor Network Theory (Latour 1987) concerned with the constitutive properties of connection, and political economy, seeking to bare the social relations of commodification (Watts 2001, 2005). The developments seek to reveal the myriad of connections that go into making a thing, product, animal or plant. Pioneer studies have examined a diversity of agri-food commodities papaya, bananas, sushi, surimi, beans, and coffee. Although the approach has been challenged for being politically conservative because it has privileged establishing connections; its potency lies, however, in its use to give connected content to politically and ethically inspired approaches. This has meant moving on from following for personal understandings, to regrouping the insights of shared stories and assembled information into articulated knowledge propositions.

This transition in methodology, according to Roe (2009: 251) conceives 'the politics of how realities are made by working off from a non-hierarchical, thus flattened world of socionatural messiness, as opposed to studying things-in-themselves or the social construction of nature. This flattened world puts nonhumans and humans on the same plane'. Roe goes on 'this relational thinking has consequences for how we understand the process of knowledge creation and activities that go with this'.

Mahmoud's study of northern Kenyan livestock marketing (Chapter 7) explores seller-seller and buyer-seller relationships as part of linking herders and bush traders in southern Somalia to urban-based livestock traders and livestock transporters. There are multiple trekking routes, those traversing territory 'owned' by specific ethnic groups, riskier but shorter routes and one used when conflict is rife. Trucks can take several highways. The chapter outlines innovations in livestock trading that spring from recognising and mobilising cattle trading partnerships and overcoming information differences of pastoral communities and the use of informal cash transfer mechanisms that help contain risk. The chapter's following-like exploration of realignments in exchange systems is an example of multiple communities deploying their assets.

The Moroccan example by Breuer and Kreuer illustrates following features, by using interviews from nearly 400 heads of households, to investigate the negotiation of connections needed to secure market relations. For sheep these dimensions are:

the rhythms of price formation, decisions about livestock type, the influence of fodder dependency on mobility and remittances from household labour working predominately in Spain. The account of following the panty (velvet antlers) in Stammler's chapter from initial exchange (goods advanced against agreed weight of antler to be supplied by the collecting enterprise), to the transition from raw material to commodity (when the panty leaves the control of the collecting enterprise), exposes creative approaches to the volatile business of antler sales.

In important ways research dedicated to documenting decades of pastoral experience creates a resource that can be accessed to provide detail for trajectory-based accounts. The sensitivity of the researchers to the ecological intra-dependencies of pastoralism and pastoral economies is noteworthy.

Approaching Spacings and Timings

The word-power of the title of the famous Massey and Allen (1984) edited collection *Geography Matters!* is of course legendary. A course reader for the Open University in the UK the book and its opening essay re-energised generations of Anglo-American geographers. Guiding metaphors have such effects. Pastoralism and pastoral economies, because of their complex mobilities and mutabilities (Law and Mol 2001) from strivings to contain exposure and build resilience in many places, present newly understood nonhuman and human materialities that give new meaning to 'geography matters'.

Geographers frequently invoke the idea of spatio-temporal relations and their constructed and constituted character that defines particular geographical and temporal contexts. But it is altogether another step to conceive of space and time in dynamic terms in relation to becoming (Doel 1996, McCormack 2009). Journeying with pastoralists and their bundles of animals-plants-community-region relations suggests some crucial aspects to the way knowledge might be produced are still to be explored. We began to ask, after re-reading the book's chapters, if the trajectory/pathway framing could lead to new directions of enquiry. We concluded it does.

First, whether we conduct empirical studies of on-the-move dimensions or sedentary moments, movement institutes contingent mixes of connection and relationship. In the worlds of pastoralists and pastoralism depicted in Wiese, Komey, Zaal, Mahmoud, Breuer and Kreuer, Janzen, Gruschke, and Stammler, herders camp, thereby actively make-do with what can be accessed in the locality, and beyond from the locality. Grazing animals forage afresh and are fed supplementary feed. Pasture that may or may not be flourishing is grazed, under heavy or gentle management by pastoralists. Movement from camp to camp is narrated, the experienced times of sedentary-spaces are part of seasonal patterns and wider biophysical variability. The duration of encampment is timed, finite, maybe or maybe not flexible. By comparison the study by Gertel, Le Heron and Le Heron on Western Australia involves very different realities: Animals on ships (a camp under market exchange rules), even when experiencing an unexceptional journey are still in a sequence of spaced times. These include, to illustrate, loading bays before loading (a space occupied for a duration, and a duration when movement of animals inhabit a new space into a feedlot), deck feedlots (the qualities are co-constituted by the materialities present, some of which may erupt into event-full episodes that become known elsewhere because surveillance and extremity combine to broadcast news) and so on. The relationality given prominence by thinking through spacings and timings indicates dimensions of old and new ways of connecting, ordering and inhabiting worlds. Kreutzmann's study on pastoralism in the Pamirs analyses market integration from the perspective of the longue durée, conceptualising spatio-temporal relations as effects of interferences with (national) boundary-making, revolutionary movements and socio-economic reforms in Central Asia.

Matters of Nonhuman and Human Materiality

The challenge, as we have been arguing, is to take advantage of the 'performative interactions between the conceptual and the empirical' (McCormack 2009: 280), so that 'theory is refigured as a conceptually charged set of mobile practices through which to apprehend the movement of an open-ended and ontogenetic world'. We are suggesting that the constitution of markets and livelihoods might be illuminated even more were we to encounter 'the world as a swarm of vibrant materials entering and leaving agentic assemblages' (Bennett 2010: 107). In particular, the assemblages understood as pastoralism and pastoral economies may not be especially able to maintain resilient relations for the actors making up the assemblages. This approach rests on conceptual shifts around exchange relations which span those of reciprocity, surplus, money and metabolism. Such conceptual shifts prioritise the mutual interdependencies that assemblage actors cannot do without and questions the boundaries of assemblage categories. Thus, as McCormack (2009: 277) holds, assemblages are dynamic and open-ended sets of relational transformations (they are becoming) that through their transformations are always constituting categories. In this interpretation, categories are provisional outcomes of change. We contend that any attempts to move categories from their provisional status is very dependent on the nature of embodying and subsequent embedding required to put new networks in place, as networks are constitutive of ethical and political engagement.

In this sense the study on trade relations between Western Australia and the Arab Gulf countries by Gertel et al. treats economic spaces as processes, an endeavour which complicates the attempt to produce fixed ontological reference points (see also Lloyd's chapter for breeding characteristics of UK sheep bodies). But this conceptualisation opens up, for example, the rethinking of value and values which intersect in constituting new materialities, such as the 'beyond (Australian)-border-die-wrong-sheep'; this constitutes a reality shared by a specific, not territorially bound community that ultimately might be able to stop live sheep exports from Australia and challenge the 'fixing' of the economy (Mitchell 1998), by demonstrating that it is a non-autonomous domain.

Embodied Risks and Embodying Uncertainties

We should not forget that any exchange relation carries both uncertainties and risks. Dramatic ruptures in pastoral commodity systems occurred, for example, with the break-down of the Soviet Union, exposing the nomadic livelihoods of millions in Central Asia to new risks. The recent world food and financial crises also threatens pastoral livelihoods and lives on a global scale, triggering international food riots in 2007/08 and in 2011, and causing the number of starving people to jump within months from 830 million to over 1 billion. Exchange relations are real: they affect human bodies and bodies ultimately die. Pastoralists and nomadic people inhabit this world. The majority of them are poor, but conversely they should not always been seen as the most vulnerable. The exchange entitlements (Sen 1981; Devereux 2007) that pastoralists are able to command (i.e. livestock, grassland and other pastoral resources), even if partly or deeply enmeshed with the world economy, still provide them with a degree of independence from the volatilities of the global financial system, particularly in relation to subsistence production.

What the book addresses is thus the development of knowledge systems that better expose the range of human and nonhuman materialities involved in the worlds of pastoralism and pastoral economies and how such complex materialities in singular, and in combination, are enmeshed in and constitutive of uncertainties and risks. This goal is far reaching, and we acknowledge the preliminary nature and partiality of the book's contributions to this end. However, by (a) creating the book's interpretive heuristic by co-framing markets and livelihoods into tension and (b) reading the book's accounts of differentiated pastoralism and pastoral economies trajectories as resources to forward-think into the contemporary literature, we have been able to mobilise this combined strategy into a knowledge production intervention.

Increasingly we find ourselves outlining and attempting to defend a shift in focus. A passionate concern with the artificiality of treating markets and livelihoods in isolation has widened to include a passionate concern that what is at stake is the identification and placing of bodies, and bodies-in-relations. Key to moving this direction of thought ahead is to admit more players into the assemblages of materialities. Through a conceptual openness, the trajectory of inquiry delineated by the book will be accelerated.

But why argue that embodiment should be prioritised? Three reasons seem especially compelling: First, unless we incorporate ourselves directly as well as indirectly into knowledge systems we lose the ability to gauge, from bodily experiences, the direction and satisfactoriness of our decisions and investments in practices and behaviours. Many chapters gesture in this direction, either through long-term in field engagement or by recognising multiple and new voices of experience and analysis. Second, perhaps it is a sign of contemporary concerns more generally, but the conceptual act of opening up is likely to see an explosion of different sorts of actors 'naming' themselves into a more open and associative assemblage. Third, these new actors will present fresh challenges to the doing of ethics and politics. These knowledge production demands engage with what Gibson-Graham and Roelvink (2009) have called an 'Economic Ethics for the Anthropocene'. There is anthropomorphic advocacy here, but aided by growing respect for all human and nonhuman enlivening forces in and through which we exist. This sensitivity to and sense of embodied mutual respect is most apparent in the ethnographically informed studies.

In re-visiting the book's chapters around the theme of exchange relations we drew three major conclusions about key influences at work. Evidence for our conclusions came from the many instances detailed in the chapters and from the overall trajectory-like experiences many contributors outlined. The insights point to the 'elephants in the room', the unspoken visible, around which there is much avoidance behaviour and about which too little is known. These influences need to be examined in terms of non-contestable criteria, which we derive from our readings of the agri-food, economic geography and development literatures that inform the book. The criteria are: the way they screen in or screen out the centrality of ecological processes, the manner in which they constrict or throw open the identification of 'relevant' effects and responses, and how they prioritise or dismiss thinking about generating and using surpluses. In a way these criteria are stating what has eluded much research in the past.

The first major influence is the resilience and productivity of pastoral economies. The authors of the chapters reveal that pastoral production and commodity systems are, in general, productive rather than resource extractive economies. The latter, for example, industrialised production of gas, oil, and minerals, increasingly compete over the same pastoral territories, as is the case in Siberia, Sudan or Chad. While century old interrelations between nomadic people, animals and grassland often fluctuate according to changing conditions and are based on flexible arrangements, they do reach long-standing balances and offer sustainable production from pastures. The enclosure of the commons, the large scale privatisation of land and the enforced neoliberal property system, however, aim to tame, standardise and prescribe production and exchange. Economic spaces of pastoral production and commodity chains thus compete over the taxonomy of (international) exchange relations, so far, from a marginal position, but offer important visions for a new order of our global commons.

The second influence is finance, and financial relationships. These enter into every kind of exchange relationship. The crux of the matter is the reduction of activities, ecological processes, community assets and so on to numbers, but especially financial numbers. Independent numbers, that is, numbers that are detached from the conditions they purportedly represent, threaten to erase actual material systems of livelihood and access to these systems. Not naming situated bodily assemblages and experiences reduces the ability to inform practical steps towards sustainable livelihood strategies. Chapter after chapter shows the precariousness of financial arrangements irrespective of the exchange relationships pastoralist make-do with. However, we acknowledge that engaging strategically and tactically in knowledge production ventures to make visible financial relations is difficult, as the few who have done so note (Rankin 2008). Whether engagement is community-centred, individual-centred or industry-centred, the regimes, models and rules of short and long-term finance must be given more attention.

The third influence we identify is reaction to disasters of exceptional magnitude and duration. These include war, drought, floods, financial crises, cyclones, fires, volcanic eruptions and so on. Disasters are moments when reassembly is prioritised. But it is the link between activities of re-assembly, potential new assembly work, and financial frameworks, which we see as the most pernicious possibility. International evidence is mounting to suggest that coping with such calamities is frequently mobilised according to existing rules, with little regard to creative options to put in place new financial relationships for particular goals. This emphasis would be a direction that the book has not covered, though Rass deals with aspects of financial models and raises issues about reconceptualising alternative financial systems and Le Heron notes that rethinking of pastoralism in New Zealand is having to work closely with doing business in the context of climate variability and extreme events. While the research was cognisant of the influences in which pastoralism and pastoral economies are embedded, the framing of consequential and constitutive transformations needs to be more actively explored.

Acknowledgements

We acknowledge the critical comments of Erena Le Heron in the development of the argument of the chapter. Jörg Gertel gratefully acknowledges the funding of the German Research Foundation (DFG) for his sabbatical leave to New Zealand in 2009/2010 and he would like to extend his gratitude to colleagues in the School of Environment, the University of Auckland, New Zealand, for a stimulating academic time. Richard Le Heron wishes to thank the New Zealand Marsden funded Biological Economies research team and colleagues at Auckland for their ongoing and stimulating engagement with agri-food questions.

References

Appadurai, A. (ed.). 1986. *The Social Life of Things: Commodities in Cultural Perspective*. Cambridge: Cambridge University Press.
Bennett, J. 2010. *Vibrant Matter. A Political Ecology of Things*. Durham: Duke University Press.
Berndt, C. and Boekler, M. 2009. Geographies of circulation and exchange: Constructions of markets. *Progress in Human Geography*, 33(4), 535-551.
Bingham, N. 2006. Bees, butterflies, and bacteria: Biotechnology and the politics of nonhuman friendship. *Environment and Planning A*, 38, 483-498.
Bingham, N. and Hinchliffe, S. 2008. Reconstituting natures: Articulating other modes of living together. *Geoforum*, 39(1), 83-87.

Callon, M. 2007. An essay on the growing contribution of economic markets to the proliferation of the social. *Theory, Culture and Society*, 24(7-8), 139-163.

Cook, I. 2004. Follow the thing: Papaya. *Antipode*, 36(4), 642-664.

Cook, I. et al. 2006. Geographies of food: Following. *Progress in Human Geography*, 30(5), 655-666.

Cook, I., Evans, J., Griffith, H., Morris, M. and Wrathmell, S. 2007. 'It's more than just what it is': Defetishising commodities, expanding fields, mobilising change... *Geoforum*, 38, 1113-1126.

Devereux, S. (ed.). 2007. *The New Famines. Why Famines Persist in an Era of Globalization*. London and New York: Routledge.

Doel, M. 1996. A hundred thousand lines of flight: A machinic introduction to the nomad thought and scrumpled geography of Gilles Deleuze and Felix Guattari. *Environment and Planning D: Society and Space*, 14(4), 421-439.

Gertel, J. and Breuer, I. 2011. *Alltagsmobilitäten. Aufbruch marokkanischer Lebenswelten*. Bielefeld: Transcript (forthcoming).

Gibson-Graham, J.-K. 2006. *A Postcapitalist Politics*. Minneapolis: University of Minnesota Press.

Gibson-Graham, J.-K. and Roelvink, G. 2009. An economic ethics for the Anthropocene. *Antipode*, 41, 320-346.

Latour, B. 1987. *Science in Action. How to Follow Scientist and Engineers Through Society*. Cambridge: Harvard University Press.

Law, J. and Mol, A. 2001. Situating technoscience: An inquiry into spatialities. *Environment and Planning D: Society and Space*, 19, 609-621.

Le Heron, E., Le Heron, R. and Lewis, N. 2011. Performing research capability building in New Zealand's social sciences: Capacity-capability insights from exploring the work of BRCSS's 'Sustainability' theme, 2004-2009. *Environment and Planning A* (forthcoming).

Li, T. 2007. Practices of assemblages and community forest management. *Economy and Society*, 36(2), 263-293.

Longhurst, R. 2009. Embodied knowing, in *International Encyclopedia of Human Geography*, edited by R. Kitchin and N. Thrift. Amsterdam: Elsevier, 429-433.

Massey, D. and Allen, J. (eds.). 1984. *Geography Matters!* Cambridge: Cambridge University Press.

McCormack, D. 2009. Becoming, in *International Encyclopedia of Human Geography*, edited by R. Kitchin and N. Thrift. Amsterdam: Elsevier, 277-281.

Mitchell, T. 1998. Fixing the economy. *Cultural Studies*, 12(1), 82-101.

Murdoch, J. 2006. *Post-Structuralist Geographies: A Guide to Relational Space*. London: Sage.

Rankin, K. 2008. Manufacturing rural finance in Vietnam: Contested governance, market societies and entrepreneurial subjects. *Geoforum*, 39, 1965-1977.

Roe, E. 2009. Human-Nonhuman, in *International Encyclopedia of Human Geography*, edited by R. Kitchin and N. Thrift. Amsterdam: Elsevier, 251-257.

Sen, A. 1981. *Poverty and Famines. An Essay on Entitlement and Deprivation*. Oxford: Oxford University Press.

Thrift, N. 2008. *Non-Representational Theory: Space, Politics, Affect*. London/ New York: Routledge.

Watts, M. 2001. Petro-violence: Nation, identity and extraction in Nigeria and Ecuador, in *Violent Environments*, edited by N. Peluso and M. Watts. New York: Cornell University Press, 182-212.

Watts, M. 2005. Commodities, in *Introducing Human Geographies*, edited by P. Cloke, P. Crang and M. Goodwin. London: Hodder Arnold.

Whatmore, S. 2006. Materialist returns: Practicing cultural geography in and for a more-than-human world. *Cultural Geographies*, 13(4), 600-609.

Index

Note: page references in **bold** refer to
figures